Electrospinning

From Basic Research to Commercialization

Soft Matter Series

Series editors:
Hans-Jürgen Butt, *Max Planck Institute for Polymer Research, Germany*
Ian W. Hamley, *University of Reading, UK*
Howard A. Stone, *Princeton University, USA*

Titles in this series:
1: Functional Molecular Gels
2: Hydrogels in Cell-based Therapies
3: Particle-stabilized Emulsions and Colloids: Formation and Applications
4: Fluid–Structure Interactions in Low-Reynolds-Number Flows
5: Non-wettable Surfaces: Theory, Preparation and Applications
6: Wormlike Micelles: Advances in Systems, Characterisation and Applications
7: Electrospinning: From Basic Research to Commercialization

How to obtain future titles on publication:
A standing order plan is available for this series. A standing order will bring delivery of each new volume immediately on publication.

For further information please contact:
Book Sales Department, Royal Society of Chemistry, Thomas Graham House, Science Park, Milton Road, Cambridge, CB4 0WF, UK
Telephone: +44 (0)1223 420066, Fax: +44 (0)1223 420247
Email: booksales@rsc.org
Visit our website at www.rsc.org/books

Electrospinning
From Basic Research to Commercialization

Edited by

Erich Kny
KEMYK, Burgenland, Austria
Email: erich.kny@kemyk.eu

Kajal Ghosal
Dr B. C. Roy College of Pharmacy and Allied Health Services, Durgapur, India
Email: kajal.ghosal@gmail.com

and

Sabu Thomas
Mahatma Gandhi University, Kottayam, India
Email: sabuchathukulam@yahoo.co.uk

ROYAL SOCIETY
OF CHEMISTRY

Soft Matter Series No. 7

Print ISBN: 978-1-78801-100-6
PDF ISBN: 978-1-78801-294-2
EPUB ISBN: 978-1-78801-491-5
Print ISSN: 2048-7681
Electronic ISSN: 2048-769X

A catalogue record for this book is available from the British Library

The Royal Society of Chemistry is a charity, registered in England and Wales, Number 207890, and a company incorporated in England by Royal Charter (Registered No. RC000524), registered office: Burlington House, Piccadilly, London W1J 0BA, UK, Telephone: +44 (0) 207 4378 6556.

For further information see our web site at www.rsc.org

Printed in the United Kingdom by CPI Group (UK) Ltd, Croydon, CR0 4YY, UK

Preface

This book revolves around the central theme of electrospinning, and discusses in greater detail the recent developments in design, process parameters, polymer selection, application areas and alternative electrospinning processes. Electrospinning produces continuous ultrafine fibers in the semi-micro to nano-diameter range from polymer solution/melt with the help of an electrostatic force. The technology was first patented by Formhals in 1931 and since then a series of technological advancements has established it as one of the most popular semi-micro to nano-fabrication processes. In the past few decades, electrospun non-woven matrices/membranes have been successfully exploited by researchers across a wide variety of applications including tissue engineering, nanocatalysis, clothing, biomedical/pharmaceutical/healthcare, filtration, optical electronics, biosensors, defense and environmental protection, just to name a few. In contrast to existing literature and books available on electrospinning and nanofibers, which are introductory and more fundamental in nature, this book focuses on recent developments in various aspects of the electrospinning process. The primary target is to encompass various approaches for successful implementation of this fabrication process towards commercialization from basic research and development.

The book has 10 chapters. In Chapter 1, the historical journey of the electrospinning process towards its establishment in laboratory practice is explained. Although fine fibers made by electrostatic force have been discussed, investigated and patented for a long time, until the 1980s potential applications for such fibers were restricted by different technological limitations. In 1995, Doshi and Reneker reworked and simplified the electrical spinning process. The first chapter details these historical developments. Chapter 2 discusses the encroachment of traditional electrospinning by process modifications. Higher fiber production rates, better alignment, and uniform diameter of fibers can

Soft Matter Series No. 7
Electrospinning: From Basic Research to Commercialization
Edited by Erich Kny, Kajal Ghosal and Sabu Thomas

Published by the Royal Society of Chemistry, www.rsc.org

only be achieved by continuous improvements of the electrospinning process. Chapter 3 deals with electrospun composites that are biocompatible and bioactive in nature. Their uses, compatibilization strategies, possible modifications, utilization and recent commercialization are discussed. Chapter 4 discusses cell electrospinning and technology transfer from lab to market scale. The authors describe methods, challenges, opportunities, current research works, key aspects and market potentials of cell electrospinning in this chapter. Chapter 5 is about health care and wound care textiles obtained through the electrospinning process. A detailed discussion of their recent developments, essential aspects, basic functions, specific characteristics, market potentials, and recent success is presented. This chapter is focused on the applications of electrospinning techniques specific to nanomedicine. Chapter 6 focuses on several general methods of bioprinting, with an emphasis on hybrid systems that involves both 3D printing and electrospinning. Future improvements in bioprinting technologies and indications on potential markets are summarized. Chapter 7 is about the industrial applications of super-hydrophobic surfaces obtained through electrospinning processes on a large scale in filtration/sorbents applications. Electrospun mats have unique properties that make them suitable for water filtration, oil clean-ups and greenhouse gas sorbents. This chapter presents various materials and methods used for this purpose and comments on the large-scale production feasibility of electrospun fibers. In Chapter 8, the authors discuss the perspectives of the electrospinning process for electronics and sensors. Chapter 9 shows that electrospinning has opened up new applications in energy applications such as proton exchange membrane fuel cells and water electrolysis, *etc.* This chapter describes the recent advancements in the field along with the materials used, methods applied, and future application possibilities. Chapter 10 discusses recent developments and the latest trends in packaging involving electrospun ultrathin fibers, focusing on their technological impact and its potential scale-up and market introduction.

We hope that we have presented our theme in a very easy, understandable and continuous manner. The contributing authors are experts in their particular fields and come from different parts of the world. To gain commercial insight not just from academia, authors from industry are also involved.

We thank and acknowledge the kind assistance of our reviewers, whose timely review and expert comments have helped us in accomplishing our goal. We would also like to thank Connor Shephard and Leanne Marle for their editorial assistance. We also take the opportunity to thank Dr Hanna Maria and Dr Aniruddha Chandra who helped to form the initial proposal. A further acknowledgment goes to the COST Networking Action MP1206 on electrospinning, which was helpful in promoting the technological basis, especially for Chapters 1, 2, 9 and 10, for describing and elucidating the latest trends in these application fields.

Erich Kny, Kajal Ghosal and Sabu Thomas

Contents

Soft Matter Series No. 7
Electrospinning: From Basic Research to Commercialization
Edited by Erich Kny, Kajal Ghosal and Sabu Thomas

Published by the Royal Society of Chemistry, www.rsc.org

CHAPTER 1

Electrical Spinning to Electrospinning: a Brief History

K. GHOSAL*[a], C. AGATEMOR[b], N. TUCKER[c], E. KNY[d] AND S. THOMAS[a]

[a]Mahatma Gandhi University, Department of Chemical Science, Priyadarshini, Kottayam, 686560, India; [b]Harvard University, Harvard John A. Paulson School of Engineering and Applied Sciences, 29 Oxford Street, Cambridge, MA 02138, USA; [c]University of Lincoln, School of Engineering, Lincoln LN6 7TS, UK; [d]CEST GmbH, 2700 Wiener Neustadt, Austria
*E-mail: kajal.ghosal@gmail.com

1.1 Introduction to Early Concepts, Phenomena and Technology Related to Electrospinning

1.1.1 Initial Identity of Triboelectric Charge and the Consequences for Electrohydrodynamic Processes

Electrospinning technology has progressed a long way, from certain fundamental concepts to substantial industrial applications. Electrospinning is an electrohydrodynamic process resulting in the movement of a fluid by means of the application of an electrostatic field. This movement of liquids by electrostatic force was first observed by William Gilbert, sometime personal physician to Queen Elizabeth I of England and president of the Royal College of Physicians. In about 1600, Gilbert[1] first recorded the movement of liquid

Soft Matter Series No. 7
Electrospinning: From Basic Research to Commercialization
Edited by Erich Kny, Kajal Ghosal and Sabu Thomas

Published by the Royal Society of Chemistry, www.rsc.org

under the influence of a triboelectric field produced by rubbing amber. Due to friction-induced charge separation, the amber is electrified. The rubbing or friction of some material with another material to produce a separated charge was widely known, having said to have been first observed by the Greek philosopher, Thales of Miletus.[2] The use of amber (in ancient Greek ἤλεκτρον or *ēlektron*) gave rise to the coinage by Gilbert of the word '*electricus*', which in turn lead to the terms 'electric' and 'electricity'. When the rubbed amber was placed in proximity to a water droplet placed on a dry surface, the water droplet distorted under the influence of electric charge into a diagnostic conical shape that later became known as the Taylor cone.[3] William Gilbert observed:

Indeed it plainly does draw the body itself in the case of a spherical drop of water standing on a dry surface; for a piece of amber applied to it at a suitable distance pulls the nearest parts out of their position and draws it up into a cone; otherwise, if it were drawn by means of the air rushing along, the whole drop would have moved.[1]

Before the invention of reliable methods of generating low-voltage, high-current electricity, a considerable amount of fundamental work on electricity using high-voltage electrostatics was undertaken by workers such as Otto von Guericke, Stephen Gray, Charles Francois Dufay and Georg Mathias Bose.

In 1749, French clergyman and physicist Jean-Antoine Nollet[4] observed another electrohydrodynamic phenomenon, namely how electrifying water can cause the formation of droplets, a process now known as electrospraying. None of this early work resulted in the development of recognizable processing technology for either electrospraying or electrospinning. The efforts of these early workers were focused on the description and understanding of these phenomena. The concept of an artificial fiber dates from this period, being proposed by the natural philosopher Robert Hooke. Hooke[5] said that to succeed in producing artificial fibers from an unnamed 'glutinous composition', "very quick ways of drawing it out into small wires for use could be found".

There were also parallel attempts to store the charge. The storing of electrostatic charge was first experimentally achieved using a glass jar (known as a Leyden jar), coated inside and out with metal foil. The Leyden jar form of the capacitor was developed by Pieter van Musschenbroek, Professor of Physics at Leiden University, but the credit for performing this experiment is often attributed to the German scientist, Ewald G. von Kleist, in 1745.[6]

In electrospinning, electrostatic charge provides the motive force for the production of fibers. The speed of a fiber in flight is commonly between 60 and 100 m s^{-1}, but the mass rate of production from a single spinneret is typically measured in mg per hour. The electrostatic charge that is engendered in the fluid by applying a high voltage causes the fluid to overcome the surface tension that holds a pendant droplet in shape. Ultimately the surface of the droplet fails and droplets or – if there is sufficient molecular cohesion – a fiber are ejected from the droplet. This form of use of electrostatic charge is applied in various fields in addition to electrospinning. Examples include photocopying, laser printing, electrostatic painting and acoustic transducers, including high-fidelity loudspeakers, particle separators, and electrostatic filters.[6]

1.1.2 The Development of Industrial Spinning Processes

The establishment of the electrohydrodynamic process of electrospinning as a commercial application is the result of work by a number of researchers and manufacturers. It should be noted that the work by the academic community has been of most use in understanding the parallel efforts of commercial companies to develop the spinning process. Electrospinning is but one of a number of artificial fiber manufacturing processes that were developed during the 19th century.

The need for large quantities of reliable fiber for the production of woven and embroidered goods during the Victorian period was an early driver for the production of artificial fibers. At that time, machine embroidery was mainly used for decorative furnishings. In 1828, Josué Heilmann developed commercial machine fiber combing and stressed the importance of improvements to profitability that this process offered for large-scale production; the machine was patented in 1829 in England. Joshua Wordsworth, the owner of a weaving mill who took out a patent on the machine, and in turn established a parallel arrangement with Manchester silk manufacturer Louis Schwabe for fabricating embroidered goods. Louis Schwabe devised a process for spinning glass fiber, but was unable to find a more suitable spinning material for fabric production,[7] although he put considerable work into "carrying out experiments which would lead to the discovery of a substance which would form a homogeneous mass possessing the quality of ductility and susceptible of being drawn out through fine holes, or otherwise, into filaments or fibers possessing suitable strength and other properties to adapt it for manufacturing purposes".

The invention of a potential spinning formulation by George Audemars[8] in 1855, of suitable strength and suitable for the manufacturing process, had been demonstrated in concept by Schwabe. Audemars dipped needles into a nitrocellulose solution, then drew it into air, obtaining a long durable thread as the solvent evaporated. The manually applied dragging force pulling the fibers from the polymer solution was low, but the process was very slow and hence only suitable for small-scale production. Cellulose was first nitrated to a high degree by the Swiss chemist Christian Schönbein, in 1845, who patented his process for the production of the explosive known as guncotton. In 1846 Louis-Nicolas Ménard and Florès Domonte dissolved cellulose nitrate in a mixture of ether (ethoxyethane) as the solvent and ethanol as a diluent to produce a solution that became known as collodion. The expedition of generating fibers continued. Count Hilaire De Chardonnet, a French engineer and industrialist, forced a solution of cellulose derivatives through small holes like a spinneret, then dried these in air and further treated the fibers with a coagulating medium: for this work he became known as the 'father of the artificial silk industry'.[9] His artificial silk was sold as 'Chardonnet silk'. Chardonnet silk was the first artificial semi-synthetic fiber. The invention was exhibited by Chardonnet in 1889, but pre-dated in 1855 by Joseph Wilson Swan, the British chemist who produced fibers for electric light filaments from 'parchmentized' cotton, and who went on by 1881 to develop and patent a method for extruding nitrocellulose to form conducting fiber filaments – this material was used to produce incandescent light bulbs by the Edison & Swan United Electric Light Company.

The next significant development was cuprammonium silk. In 1857, Eduard Schweizer investigated the solubility of cellulose in an ammoniacal solution of copper hydroxide (cuprammonium hydroxide solution). This observation was applied to the fiber production process in 1890 by L. H. Despaissis, who regenerated a solution of cellulose into a solid[10] by treating a cuprammonium solution of cellulose in a coagulating bath. Although the process was patented, it was not successfully commercialized. Two years later, in 1892, Dr Max Fremery and Johan Urban used a development of the process to make lamp filaments: the first profitable outcome of manufacturing cellulose fibers from a cuprammonium solution. At first only short discontinuous lamp filament fibers were produced. It took significant further work to produce continuous long fibers. Early in the market in 1908 with this material was J. S. Bemberg with Bemberg silk. Bemberg silk fiber was spun through a stretch spinning system developed by Dr Edmund Thiele and Emil Elsässer, in which a thick fiber was extruded, then stretched to a final thickness. Asahi Kasei in Japan is the only manufacturer still producing Bemberg fiber. Having started production of this fiber in 1931, by 2015 they had a production capacity of 17 000 tons per year. In parallel with the development of the stretch spinning system, another process for spinning the cellulose fiber known as 'viscose' was developed. This fiber-producing process was patented in 1893 by Cross and Bevan in England.

Over the last century, the basic process for producing the fibers has been evolved through many iterations. Wood pulp was reacted with sodium hydroxide solution and then the solution was treated with carbon disulfide to produce cellulose xanthate. This procedure was followed by addition of further caustic soda to dissolve cellulose xanthate into a spinnable viscous solution. The viscous solution[11] is pumped through a spinneret into a coagulation medium to produce a solid fiber. This fiber or thread is further treated with dilute sulfuric acid to decompose the xanthate and to regenerate the material into cellulose. As rayon, this regenerated cellulose fiber entered the USA textile market in 1920 and the manufactured artificial fiber is still widely used in numerous commercial applications.

1.2 The Establishment of Electrospinning for Fiber Generation

The use of electricity for spinning fibers was attempted by Professor Charles Vernon Boys in 1888. Boys is noted for his invention of a number of scientific instruments. While making a torsion balance to determine the gravitational constant he required a ligament to suspend the moving part of the instrument. His first attempt to make such a fiber was recognizably electrospinning.[12] He used a small insulated dish which was connected to a high-voltage electrical supply, and successfully spun from various melts such as beeswax, shellac, and collodion. This he described as "the old, but little known experiment of electrical spinning". Unfortunately, none were strong enough for the purpose in hand, and he eventually achieved success by attaching a silica billet to a crossbow quarrel, heating the billet to soften it, and then firing the crossbow down the length of his laboratory: this is reasonably assumed to be the first successful attempt to produce a nano-scale fiber.

In the early 20th century, John Francis Cooley, an inventor and professional electrician of Penn Yan, New York, took out a UK patent (GB 6385) on 5th April 1900. His patent was titled *Improved methods of and apparatus for electrically separating the relatively volatile liquid component from the component of relatively fixed substances of composite fluids.*[13] On 4th February 1902 and 24th November 1903 Cooley took out another two patents (US692631 and US745276). These patents covered methods and the apparatus for dispersion of fluids by means of electrostatic force. Direct current was used to charge the fluids, and spinnable fluids were dropped in between the electric field source and the grounded collector and fibers were collected by reeling onto a glass rod (Figure 1.1).

Cooley observed that when his diluent–solvent mixture (ether–ethanol) was modified with the addition of a third liquid (benzole), the spinnability of the fluid was enhanced. He also described spinneret systems for coaxial spinning, air-assisted spinning and rotating collectors (Figure 1.2).

W. J. Morton, an American physician working on the use of electricity to cure ailments, known as electrotherapy, patented a method for the dispersion of fluids (US 705691) in 1902. Dispersion of the fluid was assisted by indirect and direct electric charging methods. He did not attempt to commercialize the system, saying his "cobweb-like mass" of fibers "may be put to any industrial use". Most electrospinning methods use the direct charging method proposed by Morton.

Professor Kiyohiko Hagiwara, of Kyoto College of Industrial Arts, Japan, worked on the manufacture of artificial silk and made significant contributions to rayon manufacturing technologies. In Japan, rayon manufacturing was advanced after the expiry of the patent for the basic viscose process for producing artificial silk in Europe.[14] Hagiwara charged the colloidal liquid viscose solution before spinning the fibers. The electric charge was employed here to retain the aligned molecular structure of fibers, limiting the tendency

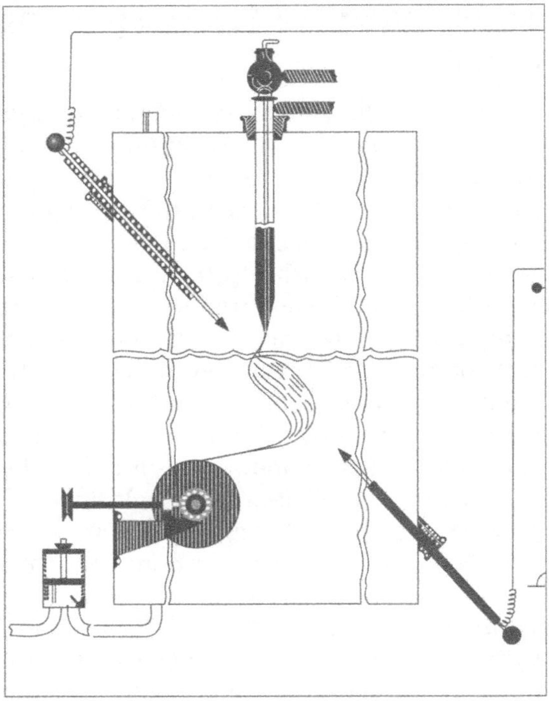

Figure 1.1 Cooley's indirect charging method. Reproduced from ref. 12 with permission from the Royal Society of Chemistry.

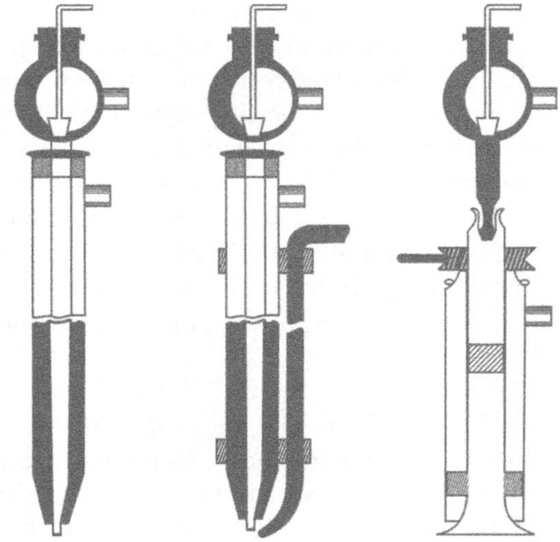

Figure 1.2 Coaxial, air-assisted and rotating emitters. Reproduced from ref. 12 with permission from the Royal Society of Chemistry.

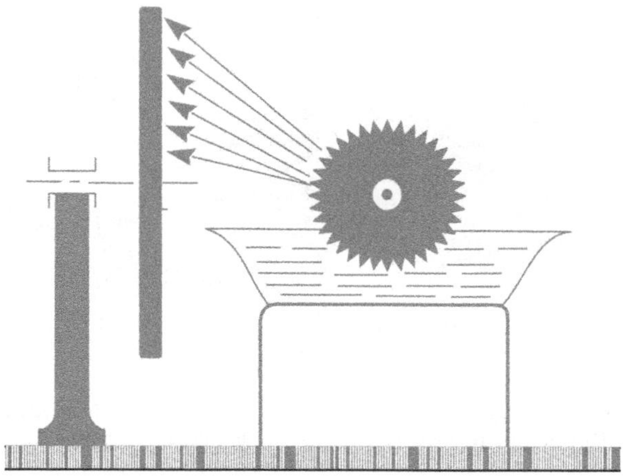

Figure 1.3 An early spinning machine with saw-tooth rotating emitter. Reproduced from ref. 12 with permission from the Royal Society of Chemistry.

towards fiber aggregation. The fiber obtained had better luster and Hagiwara patented the use of electricity during artificial production as US patent 169965 in January 1929. He also tried to increase the thickness of the fibers produced by combining electrostatic force with air pressure to increase the flow rate of spinnable fluid through the spinneret. His method used an 80 Hz oscillating voltage, presumably to avoid charge build-up in the product.

Anton Formhals made a large number of highly significant contributions to the development of electrospinning technology. He had 22 patents to his credit, taken out between 1929 and 1944, which covered numerous aspects of the electrospinning process. Formhals invented the saw-tooth emitter for distribution of spinnable fluid. The saw-tooth emitter (Figure 1.3) resembled a circular saw blade that dipped into the spinnable fluid, and emitted fibers from the saw-tooth points where the electrostatic field was strongest, and the curved surface of the liquid promoted the fracture of the droplet. The fibers were collected on a charged rotating collector.

Formhals developed a multi-head spinneret (Figure 1.4), which was filled up individually with spinnable fluids from an overhead tank and produced short fibers known as staple fibers gathered on a rotating notched-disc counter electrode. Further modifications for the collection of short fibers used an air blast. In another patent[15] he proposed a conical, tapered nozzle for distributing spinnable fluid.

Electrospun staple fiber manufacturing was also proposed by Ernest K. Gladding while he was working for E. I. du Pont de Nemours and Company. In his patent (US 2168027), Gladding proposed his method for producing staple fibers. Unlike Formhals' method, where the short fibers were collected on a charged counter electrode in the method depicted in Figure 1.5, the collecting electrode is not conducting, with the counter electrode placed behind the collector. This means that the loosely attached fibers on the collector belt

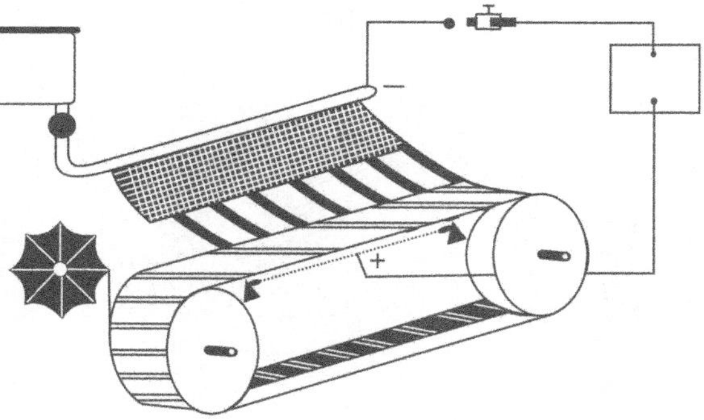

Figure 1.4 Formhals' spinning machine for short fibers. Reproduced from ref. 12 with permission from the Royal Society of Chemistry.

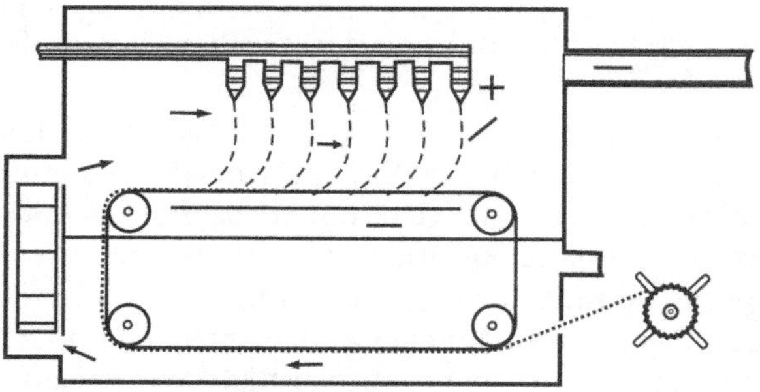

Figure 1.5 Gladdings' spinning machine for short fibers. Reproduced from ref. 12 with permission from the Royal Society of Chemistry.

are easier to remove onto reels. Here also multi-head spinnerets or nozzles were used to produce fiber strands. The mechanism for collecting fibers in a multi-head spinneret machine is shown in Figure 1.5.

Charles Ladd Norton was born in Springfield, Massachusetts, in 1870. He graduated from the Electrical Engineering and Physics Department of MIT in 1893. He worked on fire-resistant materials and the exploitation of X-rays as a medical diagnostic tool with Francis Williams, at the Boston City Hospital. In his 1936 patent (US 2048651) and his posthumous 1940 patent (US 2185417), Norton described a method and apparatus for forming filamentous or fibrous material by electrospinning from a melt (Figure 1.6). Currently, this process is attracting commercial attention for wound dressings, filtration, textiles, nanomedicine and regenerative medicines.[16] This process produces fibers without any risk of solvent carryover into the finished product.

The materials suggested for processing included gums, pitches, synthetic rosins and fused glass. The materials are held inside a refractory container, and one pole of a high-voltage electric source is connected to the container and the other pole of the source to a target screen, which

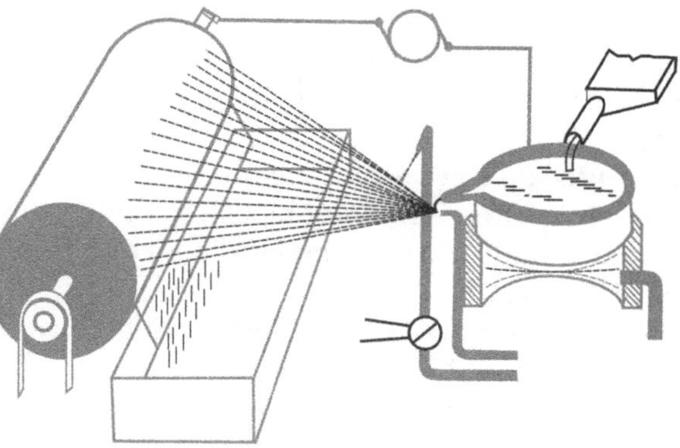

Figure 1.6 Norton's air-assisted melt spinning process. Reproduced from ref. 12 with permission from the Royal Society of Chemistry.

is a flat plate or a cylinder. The distance between collector and container was changed depending on the material to be spun. Norton maintained the distance between 1 foot and 6 feet (0.3–1.8 m) for high melting point materials and for low melting point materials such as rosins, the distance was changed to 20 feet (6 m) or more. Fibers were formed by a combination of electrostatic repulsion and air blast.[17] The material was extracted from the container through its tilted lip or spout, to encourage the formation of fine fibers rather than droplets or coarse fibers.

In the USSR the industrial production of air filters started in late 1939. The commercial production was the result of research performed in 1938. In 1938 two scientists, Igor' Vasil'evich Petryanov-Sokolov and Natalya D. Rosenblum, generated electrospun fibers from a nitrocellulose feedstock, which they developed into what came to be known as 'Petryanov filters'. This work won them the Stalin Prize, and rapidly led to the establishment of a works in Tver, manufacturing electrospun smoke filters used in gas masks. The Tver works is the first industrial-scale electrospun fiber factory. The material was spun from cellulose acetate in a solvent mixture of dichloroethane and ethanol.[18] These filters are for protection against aerosols and airborne particulates.[19] The USSR production capacity[20] of this filter reached 6.5 kg $hours^{-1}$. This work is particularly notable because it marks the first use of electrospun fiber as a non-woven, thus doing away with the difficulties associated with handling near invisible and highly charged fibers in further production processes.

F. W. Manning, a Canadian resident in the USA, took out a patent in 1943[21] on producing beds of randomly orientated fibers described as "by the deposition and induration of disrupted filaments in a promiscuous and intersected condition". His method was not intended to produces yarns but rather non-woven fabrics for uses as "sanitary napkins, surgical dressings, filtering purposes, leather substitutes, blankets, draperies, rugs, upholstering, insulating and most purposes for which woven fabrics are now used".

This neatly sidesteps the difficulties of handling highly charged nanoscale fibers for manufacture into handleable yarns. Manning's design used Cooley-type indirect charging, with the fibers cast onto the collector belt through a venturi.

H. R. Child patented an electrospinning process in 1944 while working for Eastman Kodak.[22] To address spinneret blocking, Child proposed the use of glass to make a non-conducting spinneret. Fiber collection is assisted by a directed current of warm air.

W. C. Heubner of Dayton, Ohio, USA, suggested preheating the spinning dope to speed up the solvent losses in flight.[23] He also suggested a modification to the spinneret to produce hollow fibers. He used a teardrop-shaped electrode inside his rotating drum collector to concentrate the electrostatic field to ease fiber removal.

In 1966, H. L. Simons, working for the Kendall Company in the USA, described in a patent the production of aligned fibers,[24] claiming that previous textile non-wovens are limited in application because they must be robust enough to cope with the bonding stage required to make the finished product, but by implying a minimum areal weight of "12 to 16 grams per square yard" he proposed to better this level by direct fiber deposition. The patterned deposition characteristic of his process is due to the use of a 'segmented receiver' with areas of different electric potentials. This is achieved using a rotating drum with a perforated surface as a collector – the fiber landing on the metal rather than the voids.

J. E. Owens and S. P. Scheinberg, working for du Pont de Nemours and Company, proposed a method called 'flash spinning' for making electrospun fibers.[25] The liquid raw material is heated "far above the boiling point of the solvent and at a pressure usually near the critical pressure of the solvent". When the liquid leaves the spinneret, the solvent evaporates very rapidly to form what Owens and Scheinberg call a "plexifilamentary" non-directional fiber. Their adjustable spinning distance collector system is held at a non-zero controlled potential above earth. Oscillating baffles are deployed to evenly disperse the fiber by influencing an electrostatic field in the fiber flight path. They also seek to minimize corona discharge, to help achieve a uniform mat deposition. In the early 1970s this work was extended by Hollberg and Owens[26] by using the electrostatic wind from the spinning nozzle and a variation of the oscillating baffle system to improve the uniformity of mat deposition.

W. Simm, working for Bayer Aktiengesellschaft in Germany, devised a multiple spinning zone apparatus for making Petryanov-type filters, offsetting the position of the spinnerets to even out fiber deposition.[27] Multiple spinneret systems sometimes do not give an even deposition of fibers due to mutual repulsion between the fiber streams. The spinnerets are in the form of a rotating ring dipping into the spinnable dope. A curved liquid surface is required to initiate fiber production; the curve puts the liquid surface under tension, and in addition maximizes the electrostatic field strength, causing a stress concentration that results in rupture of the surface. Simm suggests that a "free flowing material" powder can be added to the surface of the fiber.

A group led by G. E. Martin at Imperial Chemical Industries in the UK patented a method for making prosthetic devices "such as an internal organ" using an earthed spinneret and a collector belt connected to a Van de Graaf triboelectric generator.[28] The method allows a number of fiber collection methods to make a layered structure typical of human tissues. Martin suggests the employment of shaped collector substrates, such as a face mask to produce finished articles, with the advantage of reducing difficulties associated with post-deposition fiber handling.

A. Bornat, also working for Imperial Chemical Industries, and the University of Liverpool, has proposed a method to electrospin prosthetic components.[29] The method uses the earthed spinneret technique previously described and a number of spinnerets to deposit fiber onto a rotating drum collector.

The collector has a rotating internal metal rod. The outer layer is collapsible to allow removal of the fiber product. The fiber can then be sintered to improve the strength of the prosthesis. Bornat later proposed that the fiber tube can be curved while still on the outer sheath and before sintering to fix the shape.

C. Guignard, working at the Batelle Memorial Institute in Geneva, Switzerland, devised a melt spinning method.[30] The solid polymer is deposited on a charged belt and passed through a heater. The polymer melt then passes under a secondary earthed belt, and the liquid polymer is drawn up as fiber onto another belt acting as the collector.

B. Kliemann and M. Stoll, working for Rhodia AG in Germany, proposed the extrusion of a polymer melt as a feed material for electrospinning.[31] The molten strand is charged by a wire electrode inside the extrusion nozzle. The patent notes the charge on the fiber persists for up to 300 days after manufacture. It is likely that this persistent charge is due to the formation of an electret as the polymer change state while under the influence of a strong electrical field.

T. V. How, working with the University of Liverpool and Ethicon Inc. USA,[32] suggests varying the rotation rate of a drum collector between 2000 and 20 000 rev min^{-1} to control orientation of the fiber product. This is presumably to match the surface speed of the drum to that of the terminal speed of the fiber (typically ranging from 60 to 100 m s^{-1}). The spinneret scans along the mandrel at speeds between 2 and 40 cm s^{-1} to lay the fiber lay in a controlled fashion to vary the mechanical properties of the resulting tubular structure.

Bornat suggests controlling alignment of fibers in tubular structures[33] by means of an electrode array made of parallel bars placed at a small displacement from the collecting drum. This array is set at a voltage 4–5 kV less than the drum, the drum being the charged component and the spinneret is earthed.

Berry[34] describes a method that produces larger diameter fibers with a degree of orientation around the circumference of the cylinder, and embedded in a random orientated mat of smaller diameter fibers. He does this by means of two additional electrodes intended to bifurcate the fiber strand, meaning that one strand of the fiber has a longer flight path than the other, and will hence be drawn thinner. The bimodal fiber thicknesses result in a tube that resists buckling.

Doshi and Reneker's historic paper[35] marks a step-change in the rate of academic publishing and patenting. The process has gradually gained momentum in the USA, as well as Europe and the Pacific Rim. According to Barhate *et al.*,[36] more than 20 industries are currently engaged in the production of electrospun fiber-based filter production processes alone.

1.3 Current Status of the Electrospinning Process in Materials Science and Engineering

Today, electrospinning is a vibrant research field in materials science and engineering and is a promising lab- and industrial-scale technique to spin micro- and nano-scale materials. A SciFinder database search on 19 November 2017 found that over 4000 studies had been published since January 2017, and that there had been a significant increase (~272 times) in the number of published reports between 2000 and 2017 (Figure 1.7). The outcome of these research studies is an expanding frontier that provides numerous

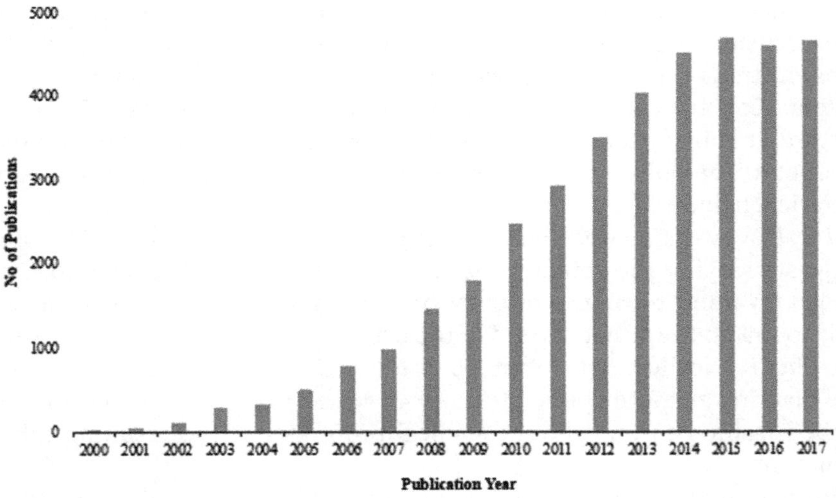

Figure 1.7 Research on electrospinning progressively increased between 2000 and 2017, as shown by the increase in the number of publications (SciFinder database, 18 November 2017).

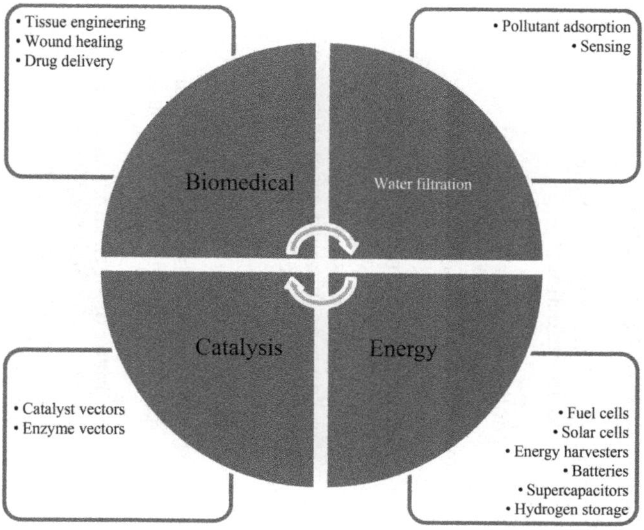

Figure 1.8 Functional applications of electrospun materials.

options regarding spinning technologies, materials to spin, and applications. Depending on the target application, polymers, ceramics, or metals, as well as their composites, can be spun from the melt, solution, or emulsion, with each technique providing a distinct advantage. Melt electrospinning, for instance, ensures safety and high throughput, while solution electrospinning is more versatile with regards to electrospun materials and is more economical with respect to energy consumption. Compared to melt electrospinning, the solution process affords fibers with better optoelectronic and mechanical properties. Researchers usually resort to emulsion electrospinning if high melting point materials or composites with immiscible components are to be spun. Also, recent innovations in the field now enable encapsulation of functional materials into spun fibers to design nano-scale materials with wide-ranging implications for biomedicine, catalysis, the environment, and energy (Figure 1.8 and Table 1.1).

Currently, materials science-orientated research into electrospinning has evolved beyond fabricating textile yarns into diverse fields (Figure 1.8). In the biomaterials field, electrospinning is now explored to fabricate vectors for drug delivery,[37–41] platforms to heal wounds,[41–45] and scaffolds for tissue engineering.[46–50] Polymers including poly(lactic acid),[51–54] poly(lactic-*co*-glycolic acid),[55–60] poly(vinyl alcohol),[61,62] poly(vinylpyrrolidone),[63,64] polycaprolactone,[65–67] polyurethane,[68,69] and poly(3-hydroxybutyrate)[70–72] are classic materials that are electrospun into scaffolds for diverse biomedical applications. It is now feasible to encapsulate cells, growth factors or drugs into these polymers and electrospin the resulting composite into nano-scale scaffolds for tissue engineering. Recently, Ramakrishna

Table 1.1 Selected examples of functional applications of electrospun materials.

Electrospun material	Precursors	Applications	Comment
SiO_2 nanofiber	Polyvinylpyrrolidone, tetraethyl orthosilicate, and nickel(II) nitrate hexahydrate	Catalysis	Encapsulated Ni catalyst promotes CO_2 re-forming of methane[79]
Carbon nanofiber	Polyacrylonitrile, multi-walled carbon nanotube, and palladium acetylacetone	Catalysis	Encapsulated Pd catalyst promotes nitrite reduction[80]
NiO nanofiber	Poly(vinyl alcohol) and nickel(II) acetate	Sensing	Nanofibers sense ammonia and hydrogen gases[81]
Poly(vinyl fluoride)/graphene nanofiber	Poly(vinylidene fluoride), graphene, and cerium sulfate dihydrate	Sensing	Piezoelectric property of nanofiber enables sensing of pressure[82]
Polyacrylonitrile/halloysite nanofibrous membrane	Polyacrylonitrile, and halloysite	Water purification	The nanofibrous membrane separates 99.5% oil from and adsorbs heavy metal ions from water[83]
Ethyl cellulose nanofiber	Ethyl cellulose and ketoprofen	Drug release	Nanofibers enable zero-order release of ketoprofen over 20 hours[84]
Chitosan/phospholipids nanofiber	Chitosan, and azolectin with vitamin B12, curcumin, or diclofenac	Drug release	Nanofiber ensures release of vitamin B12, curcumin, or diclofenac[85]
Polycaprolactone/gelatin nanofiber	Polycaprolactone, gelatin, and 6-aminopenicillanic acid-coated gold nanoparticles	Wound healing	Nanofibers assist healing of multidrug-resistant bacterial infected wounds[86]
Poly(vinylidene fluoride-trifluoroethylene) nanofiber	Poly(vinylidene fluoride-trifluoroethylene)	Energy harvesting	The dimension of the fiber tunes the electromechanical property with smaller fibers yielding greater electrical output[87]
Poly(vinylidene fluoride) nanofiber	Poly(vinylidene fluoride)	Energy harvesting	The fiber converts mechanical to electrical energy, sufficient to power Peltier cooler[88,89]

demonstrated this concept, electrospinning a polycaprolactone-based core–shell nanofiber scaffold that encapsulates nerve growth factor, or bovine serum albumin for neural tissue engineering.[73] Ramakrishna and coworkers employed an emulsion electrospinning technique to design this random or aligned nanofibrous scaffold that guides the cultured cells to grow in a predetermined orientation. Release of the nerve growth factor from the scaffold enables rat pheochromocytoma cells to grow, differentiate, and express critical biomarkers. Characteristically, biodegradable polymers such as poly(lactic acid) and polycaprolactone are poor platform technologies for tissue engineering because they lack cell recognition sites, a property that limits cell affinity and adhesion. Recent innovative developments in electrospinning now endow these polymers with improved cell affinity and adhesion properties, positioning them as promising nanofibrous scaffolds for tissue engineering. An example is the enhanced mineralization, cell adhesion, and cell differentiation properties observed in poly(lactic acid) or polycaprolactone electrospun fibers when an ossein–hydroxyapatite complex, a natural bone extracellular matrix mimic, is incorporated into the polymer matrix before electrospinning the resulting composite.[74] It was shown that composites containing a bone extracellular matrix enhances the mechanical properties, for instance, the initial modulus and breaking stress, of the electrospun nanofiber.[75,76] A typical example is the incorporation of hydroxyapatite into tussah silk fibroin to form a composite, which is spun into nanofibers with improved initial modulus and breaking stress compared to neat tussah silk fiber.[76] Also, the nanofibers from the composite are more biocompatible towards osteoblast-like MG-63, and are more efficient in promoting cell adhesion, proliferation, and biomineralization than natural tussah silk fiber.[75] Also, it is now possible to tune critical properties such as morphology, Young's modulus and tensile strength *via* developing composites of polymers and other materials such as ceramics as precursors to electrospun biomaterial scaffolds. A recent report proved that ultrafine nanofiber spun from a composite of poly(L-lactic-*co*-glycolic acid), 10% tussah silk, and 1% graphene oxide has a smaller diameter and higher Young's modulus and tensile strength than that from poly(L-lactic-*co*-glycolic acid).[75] Also, the presence of the tussah silk and graphene oxide endows the scaffold with functionalities that promote mouse mesenchymal stem cell adhesion and proliferation as well as enabling biomineralization and biodeposition.

In the energy landscape, electrospinning is increasingly explored to fabricate materials that efficiently harness energy. At the front line of this enterprise is the exploration of the electrospinning process to fabricate electrodes for solar cells. As an example, Wali *et al.*[77] incorporated tin(IV) oxide particles – an n-type semiconductor – into poly(vinylpyrrolidone), forming a polymer–inorganic oxide nanocomposite. By electrospinning this nanocomposite, Jose and coworkers obtained one-dimensional multiporous nanofibers, porous nanofiber, or nanowires, which they fabricated into a photoanode for dye-sensitized solar cells. The electrospun multiporous nanofibers exhibit a larger surface area than the other nanostructured fibers, and its dye-sensitized solar cell features 80% incident photon-to-current conversion efficiency, which is 22% higher than that of tin(IV) oxide particles. Also, the electron lifetime and electron diffusion length of dye-sensitized solar cells fabricated from the multiporous nanofiber is three times higher than that from pristine tin(IV) oxide particles.

Electrospinning has grown into a versatile technique to fabricate a broad range of materials for different applications. Titanium(IV) oxide, a semiconductor, was recently incorporated into polyvinylpyrrolidone to form a sol, which was electrospun into one-dimensional nanowires.[78] The use of these nanowires as the photoanode in a dye-sensitized solar cell resulted in increased cell performance compared to titanium(IV) oxide particles (P25). Specifically, the short-circuit current density (J_{sc}) and efficiency (η) are higher in the dye-sensitized solar cell fabricated with the nanowire (J_{sc} = 8.57 mA cm^{-2} and η = 3.45%) than that with P25 (J_{sc} = 6.76 mA cm^{-2} and η = 2.88%).[77] Another recent demonstration of the versatility of electrospinning is the incorporation of the semiconductor, kesterite (Cu_2ZnSnS_4), into polyvinylpyrrolidone or cellulose acetate to form a polymer–inorganic composite that is electrospun into nanofibers and used as counter electrodes in dye-sensitized solar cells.[78] Of fundamental importance is the effect of the polymer on the morphology of the electrospun fiber. Whereas poly(vinylpyrrolidone)-based fibers are smooth, single crystalline materials with diameters in the range of 100–150 nm, the cellulose acetate-based fibers are polycrystalline with diameters of 10–20 nm. Compared to a conventional platinum/fluorine-doped tin oxide counter electrode with power conversion efficiency (η) of 1.72%, those derived from the poly(vinylpyrrolidone)-based and cellulose acetate-based fibers are more efficient, as revealed by their respective η values of 3.10 and 3.90%.

Electrospinning is a versatile approach to the fabrication of functional materials with broad-ranging implications for diverse human needs (Table 1.1). Many of these functional materials are being harnessed for their commercial value and several companies, including academic spin-offs and small, medium, and large companies across the world, are at the forefront in driving this translation.

1.4 Conclusion

This next phase of the electrospinning story is beyond the scope of this chapter, but it is hoped that reading this chapter will give an indication to the reader of the breadth and depth of the development of the process. The following chapters will provide an insight into current developments in diverse electrospinning applications and will provide an overview of expected future developments in industrial applications.

References

1. W. Gilbert, De MagneteMagnetcisqueCorporibus, et de MagnoMagnete-Tellure (On the Magnet, Magnetick Bodies also, and on the Great Magnet the Earth; a new Physiology, demonstrated by many arguments & Experiments), *Translated 1893 from Latin to English by Paul FleuryMottelay*, Dover Publications, New York, 1991.
2. P. Iversen and D. J. Lacks, A life of its own: The tenuous connection between Thales of Miletus and the study of electrostatic charging, *J. Electrost.*, 2012, **70**(3), 309–311.
3. G. Taylor, Disintegration of Water Drops in an Electric Field, *Proc. R. Soc. A*, 1964, **280**(1382), 383–397.
4. A. Nollet, *Recherches Sur Les Causes Particulieres des PhenomenesElectriques*, les Freres Guerin, Paris, 1749.
5. R. Hooke, *Micrographia - Some Physiological Descriptions of Minute Bodies Made by Magnifying Glasses with Observations and Inquiries Thereupon*, John Martin and James Allestry, printers to the Royal Society, London, 1665.
6. T. B. Greenslade Jr, Nineteenth Century Textbook Illustrations LV, Discovery of the Leiden Jar, *Phys. Teach.*, 1994, **32**, 536–537.
7. N. Tucker, J. Stanger, M. Staiger, H. Razzaq and K. Hofman, The history of the science and technology of electrospinning from 1600 to 1995, *J. Eng. Fiber. Fabr.*, 2012, **7**, 63–73.
8. J. Gordon-Cook, *Handbook of Textile Fibres*, Woodhead Publishing Ltd, Cambridge, 2003.
9. J. L. Parsons, Count Hilaire De Chardonnet, Scientist and Inventor, *Ind. Eng. Chem.*, 1925, **17**, 754–755.
10. M. G. Luft, Artificial Silk.-With Special Reference to the Viscose Process, *Ind. Eng. Chem.*, 1925, **17**, 1037–1042.
11. *Regenerated Cellulose Fibres*, ed. C. Woodings, Woodhead Publishing, 2001, vol. 18.
12. N. Tucker, The development of electrospinning technologies for commercial application, in *Electrospinning*, 2015, pp. 34–56.
13. J. F. Cooley, Improved methods of and apparatus for electrically separating the relatively volatile liquid component from the component of relatively fixed substances of composite fluids, UK, Patent No. 6385, 1900.
14. B. Z. Tang, *Electrospinning: Principles, Practice and Possibilities*, Royal Society of Chemistry, 2015.
15. A. Formhals, Production of Artificial Fibres, USA, Patent No. 2077373, 1937.
16. L. H. Zhang, X. P. Duan, X. Yan, M. Yu, X. Ning, Y. Zhao and Y. Z. Long, Recent advances in melt electrospinning, *RSC Adv.*, 2016, **6**(58), 53400–53414.
17. C. L. Norton, Method and apparatus for producing fibrous material, U.S. Patent 2,219,346, issued October 29, 1940.

18. Y. Filatov, A. Budyka and V. Kirichenko, *Electrospinning of Micro and Nanofibers: Fundamentals and Applications in Separation and Filtration Processes*, Begell House Inc, New York, USA, 2007.
19. M. Botes and T. Eugene Cloete, The potential of nanofibres and nanobiocides in water purification, *Crit. Rev. Microbiol.*, 2010, **36**(1), 68–81.
20. C. J. Luo, S. D. Stoyanov, E. Stride, E. Pelanb and M. Edirisinghe, *Chem. Soc. Rev.*, 2012, **41**, 4708.
21. F. W. Manning, Method for the production of filaments and fabrics from fluids, U.S. Patent 2,433,000, 1947.
22. H. R. Childs, Process of electrostatic spinning, USA, Patent No. 2338570, 1944.
23. W. C. Huebner, Method and means of producing threads or filaments electrically, USA, Patent No. 2636216, 1953.
24. H. L. Simons, Process and apparatus for producing patterned non-woven fabrics, USA, Patent No. 3280229, 1966.
25. J. E. Owens, and S. P. Scheinberg, Apparatus for collecting fibrous material in sheet form, USA, Patent No. 3490115, 1970.
26. H. J. Hollberg and J. E. Owens, Process for forming a non-woven web, USA, Patent No. 3689608, 1972.
27. W. Simm, Apparatus for the production of filters by electrostatic fiber spinning, USA, Patent No. 3994258, 1976.
28. G. E. Martin, I. D. Cockshott and F. J. T. Fildes, Fibrillar lining for prosthetic device, USA, Patent No. 4044404, 1977.
29. A. Bornat, Electrostatic spinning of tubular products, USA, Patent No. 4323525, 1982.
30. C. Guignard, Process for the manufacture of a plurality of filaments, USA, Patent No. 4230650, 1980.
31. B. Kliemann, and M. Stoll, Process and apparatus for the preparation of electret filaments, textile fibres and similar articles, USA, Patent No. 4486365, 1984.
32. T. V. How, Synthetic vascular grafts, and methods of manufacturing such grafts, USA, Patent No. 4552707, 1985.
33. A. Bornat, Production of electrostatically spun products, USA, Patent No. 4689186, 1987.
34. J. P. Berry, Electrostatically produced structures and methods of manufacturing, USA, Patent No. 4965110, 1990.
35. J. Doshi and D. H. Reneker, Electrospinning process and applications of electrospun fibers, *J. Electrost.*, 1995, **35**, 9.
36. R. S. Barhate and S. Ramakrishna, Nanofibrous filtering media, Filtration problems and solutions from tiny materials, *J. Membr. Sci.*, 2007, **296**, 1–8.
37. T. J. Sill and H. A. von Recum, Electrospinning: applications in drug delivery and tissue engineering, *Biomaterials*, 2008, **29**, 1989–2006.
38. J. Zeng, X. Xu, X. Chen, Q. Liang, X. Bian, L. Yang and X. Jing, Biodegradable electrospun fibers for drug delivery, *J. Controlled Release*, 2003, **92**, 227–231.

39. H. S. Yoo, T. G. Kim and T. G. Park, Surface-functionalized electrospun nanofibers for tissue engineering and drug delivery, *Adv. Drug Delivery Rev.*, 2009, **61**, 1033–1042.
40. X. Hu, S. Liu, G. Zhou, Y. Huang, Z. Xie and X. Jing, Electrospinning of polymeric nanofibers for drug delivery applications, *J. Controlled Release*, 2014, **185**, 12–21.
41. R. Sridhar, R. Lakshminarayanan, K. Madhaiyan, V. A. Barathi, K. H. C. Lim and S. Ramakrishna, Electrosprayed nanoparticles and electrospun nanofibers based on natural materials: applications in tissue regeneration, drug delivery and pharmaceuticals, *Chem. Soc. Rev.*, 2015, **44**, 790–814.
42. H. Lai, C. Kuan, H. Wu, J. Tsai, T. Chen, D. Hsieh and T. Wang, Tailored design of electrospun composite nanofibers with staged release of multiple angiogenic growth factors for chronic wound healing, *Acta Biomater.*, 2014, **10**, 4156–4166.
43. K. A. Rieger, N. P. Birch and J. D. Schiffman, Designing electrospun nanofiber mats to promote wound healing–a review, *J. Mater. Chem. B*, 2013, **1**, 4531–4541.
44. A. Lowe, J. Bills, R. Verma, L. Lavery, K. Davis and K. J. Balkus, Electrospun nitric oxide releasing bandage with enhanced wound healing, *Acta Biomater.*, 2015, **13**, 121–130.
45. R. Augustine, E. A. Dominic, I. Reju, B. Kaimal, N. Kalarikkal and S. Thomas, Electrospun polycaprolactone membranes incorporated with ZnO nanoparticles as skin substitutes with enhanced fibroblast proliferation and wound healing, *RSC Adv.*, 2014, **4**, 24777–24785.
46. K. Ghosal, S. Thomas, N. Kalarikkal and A. Gnanamani, Collagen coated electrospun polycaprolactone (PCL) with titanium dioxide (TiO2) from an environmentally benign solvent: preliminary physico-chemical studies for skin substitute, *J. Polym. Res.*, 2014, **21**(5), 410.
47. K. Ghosal, M. S. Latha and S. Thomas, Poly (ester amides)(PEAs)–Scaffold for tissue engineering applications, *Eur. Polym. J.*, 2014, **60**, 58–68.
48. K. Ghosal, A. Manakhov, L. Zajíčková and S. Thomas, Structural and surface compatibility study of modified electrospun poly (ε-caprolactone) (pcl) composites for skin tissue engineering, *AAPS PharmSciTech*, 2017, **18**(1), 72–81.
49. S. Baiguera, C. Del Gaudio, E. Lucatelli, E. Kuevda, M. Boieri, B. Mazzanti, A. Bianco and P. Macchiarini, Electrospun gelatin scaffolds incorporating rat decellularized brain extracellular matrix for neural tissue engineering, *Biomaterials*, 2014, **35**, 1205–1214.
50. H. G. Ş. Ayaz, A. Perets, H. Ayaz, K. D. Gilroy, M. Govindaraj, D. Brookstein and P. I. Lelkes, Textile-templated electrospun anisotropic scaffolds for regenerative cardiac tissue engineering, *Biomaterials*, 2014, **35**, 8540–8552.
51. J. W. Reboredo, T. Weigel, A. Steinert, L. Rackwitz, M. Rudert and H. Walles, Cartilage Tissue Engineering: Investigation of Migration and Differentiation of Human Mesenchymal Stem Cells on Five-Layered Collagenous Electrospun Scaffold Mimicking Native Cartilage Structure, *Adv. Healthcare Mater.*, 2016, **5**, 2304.

52. T. Valente, D. M. Silva, P. S. Gomes, M. H. Fernandes, J. D. Santos and V. Sencadas, Effect of sterilization methods on electrospun poly (lactic acid) (PLA) fiber alignment for biomedical applications, *ACS Appl. Mater. Interfaces*, 2016, **8**, 3241–3249.
53. G. Narayanan, V. N. Vernekar, E. L. Kuyinu and C. T. Laurencin, Poly (lactic acid)-based biomaterials for orthopaedic regenerative engineering, *Adv. Drug Delivery Rev.*, 2016, **107**, 247–276.
54. T. V. Toniatto, B. Rodrigues, T. Marsi, R. Ricci, F. R. Marciano, T. J. Webster and A. O. Lobo, Nanostructured poly(lactic acid) electrospun fiber with high loadings of TiO_2 nanoparticles: Insights into bactericidal activity and cell viability, *Mater. Sci. Eng., C*, 2017, **71**, 381–385.
55. P. I. Siafaka, P. Barmbalexis and D. N. Bikiaris, Novel electrospun nanofibrous matrices prepared from poly(lactic acid)/poly(butylene adipate) blends for controlled release formulations of an anti-rheumatoid agent, *Eur. J. Pharm. Sci.*, 2016, **88**, 12–25.
56. D. R. Marques, L. A. dos Santos, M. A. O'Brien, S. H. Cartmell and J. E. Gough, In vitro evaluation of poly(lactic-co-glycolic acid)/polyisoprene fibers for soft tissue engineering, *J. Biomed. Mater. Res., Part B*, 2017, **105**, 2581–2591.
57. X. Liu, J. Aho, S. Baldursdottir, A. Bohr, H. Qu, L. P. Christensen, J. Rantanen and M. Yang, The effect of poly(lactic-co-glycolic) acid composition on the mechanical properties of electrospun fibrous mats, *Int. J. Pharm.*, 2017, **529**, 371–380.
58. W. Di, R. S. Czarny, N. A. Fletcher, M. D. Krebs and H. A. Clark, Comparative study of poly(ε-Caprolactone) and poly(lactic-co-glycolic acid)-based nanofiber scaffolds for pH-sensing, *Pharm. Res.*, 2016, **33**, 2433–2444.
59. X. Liu, L. H. Nielsen, S. N. Kłodzińska, H. M. Nielsen, H. Qu, L. P. Christensen, J. Rantanen and M. Yang, Ciprofloxacin-loaded sodium alginate/poly (lactic-co-glycolic acid) electrospun fibrous mats for wound healing, *Eur. J. Pharm. Biopharm.*, 2018, **123**, 42–49.
60. S. E. Aniagyei, L. B. Sims, D. A. Malik, K. M. Tyo, K. C. Curry, W. Kim, D. A. Hodge, J. Duan and J. M. Steinbach-Rankins, Evaluation of poly(lactic-co-glycolic acid) and poly(D,L-lactide-*co*-ε-caprolactone) electrospun fibers for the treatment of HSV-2 infection, *Mater. Sci. Eng., C*, 2017, **72**, 238–251.
61. Z. Gu, H. Yin, J. Wang, L. Ma, Y. Morsi and X. Mo, Fabrication and Characterization of TGF-β1-loaded Electrospun Poly (lactic-co-glycolic acid) Core-sheath Sutures, *Colloids Surf., B*, 2018, **161**, 331–338.
62. S. S. Ray, S. Chen, N. C. Nguyen, H. Hsu, H. T. Nguyen and C. Chang, Poly(vinyl alcohol) incorporated with surfactant based electrospun nanofibrous layer onto polypropylene mat for improved desalination by using membrane distillation, *Desalination*, 2017, **414**, 18–27.
63. A. Shirole, J. Sapkota, E. J. Foster and C. Weder, Shape memory composites based on electrospun poly(vinyl alcohol) fibers and a thermoplastic

polyether block amide elastomer, *ACS Appl. Mater. Interfaces*, 2016, **8**, 6701–6708.
64. S. Montolio, G. Abarca, R. Porcar, J. Dupont, M. I. Burguete, E. Garca-Verdugo and S. V. Luis, Hierarchically structured polymeric ionic liquids and polyvinylpyrrolidone mat-fibers fabricated by electrospinning, *J. Mater. Chem. A*, 2017, **5**, 9733–9744.
65. P. B. Tsekova, M. G. Spasova, N. E. Manolova, N. D. Markova and I. B. Rashkov, Electrospun curcumin-loaded cellulose acetate/polyvinylpyrrolidone fibrous materials with complex architecture and antibacterial activity, *Mater. Sci. Eng., C*, 2017, **73**, 206–214.
66. K. Sun, Z. Liu, C. Liu, T. Yu, M. Zhou, C. Liu, F. Ran, L. Pan and H. Zhang, In vivo study of alginate hydrogel conglutinating cells to polycaprolactone vascular scaffolds fabricated by electrospinning, *J. Biomed. Mater. Res., Part B*, 2017, **105**, 2443–2454.
67. S. P. Miguel, M. P. Ribeiro, P. Coutinho and I. J. Correia, Electrospun Polycaprolactone/Aloe Vera_Chitosan Nanofibrous Asymmetric Membranes Aimed for Wound Healing Applications, *Polymers*, 2017, **9**, 183.
68. S. N. Gorodzha, A. R. Muslimov, D. S. Syromotina, A. S. Timin, N. Y. Tcvetkov, K. V. Lepik, A. V. Petrova, M. A. Surmeneva, D. A. Gorin and G. B. Sukhorukov, A comparison study between electrospun polycaprolactone and piezoelectric poly(3-hydroxybutyrate-co-3-hydroxyvalerate) scaffolds for bone tissue engineering, *Colloids Surf., B*, 2017, **160**, 48–59.
69. M. E. Wright, I. C. Parrag, M. Yang and J. P. Santerre, Electrospun polyurethane nanofiber scaffolds with ciprofloxacin oligomer *versus* free ciprofloxacin: Effect on drug release and cell attachment, *J. Controlled Release*, 2017, **250**, 107–115.
70. Y. Wang, P. Li, P. Xiang, J. Lu, J. Yuan and J. Shen, Electrospun polyurethane/keratin/AgNP biocomposite mats for biocompatible and antibacterial wound dressings, *J. Mater. Chem. B*, 2016, **4**, 635–648.
71. W. Chen, C. Chen, H. Tseng, Y. Liu, Y. Chen, C. Lee, Y. Kuo, C. Hsu and Y. Sun, Surface functionalized electrospun fibrous poly (3-hydroxybutyrate) membranes and sleeves: a novel approach for fixation in anterior cruciate ligament reconstruction, *J. Mater. Chem. B*, 2017, **5**, 553–564.
72. L. Gong, D. B. Chase, I. Noda, C. A. Marcott, J. Liu, D. C. Martin, C. Ni and J. F. Rabolt, Polymorphic Distribution in Individual Electrospun Poly[(R)-3-hydroxybutyrate-co-(R)-3-hydroxyhexanoate](PHBHx) Nanofibers, *Macromolecules*, 2017, **50**, 5510–5517.
73. J. Hu, L. Tian, M. P. Prabhakaran, X. Ding and S. Ramakrishna, Fabrication of nerve growth factor encapsulated aligned poly (ε-caprolactone) nanofibers and their assessment as a potential neural tissue engineering scaffold, *Polymers*, 2016, **8**, 54.
74. I. Rajzer, E. Menaszek and O. Castano, Electrospun polymer scaffolds modified with drugs for tissue engineering, *Mater. Sci. Eng., C*, 2017, **77**, 493–499.

75. W. Shao, J. He, F. Sang, Q. Wang, L. Chen, S. Cui and B. Ding, Enhanced bone formation in electrospun poly (L-lactic-co-glycolic acid)–tussah silk fibroin ultrafine nanofiber scaffolds incorporated with graphene oxide, *Mater. Sci. Eng., C*, 2016, **62**, 823–834.
76. W. Shao, J. He, F. Sang, B. Ding, L. Chen, S. Cui, K. Li, Q. Han and W. Tan, Coaxial electrospun aligned tussah silk fibroin nanostructured fiber scaffolds embedded with hydroxyapatite–tussah silk fibroin nanoparticles for bone tissue engineering, *Mater. Sci. Eng., C*, 2016, **58**, 342–351.
77. Q. Wali, A. Fakharuddin, I. Ahmed, M. H. Ab Rahim, J. Ismail and R. Jose, Multiporous nanofibers of SnO 2 by electrospinning for high efficiency dye-sensitized solar cells, *J. Mater. Chem. A*, 2014, **2**, 17427–17434.
78. C. Brundha, R. Govindaraj, N. Santhosh, M. S. Pandian, P. Ramasamy and S. Karuppuchamy, Preparation of one dimensional titanium dioxide nanowires using electrospinning process for dye-sensitized solar cells, *J. Mater. Sci.: Mater. Electron.*, 2017, **28**, 1–6.
79. S. S. Mali, P. S. Patil and C. K. Hong, Low-cost electrospun highly crystalline kesterite Cu2ZnSnS4 nanofiber counter electrodes for efficient dye-sensitized solar cells, *ACS Appl. Mater. Interfaces*, 2014, **6**, 1688–1696.
80. S. Wen, M. Liang, J. Zou, S. Wang, X. Zhu, L. Liu and Z. Wang, Synthesis of a SiO 2 nanofibre confined Ni catalyst by electrospinning for the CO 2 reforming of methane, *J. Mater. Chem. A*, 2015, **3**, 13299–13307.
81. T. Ye, D. P. Durkin, M. Hu, X. Wang, N. A. Banek, M. J. Wagner and D. Shuai, Enhancement of nitrite reduction kinetics on electrospun Pd-carbon nanomaterial catalysts for water purification, *ACS Appl. Mater. Interfaces*, 2016, **8**, 17739–17744.
82. A. Khalil, J. J. Kim, H. L. Tuller, G. C. Rutledge and R. Hashaikeh, Gas sensing behavior of electrospun nickel oxide nanofibers: effect of morphology and microstructure, *Sens. Actuators, B*, 2016, **227**, 54–64.
83. S. Garain, S. Jana, T. K. Sinha and D. Mandal, Design of *in situ* poled Ce3 -Doped electrospun PVDF/Graphene composite nanofibers for fabrication of nanopressure sensor and ultrasensitive acoustic nanogenerator, *ACS Appl. Mater. Interfaces*, 2016, **8**, 4532–4540.
84. M. Makaremi, R. T. De Silva and P. Pasbakhsh, Electrospun nanofibrous membranes of polyacrylonitrile/halloysite with superior water filtration ability, *J. Phys. Chem. C*, 2015, **119**, 7949–7958.
85. D. Yu, X. Li, X. Wang, J. Yang, S. A. Bligh and G. R. Williams, Nanofibers fabricated using triaxial electrospinning as zero order drug delivery systems, *ACS Appl. Mater. Interfaces*, 2015, **7**, 18891–18897.
86. A. C. Mendes, C. Gorzelanny, N. Halter, S. W. Schneider and I. S. Chronakis, Hybrid electrospun chitosan-phospholipids nanofibers for transdermal drug delivery, *Int. J. Pharm.*, 2016, **510**, 48–56.
87. X. Yang, J. Yang, L. Wang, B. Ran, Y. Jia, L. Zhang, G. Yang, H. Shao and X. Jiang, Pharmaceutical Intermediate-Modified Gold Nanoparticles: Against Multidrug-Resistant Bacteria and Wound-Healing Application *via* Electrospun Scaffold, *ACS Nano*, 2017, **11**, 5737–5745.

88. G. Ico, A. Showalter, W. Bosze, S. C. Gott, B. S. Kim, M. P. Rao, N. V. Myung and J. Nam, Size-dependent piezoelectric and mechanical properties of electrospun P (VDF-TrFE) nanofibers for enhanced energy harvesting, *J. Mater. Chem. A*, 2016, **4**, 2293–2304.
89. J. Fang, H. Niu, H. Wang, X. Wang and T. Lin, Enhanced mechanical energy harvesting using needleless electrospun poly (vinylidene fluoride) nanofibre webs, *Energy Environ. Sci.*, 2013, **6**, 2196–2202.

CHAPTER 2

Encroachment of Traditional Electrospinning

I. SAVVA* AND T. KRASIA-CHRISTOFOROU*

University of Cyprus, Department of Mechanical and Manufacturing Engineering, 75 Kallipoleos Avenue, PO Box 20537, 1678 Nicosia, Cyprus
*E-mail: joan.savva@gmail.com, krasia@ucy.ac.cy

2.1 Introduction

During the last two decades there has been tremendous interest in the advancement of electrospun nanofibers. These 1D nanomaterials show great promise in the development of new products applicable in various fields such as biomedicine, catalysis, energy, the environment, sensing, *etc.* This is attributed to their unique properties (including their extremely long length, high surface-to-volume ratio, ease of functionalization and versatility in chemical composition), the possibility of generating nanofibers exhibiting specific orientation, porosity, and topography, their tunability in terms of fiber morphology (core–shell, hollow, helical, necklace-like, *etc.*) and most importantly the high potential for their mass production. Furthermore, the flexibility of electrospinning in the processing of solutions, melts, emulsions, and suspensions has increased the number of nanofibrous products.

Traditional electrospinning uses a single-nozzle spinneret where the produced fibers are deposited on a flat collector at a very low production rate (less than 0.3 g h^{-1}) in a random orientation. Encroachment of traditional electrospinning is being realized *via* the synergistic contributions of academia and

Soft Matter Series No. 7
Electrospinning: From Basic Research to Commercialization
Edited by Erich Kny, Kajal Ghosal and Sabu Thomas

Published by the Royal Society of Chemistry, www.rsc.org

Table 2.1 Names and websites of representative companies involved in nanofiber processing using different electrospinning processes.

Company name	Electrospinning process	Website
DIENES Apparatebau GmbH	Nozzle centrifugal electrospinning	http://www.dienes.net
Donaldson	Electrospinning	https://www.donaldson.com/en-be
Elmarco	Nozzle-less electrospinning	http://www.elmarco.com
Bioinicia	Needle-based electrospinning	http://bioinicia.com
Revolution fibres	Sonic electrospinning	https://www.revolutionfibres.com
Arsenal Medical	Slit-surface electrospinning	http://www.arsenalmedical.com

the industrial sectors. This synergy has led to significant advancements in the electrospinning process, with some of these further translated to industry. Recognition of the increasing interest in nanofibers, reflected by the continuously increasing value of the nanofiber market ($43 million in 2006; $101 million in 2010; projected to reach $2.2 billion by 2020)[1,2] has led to an increase in entrepreneurial activities and new investments from several companies. To date, there are companies worldwide that supply laboratory and large-scale electrospinning equipment, whereas some are actively involved in the production and trading of electrospun nanofibrous products.

Table 2.1 provides a list of representative companies involved in nanofiber processing using electrospinning.[3,4]

As seen in Table 2.1, most companies use non-conventional electrospinning processes including free-surface, roller, sonic and slit-surface electrospinning. This, together with the fact that industrial-scale electrospinning exhibits significant differences compared to the laboratory-scale traditional electrospinning set-ups, such as much higher applied voltages (30–120 kV *vs* 5–20 kV), longer distances (20–50 cm *vs* 10–15 cm) and much higher production rates, emphasize the necessity of adapting the academic research outcomes to industrial-scale electrospinning requirements.

Although the commercially available electrospun products developed so far are mainly used in filtration and purification processes, it is expected that more products applicable to various fields including energy (photovoltaics, batteries, *etc.*), automotive, electronics, pharmaceutics, sensing and so on, will appear in the near future. Through the effective communication technology and knowledge transfer between academia and industry, existing industrial problems will be overcome. These include, among others, the large-scale and controlled production of high-quality nanofibers, the decrease in sample loss and the possibility of producing nanofibers with controlled orientation. In line with the latter, the design and construction of new modified electrospinning set-ups enabled the production of aligned electrospun nanofibrous mats and patterned structures, as well as 3D nanofibrous architectures. The adaptation of these new controlled fiber deposition concepts on an industrial scale in the future will certainly create new prospects in the production and commercialization of advanced nanofibrous products.

This chapter focuses on the main directions of modification/innovation of the traditional electrospinning equipment in order to: (a) increase productivity, and (b) control the orientation, deposition, and morphology of the electrospun nanofibers. The first part (Section 2.2) describes the methodologies developed for targeting the translation of electrospinning from laboratory-scale to commercialization, whereas Section 2.3 provides an overview of the electrospinning advancements that have promoted controlled nanofiber alignment/deposition, the generation of fibers with unusual morphologies and the fabrication of 3D nanofibrous electrospun materials.

2.2 Encroachment of Traditional Electrospinning Towards Commercialization

Electrospinning is a well-known and highly versatile technique applied for the generation of ultrafine continuous micro- and nanofibers from a wide variety of materials.[5] In conventional electrospinning, a polymer solution or melt is placed in a syringe and subjected to a very high electrostatic force, resulting in the rapid ejection of the solution/melt from a nozzle or a spinneret, and deposition of the produced fibers on an electrically grounded collector.[6]

A simple electrospinning set-up consists of four major components: a high-voltage power supply; a syringe with a metallic needle, known as a spinneret; a syringe pump used for delivering the solution or melt through the spinneret at a constant and controllable rate; and a grounded conductive collector, on which the produced fibers are deposited.

Fibrous materials deriving from electrospinning are desirable due to the fact that the diameters of the produced fibers are consistent and controllable, the fiber surfaces are defect-free, and continuous single fibers are obtained. However, research carried out over the years has shown that the above-mentioned targets are by no means easily achievable.[7,8] The success of the whole process is governed by different parameters which are classified in terms of the solution properties,[7,9–11] process,[7,12] and ambient parameters.[9,11–13] Consequently, dissimilar results may be obtained using the same material and electrospinning set-up. Thus, it is difficult to provide quantitative relationships that can be applied across a broad range of solution systems. However, there are general trends which are useful when determining the optimum conditions for a certain system.

After vital research on the basics of the electrospinning process, researchers investigated the relationship between processing and solution parameters and the obtained fiber morphology. The major challenge of the electrospinning process lies in the optimization of all the above-mentioned parameters to achieve nanofibers with desirable morphological characteristics and thus properties.[14]

Although electrospinning has a wide variety of applications, the limited throughput was a major issue in applying the technique to large-scale production. Currently, researchers focus on the enhancement of the electrospinning production rate on an industrial level, developing and exploring several methods for large-scale production.[15] Different set-ups have been developed, including multi-needle, needleless, and gas-assisted systems, providing significant improvements in the mass production of electrospun nanofibers. These set-ups are classified according to the fluid-feed mode. In the confined set-ups, fluid ejection takes place through a confined capillary such as a needle, a porous tube, a plastic tip, *etc.*[16] The use of unconfined systems that enable the production of multiple jets in the absence of capillaries allows electrospinning to occur from a free liquid surface. Although the confined electrospinning systems provide better control over the nanofiber diameters and diameter distributions through the restricted flow rate, they suffer from clogging phenomena and hence incur high maintenance costs.

A multi-jet technology was developed, aiming to increase nanofiber production, based on the increase of the jets in different structures, giving the opportunity to scale up and increase the production of electrospun nanofibers.[17] Although this was an innovative idea, the process could not be efficiently scaled up. Many jets became inactive due to the strong repulsion from adjacent jets. This led to the low productivity of non-uniform nanofibers. In order to overcome this problem, the jets were set at an appropriate distance, but this led to a requirement for large spacing and an often structurally complex arrangement. In addition, as the number of nozzles increases the whole process becomes more expensive.[18–23] Another drawback applied in the case of the needle-based multi-jet approach, which is the need to use a cleaning device to prevent clogging of the electrospinning nozzles. Moreover, the influence of the electric field around each needle requires the use of a higher voltage to induce jet initiation. All the above limitations have prompted many researchers to find alternative multi-jet electrospinning solutions.[24,25]

Another pathway followed for higher throughput nanofiber production compared to a single-nozzle set-up was gas-assisted electrospinning, where an air jacket is used to improve the solution flow rate. However, this method had a negative effect on nanofiber morphology and fineness.[20]

The free liquid surface nanofiber production method has gained considerable attention since its first inception. The nozzle-less electrospinning set-up solves most of the aforementioned problems and it is the most prevalent technique used for large-scale nanofiber production. Needle-free electrospinning set-ups involving both rotating and stationary spinnerets produce nanofibers directly from the jets generated from the exposed surface area of the fluid.[19]

2.2.1 Syringes Connected in Series

One of many classifications of electrospinning set-ups is based on the number of nozzles used, *i.e.* single-nozzle or multi-nozzle. Multi-nozzles leading to a multi-jet set-up promotes a straightforward way to increase nanofiber

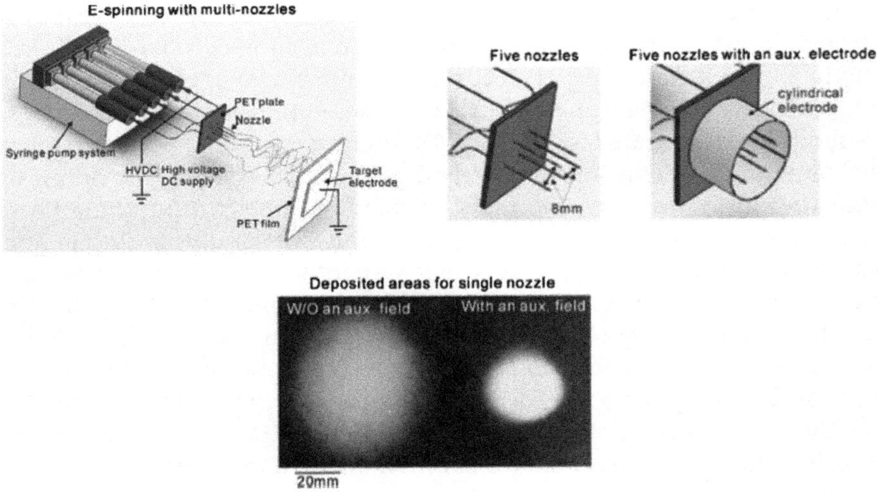

Figure 2.1 An experimental set-up for multi-jet electrospinning. Reprinted from *European Polymer Journal*, **42**, G. H. Kim, Y.-S. Cho, W. D. Kim, Stability analysis for multi-jets electrospinning process with a cylindrical electrode, 2031–2038, Copyright 2006, with permission from Elsevier.

production. As already mentioned, the increase in the number of nozzles constitutes the main problem, because the whole process becomes more difficult, less controllable and less efficient. Even though the use of multiple nozzles connected in series exhibits certain limitations, researchers have been focusing on the optimization of this approach. In order to overcome the repulsion phenomena existing between neighboring jets, in 2006 G. Kim *et al.*[26] used an extra-cylindrical electrode connected with a five-nozzle electrospinning system (Figure 2.1) aiming to increase the stability of the jets. Their results suggested that the introduction of a cylindrical electrode connected with multiple nozzles contributed to the generation of a stable jet-motion of the initial stream line and the nanofibers' jets, without interruptions from neighboring charged jets, the airflow, and interferences from near dielectric/conductive materials. Moreover, the authors introduced an electric field concentration factor in order to predict the stability of the electrospun jets.

The performance of multi-jet electrospinning systems was also examined by A. Varesano and his coworkers.[27] They carried out a series of experiments by changing various factors, including the process parameters, the jet–jet interaction (the number of nozzles was varied between 2 and 16), and the size of the deposition area.

In 2004, B. Ding and his team used a multi-jet electrospinning set-up for the fabrication of multi-component biodegradable nanofibrous mats consisting of polyvinyl alcohol (PVA) and cellulose acetate (CA). In order to obtain a uniform textile, a movable multi-jet and rotatable-grounded tubular collector were employed (Figure 2.2),[28] thus allowing for the alternative and

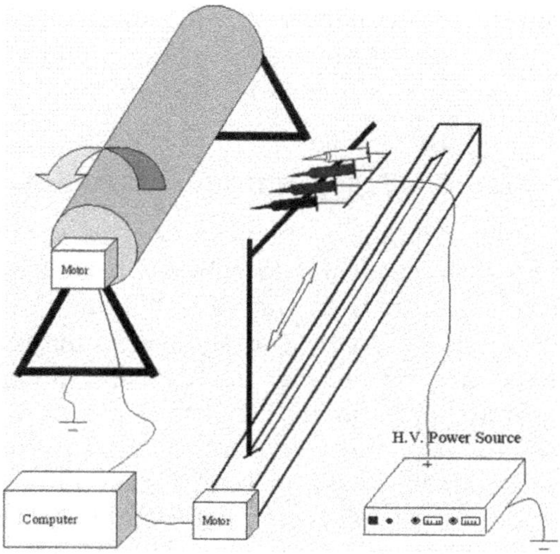

Figure 2.2 Schematic of the electrospinning process. Reprinted from *Polymer*, **45**, B. Ding, E. Kimura, T. Sato, S. Fujita, S. Shiratori, Fabrication of blend biodegradable nanofibrous nonwoven mats *via* multi-jet electrospinning, 1895–1902, Copyright 2004, with permission from Elsevier.

continuous deposition of the PVA and CA nanofibers on the collector and their homogeneous dispersion into each other. The uniform intermixing of the two fiber types allowed for the modulation of the mechanical properties of the resulting blended nanofibrous mats. More precisely it was found that the mechanical properties of the resulting mats were improved by increasing the PVA content.

In 2005, W. Tomaszewski and M. Szadkowski demonstrated high-throughput nanofiber generation by using three different types of electrospinning heads: series, elliptic, and concentric. Their study proved that the concentric electrospinning head was more efficient compared to the other two types, while in the linear-nozzle arrangement the central jets were inactive, leading to lower production efficiency.[29] In order to further improve the above set-up, in 2010 A. Varesano and coworkers introduced a nine-nozzle arrangement. Their results showed that the insertion of a non-conductive textile between the spinneret and the collector influences the electrospinning process.[30]

As previously mentioned, the number and structure of the nozzles used in multi-nozzle electrospinning are limiting factors. However, in 2015 I. G. Kim *et al.*[31] demonstrated the mass production of nanofibrous mats using a cylindrical-type multi-nozzle system. The electric field distribution was analyzed upward of the spinneret (distance: 1 m), showing the high electric field intensity at the nozzle tip, as well as the influence of the nozzle-collector distance.

Issues with the multi-nozzle electrospinning set-up include the influence of the electric field distribution on the electrospinning process and the morphology of the generated nanofibers. Thus, many studies have focused on the effects of the applied electric field. C. J. Angammana compared a single-needle system with a multi-jet set-up, showing the weakness of the electric field in the case of the multi-jet arrangement.[23] Similar results were also found by S. Xie and Y. Zeng through experimental and simulation studies.[32]

2.2.2 Gas-assisted Electrospinning/Electroblowing

Gas-assisted spinning technology (also known as spin-blowing), which combines spinning with high-velocity air flow, is well established in the industrial production of ultrafine nanofibers. Through the combination of gas-assisted spinning and electrospinning, a new method called 'electroblowing/gas-assisted electrospinning' has emerged for obtaining nanofibers from both solutions and melts.[33,34] In this process, a high-speed gas stream is coupled to the spinneret from which the electrified polymer solution or melt is ejected, resulting in the generation of ultrathin (less than 100 nm) nanofibers. This synergy between air flow and electrostatic forces enables the production of nanofibers from high-viscosity polymer melts and high surface tension polymer solutions, which are otherwise non-electrospinnable.

An inexpensive gas-assisted electrospinning method dealing with large-scale production was proposed by J. S. Varabhas *et al.*[24] A new device for launching multiple jets was constructed, enabling the simultaneous ejection of many electrospinning jets from the surface of a porous tube with the aid of low air pressure flow through the tube (Figure 2.3). By altering the tube length and the number of holes, the nanofiber production rate could be monitored.

Bubble electrospinning is another sub-category of gas-assisted electrospinning, involving the use of aerated solutions with compressed air in an electric field, which results in multiple jets rising from the free surface of the aerated polymer solution.[35–37]

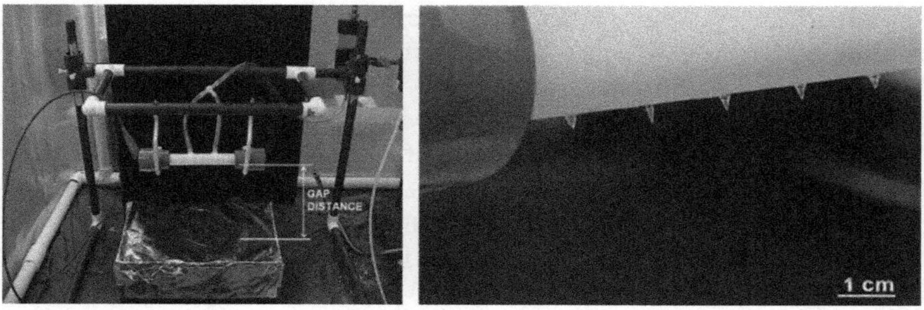

Figure 2.3 Left: Framework to hold the porous tube and aluminum foil collector. The gap distance is measured from the bottom of the white tube with end caps to the foil collector. Right: Close-up photo during electrospinning from the conical drops hanging at the locations of the drilled holes on the porous tube. Reprinted from *Polymer*, **49**, J. S. Varabhas, G. G. Chase, D. H. Reneker, Electrospun nanofibers from a porous hollow tube, 4226–4229, Copyright 2008, with permission from Elsevier.

In bubble electrospinning the size of the produced bubbles is interconnected to their surface tension. Therefore, upon decreasing the bubble size, a smaller electrostatic force is required for the production of nanofibers.[38] J. He and coworkers[39] reported the fabrication of polyvinyl alcohol (PVA) nanofibers from an aqueous solution with diameters within 100 nm, deriving from a single bubble generated above the surface of the polymer solution using a small tube (11.5 mm in diameter). In another example, a multi-jet electrospinning set-up was modified by applying a sheath-gas stream, in order to reduce the electrostatic repulsion effects appearing in such systems.[17] More precisely, the authors demonstrated that the deposition area of poly-L-lactic acid nanofibers produced by blow-assisted multi-jet electrospinning was significantly lower (~30–40%) compared to that obtained in the absence of compressed air, thus resulting in a decrease in product loss.

2.2.3 Needleless Electrospinning

As mentioned previously, in conventional electrospinning a metallic needle is used as the spinning nozzle, generating only one polymer jet, resulting in very limited fiber productivity. In order to avoid the use of capillaries and needles and their related limitations, needleless electrospinning devices have been widely used for the production of fine nanofibers on a much larger scale. This technology constitutes one extremely convenient pathway for high-throughput nanofiber production. However, one limitation of this method is the fact that the flow rate cannot be precisely controlled.[40]

In needleless electrospinning, nanofibers are generated from the open surface (planar or cylindrical) of a conductive liquid through the simultaneous formation of multiple jets. The spinneret may be either stationary or rotating. In the former, an external force is usually required to initiate the electrospinning process, such as a magnetic field, gravitational force or compressed air. In rotating spinnerets where the spinneret geometry may vary (cylinder, ball, disk, coil, beaded chain), mechanical vibration is exerted on the polymer solution.

In 2004, A. L. Yarin and E. Zussman[41] introduced a new approach enabling the mass production of nanofibers, based on the combination of magnetic and electric fields acting on a two-layer system. This mutual electromagnetic influence resulted in the induction of perturbations on the free surface and the generation of multiple jets. Since then, many researchers have focused on various types of needleless electrospinning spinnerets. Due to the fact that the electric field applied is an important influencing factor in this process,[42–44] significant research efforts focus on the generation of nanofibers with narrow diameter distributions by controlling the electric field strength and the spinneret's geometry.[45]

A patent based on needleless electrospinning came out in 2005 by O. Jirsak *et al.*,[46] involving the use of a rotating roller as a spinneret. X. Wang *et al.*[42] reported on the fabrication of polyvinyl alcohol (PVA) nanofibers using

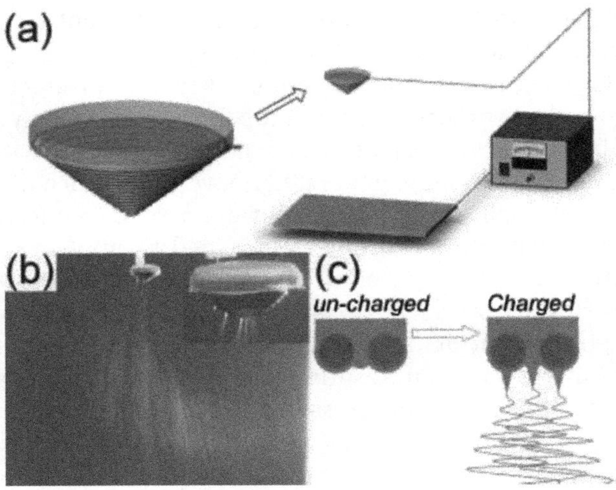

Figure 2.4 (a) Schematics of wire coil electrospinning set-up. (b) Photograph of the electrospinning process (inset: multiple jets). (c) Illustration of jet formation on the coil surface. Reproduced with permission from ref. 42, © 2009 Society of Plastics Engineers.

a cone-shaped metal wire coil as a spinneret (Figure 2.4). This spinneret enabled the processing of more viscous fluids, while a higher electric potential could be exerted on the outer surface of the cone. This novel concept resulted in nanofibers with smaller diameters and at the same time improved productivity compared with the conventional electrospinning set-up. The same group also examined improvements in electrospinning productivity, by using a rotating spiral wire coil as a spinneret (Figure 2.5). This innovative needleless electrospinning set-up, which led to a new patent,[47] exhibited a high fiber productivity rate upon increasing the coil length or coil diameter, or upon reducing the spiral distance or the wire's diameter.[22]

In 2010 B. Lu and coworkers contributed to the research relevant to the enhancement of nanofiber output, by using an electrified rotating cone as the spinneret. The nanofiber production rate reached approximately 10 g min^{-1}, which is about 1000 times higher compared to the rate corresponding to the single-nozzle electrospinning set-up. Furthermore, through their study the authors showed that the electrospinning rate did not affect fiber diameter, while the applied voltage was a key factor for the generation of thinner and uniform nanofibers.[48]

Another study focusing on mass nanofiber production was reported by H. Niu and coworkers in 2009, who used a rotating disk and a cylinder (Figure 2.6) as a spinneret in needleless electrospinning and examined the influence of the nozzle shape on fiber morphology. Comparing the two different nozzles, the disk required a relatively low applied voltage while the polymer concentration had a small influence on the process. However, in the case where the cylinder was used, a higher dependence on the applied voltage and polymer solution concentration was observed.[20]

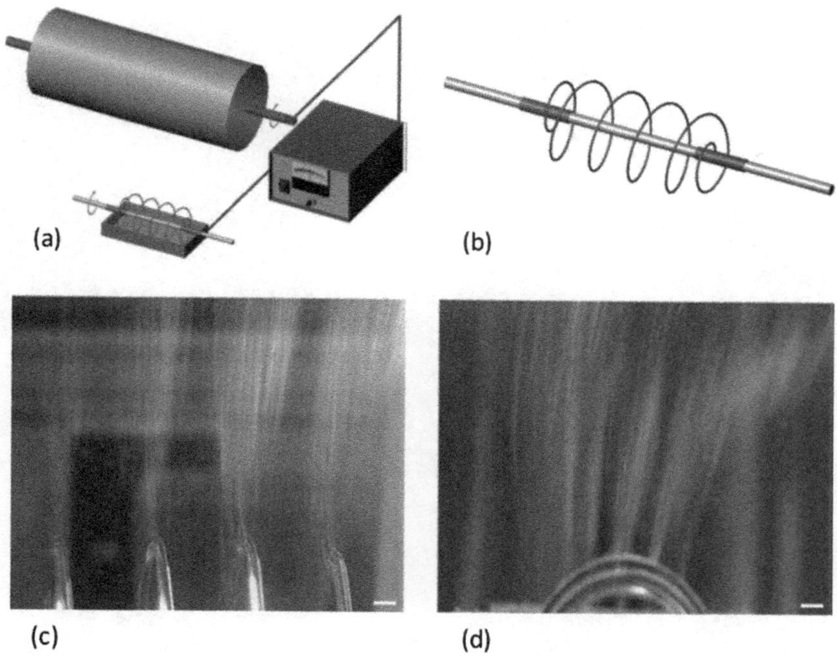

Figure 2.5 (a) Schematics of spiral coil electrospinning set-up; (b) magnified view of the coil; photos of spiral coil spinning processes; (c) front view and (d) side view. Bar = 1 cm. Reproduced from ref. 22, http://dx.doi.org/10.1155/2012/785920, Copyright The Authors. Published under the terms of the CC BY 3.0 license, https://creativecommons.org/licenses/by/3.0/.

Another example of a free-surface electrospinning apparatus developed for increasing nanofiber production rate was based on the use of multiple ring spinnerets.[49] Multiple polymer jets were formed on the top of each ring of the spinneret, generating thin and uniform nanofibers. Examination of crucial parameters, including the applied voltage and the polymer solution concentration, showed that a higher production rate was achieved upon increasing the number of rings in the spinneret, as well as the applied voltage. At the same time, higher polymer solution concentration resulted in larger fiber diameters.

Besides the introduction of cylinders, disks and rings as spinnerets in needleless electrospinning, further designs have also proved effective in increasing the production efficiency of nanofibers. Recently, S. L. Liu *et al.*[50] reported on the use of a bead wire electrode in a needleless electrospinning set-up (Figure 2.7). Their experimental results showed that this modification could also overcome several weaknesses of a conventional electrospinning system, including low production efficiency and frequent needle blockage.

In 2010, a simple unconfined feed-based electrospinning system was proposed by N. M. Thoppey *et al.*[16] The system consisted of an electrified flat plate onto which the polymer solution was placed as droplets. A different

Figure 2.6 Apparatus for disk and cylinder electrospinning and photos of the two electrospinning processes. Reproduced with permission from ref. 20, © 2009 Wiley Periodicals, Inc.

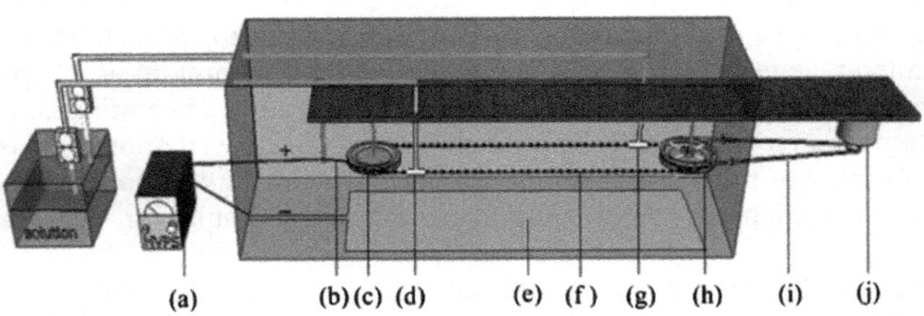

Figure 2.7 Schematic drawing of experimental apparatus: (a) HVPS; (b) electric brush; (c) single groove pulley; (d) left solution brush; (e) grounded aluminum foil; (f) bead-wire electrode; (g) right solution brush; (h) double groove pulley; (i) belt; (j) DC motor. Reproduced from S. L. Liu, Y. Y. Hung, H. D. Zhang, B. Sun, J. C. Zhang and Y. Z. Long, Needleless electrospinning for large scale production of ultrathin polymer fibers, *Mater. Res. Innovations*, 2014, **18**, 833, reprinted by permission of Taylor and Francis Ltd.

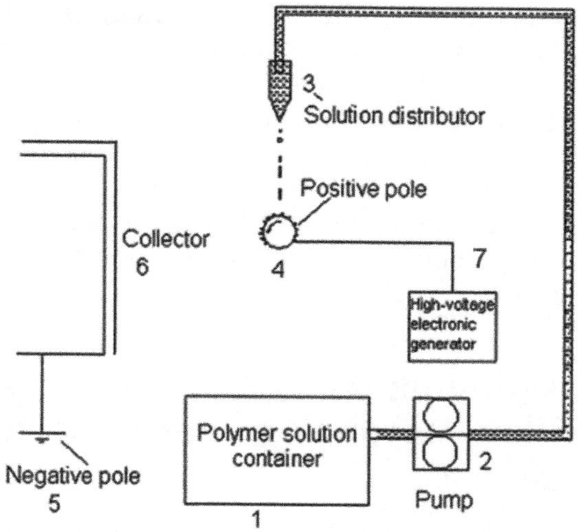

Figure 2.8 Apparatus for the splashing needleless electrospinning system. Reproduced with permission from ref. 51, © 2010 Society of Plastics Engineers.

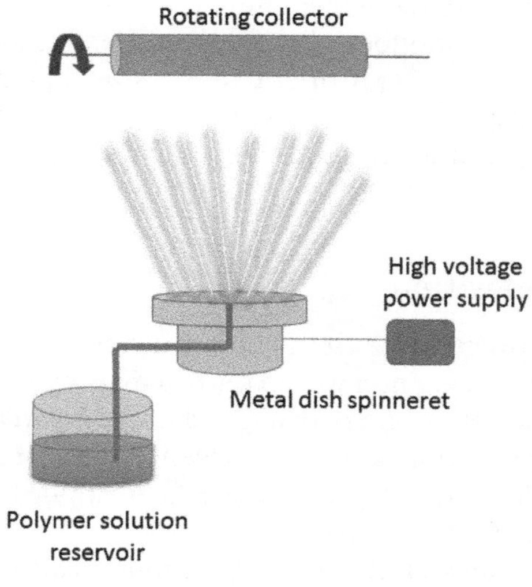

Figure 2.9 Schematic diagram of a needleless electrospinning apparatus using a metal dish as a spinneret.[45]

approach to needleless electrospinning was proposed by S. Tang and coworkers[51] who introduced the splashing method (Figure 2.8). In this modified design, the polymer solution is placed onto the surface of a metal roller spinneret, and many spinning jets are formed upon applying voltage.

In 2016, L. Wei *et al.*[45] used a metal dish spinneret in which the electric field was symmetrically distributed in three-dimensional space (Figure 2.9), with the aim of producing nanofibers with narrow diameter distributions.

Through their study, the authors investigated the effects of different process parameters including applied voltage, polymer solution concentration and distance between the metal dish and the collector on morphology, diameter, and productivity. Their results showed that the throughput increases upon increasing the polymer solution concentration and the applied voltage and decreases when the collection distance increases. Moreover, the lower polymer solution concentration resulted in bead-shaped fibers, whereas the higher applied voltage and longer collection distance resulted in smaller nanofiber diameters.

2.3 Electrospinning Encroachment in Fiber Deposition, Morphology and 3D Structure Development

The controlled deposition of nanofibers, the tuning of their morphological characteristics, as well as the generation of 3D architectures are considered very important for the development of industrial electrospun products with unique properties and functions applicable in various fields. The following sections provide an overview of the different modification approaches developed to promote: (a) nanofiber alignment and patterned nanofiber structures (contrary to the randomly oriented nanofibers produced by means of conventional electrospinning); (b) generation of fibers with non-conventional surface morphologies and structures; and (c) production of 3D nanofibrous materials.

2.3.1 Fiber Deposition

In traditional electrospinning the bending instability of the produced jet results in the production of non-woven fibrous mats with random fiber orientation. However, the development of aligned fibers is considered to be highly important for various applications.[52] These include tissue engineering,[53,54] sensing,[55–57] optoelectronics[58] and energy-related applications including fuel cell[59–61] and solar cell[62–64] technologies.

Many researchers have been focusing on modification of the traditional electrospinning set-up to produce aligned nanofibers with specific orientations.[54] Fiber alignment *via* electrospinning can be achieved by (a) modifying the geometry of the static collector resulting in adjustment of the electric field or (b) employing a mechanical rotating collector.[65] Good comprehensive reviews of the different approaches to modifying the traditional electrospinning set-up developed to induce fiber alignment have been provided by Teo and Ramakrishna.[66,67]

In 2003, Xia and coworkers[68] pioneered the production of aligned fibers by using parallel conductive silicon stripes/plates (electrodes) separated by an insulating gap. In this case, fiber orientation results from the interaction of the electrostastic force deriving from the electric field splitting due to the presence of the insulating gap between the two electrodes, with that originating from coulombic interactions. The same group has also studied the effect of the area and the geometric characteristics of the insulating gap on fiber deposition and alignment.[69] Their work stimulated the development of new methodologies, focusing on the upscaling of the production of highly aligned fibers by substituting the static collectors with rotating ones employed in various geometries such as mandrels,[70–72] disks,[73–75] and conveyors.[76,77]

Recently we have proposed a very simple concept for producing aligned electrospun fibers based on modification of the electric field profile by replacing the flat metallic collector with a concave one. Quantification of fiber alignment carried out by a Fourier transform method on different deposition sites showed up to 70% fiber alignment in the central area of the concave collector.[78]

Table 2.2 provides a few examples of fiber collectors (either static or dynamic/rotating) developed over the years to produce uniaxial aligned electrospun fibers and summarizes the main advantages/disadvantages of each collector type.[8,15]

To achieve a high degree of alignment and control over fiber diameter while providing the potential for high-scale production, Jafari and coworkers have reported the design and fabrication of a three-pole electrospinning set-up integrated with a blade-cage collector.[82] In this modified electrospinning device, the electric field can be accurately controlled, providing excellent control over fiber diameter and the possibility of mass production.

In another example in which the set-up modification also resulted in the elimination of jet instabilities caused by whipping and bending processes, highly aligned electrospun micro- and nanoscale fibers and pseudo-woven mats with 30 layers (alternating 0°/90° layup) (Figure 2.10) were produced by incorporating an auxiliary counter electrode, which led to the generation of an electric field of controlled geometry and magnitude.[83]

High-precision deposition electrospinning (HPDE) is another alternative approach developed based on the die-counter electrode set-up used in traditional electrospinning. In the case of HPDE, the distance of the electrodes is reduced from centimeters, typically applied in a conventional electrospinning process, down to millimeters and below.[84] This slight modification suppresses the bending instability, thus resulting in the deposition of aligned nanofibers onto the substrate, whereby the deposition and displacement speeds are similar.[84]

In order to generate well-aligned electrospun fibers over large areas new methods have also been developed, namely *magnetic electrospinning*, *near-field electrospinning*, *scanning-tip electrospinning* and *centrifugal electrospinning*.

Magnetically-triggered fiber alignment (*magnetic electrospinning*) is realized by using essentially the same experimental set-up as that of conventional electrospinning, the only difference being the introduction of two magnets at the sides of the conductive flat electrospinning collector, resulting in the generation of a magnetic field.[85,86]

Table 2.2 Representative examples of static and dynamic fiber collectors, developed for the production of aligned electrospun fibers and summary of their main advantages and limitations.

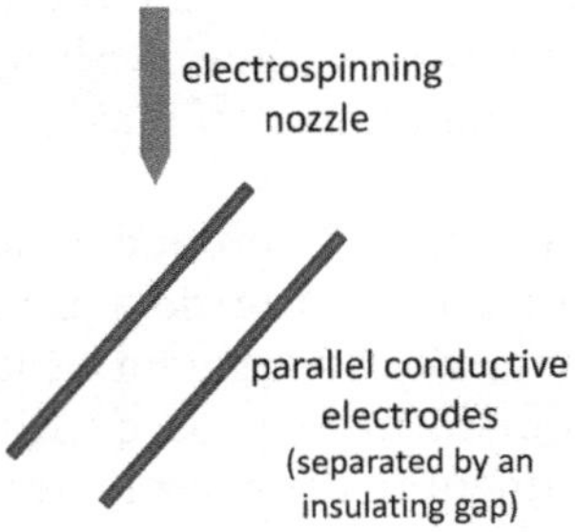

Parallel electrodes[68]

Advantages

- Simple set-up and modification
- Highly aligned fibers
- Easy transfer of the fibers to other substrates

Disadvantages

- Limitations on obtaining a fibrous mat of high thickness comprised of aligned fibers
- Limitations on the length of the fibers aligned between the two electrodes
- Fiber alignment is possible only in small areas

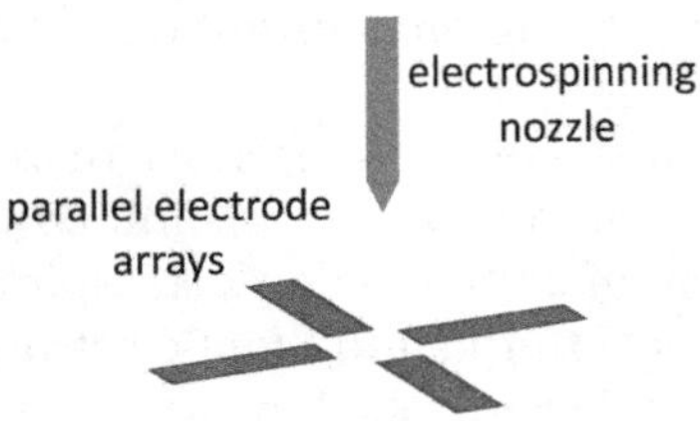

Parallel electrode arrays[79]

Advantages

- Simple set-up and modification
- Determination of the spatial orientation and position of the fibers by controlling the location and configuration of the parallel electrode arrays
- Possibility of depositing multi-layered mats of variable orientation

Disadvantages

- Very difficult to achieve large area (>1 mm^2) of aligned fibers
- Limitations on obtaining a fibrous mat of high thickness comprised of aligned fibers
- Limitations on the distance that can be set between the electrodes

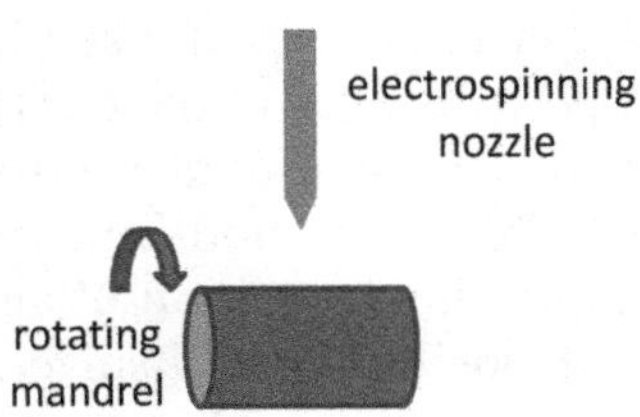

Rotating mandrel[66,80]

Advantages

- Simple set-up and modification
- Collection of aligned fibers over a large surface area
- Can be coupled with a translatable nozzle enabling control over fiber alignment

Disadvantages

- Difficulties in the generation of fibers with high degree of alignment
- Degree of alignment depends on rotational speeds – very high speeds may result in fiber fracture
- Energy-intensive
- As the mat thickness increases, fiber orientation is lost

(*continued*)

Table 2.2 (*continued*)

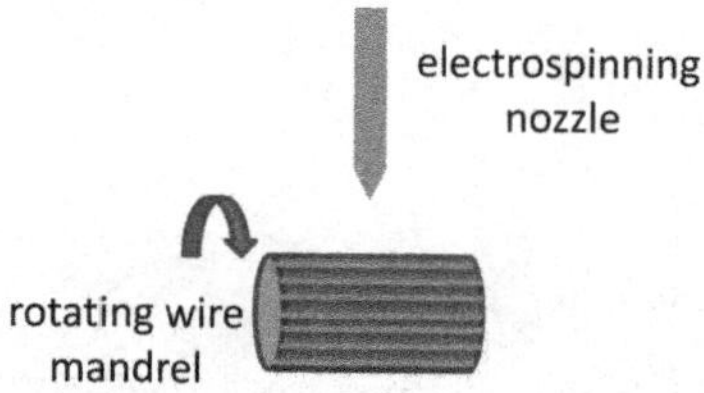	**Rotating wire mandrel**[71,81] **Advantages** • Simple set-up and modification • Collection of fibers with high degree of alignment **Disadvantages** • Difficulties with fiber detachment from the collector
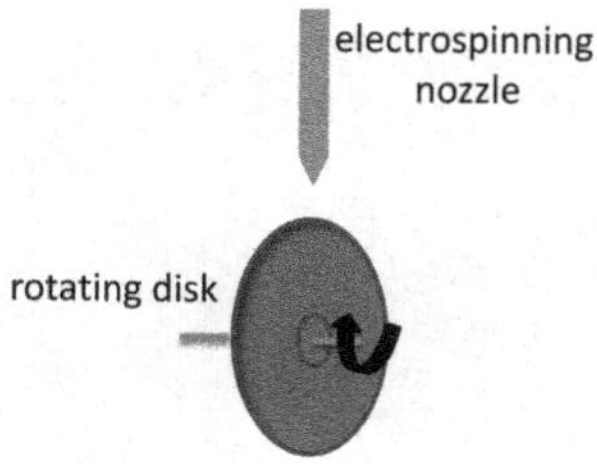	**Rotating disk with a sharp edge**[74] **Advantages** • Simple set-up • High degree of alignment • Continuous production of fibers **Disadvantages** • Loss of alignment when moving to 3D meshes • Small area of deposited aligned fibers
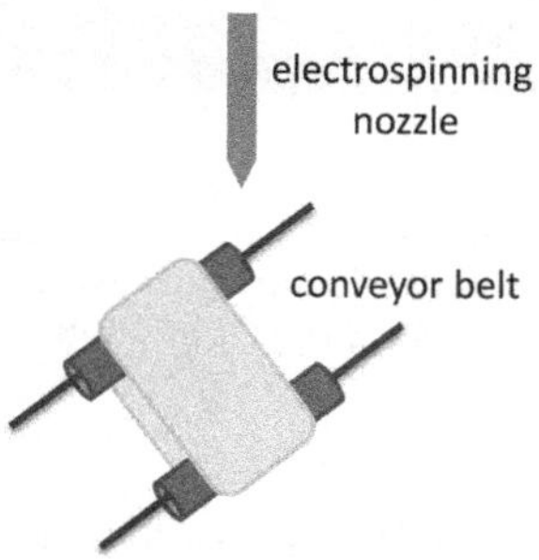	**Conveyor belts**[77] **Advantages** • Possibility for the production of aligned fibers on a large scale • Continuous operation processing over various substrates may be feasible *via* the integration of conveyor guides

The latter causes the alignment of magnetically modified fibers along the magnetic field line. Besides fiber alignment over large surface areas, advantages of this process include, among others, its simplicity in regards to set-up modification, as well as the possibility of producing more complex nanofiber patterns by rotating the collecting substrate. Furthermore, J. A. Ajao and coworkers demonstrated that fiber alignment can be realized on only one face of a box made of silicon wafers having a cylindrical magnet inside, while on the other faces random fiber orientation was observed.[87]

In *near-field electrospinning (NFES)*, the spinneret-to-collector distance is very small, ranging from a few millimeters down to hundreds of micrometers.[88,89] This technique requires that the applied voltage is significantly reduced, down to a few kV (compared to that used in conventional electrospinning, typically 10–40 kV), in order to avoid electric breakdown of the liquid jet. Because the distance between the spinneret and the collector is very small, the jet flies in the stable region, *i.e.* it does not undergo any bending/whipping processes. As a consequence, near-field electrospinning

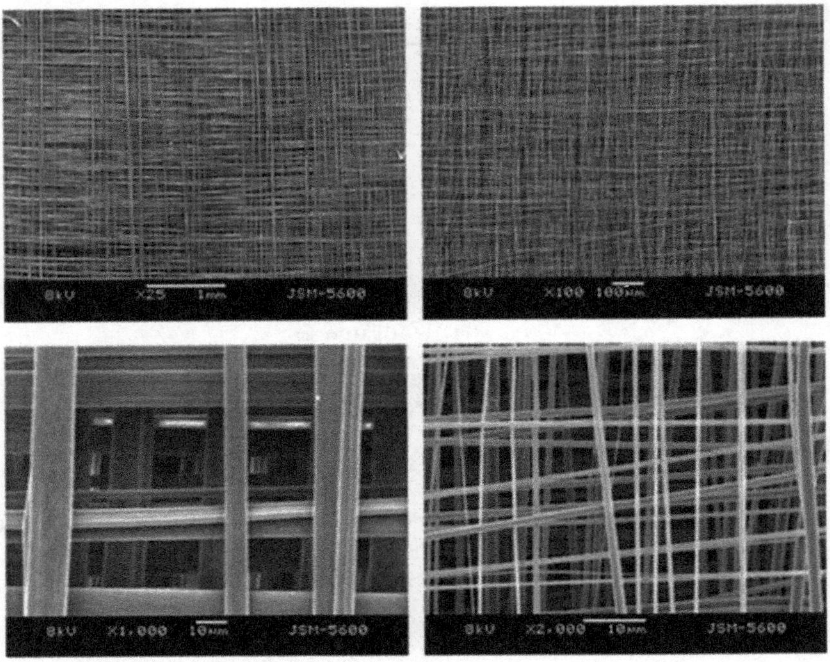

Figure 2.10 Electrospun pseudo-woven mats with 30 layers alternating 0°/90° layup. Reprinted (adapted) with permission from L. S. Carnell, E. J. Siochi, N. M. Holloway, R. M. Stephens, C. Rhim, L. E. Niklason and R. L. Clark, *Macromolecules*, 2008, **41**, 5345, Copyright 2008 American Chemical Society.

can be employed in depositing straight fibers following a predesigned pattern.[90,91] Further improvements include the use of dual-nozzle[92] or a needleless multi-spinneret NFES set-up to achieve higher production efficiency of continuous and well-ordered fibers.[93]

Scanning-tip electrospinning involves the immersion of a microfabricated scanned tip in a polymer solution followed by electrospinning, therefore acting itself as an electrospinning spinneret. By controlling the motion of the tip relative to the substrate, oriented nanofibers can be deposited and integrated with microfabricated surface structures.[94]

Both scanning-tip and NFES use the polymer droplet approach (*i.e.* involving dipping of a conductive tip into a polymer solution) which results in non-uniformity of fiber thickness and limits the deposition area.[95] Furthermore, the above processes are characterized by low throughput, thus restricting their industrial-scale applicability.

Centrifugal electrospinning leads to the production of highly aligned fibers over a large area.[96,97] The centrifugal electrospinning set-up consists of a rotating hub in which a spinneret is accommodated, surrounded by a number of grounded plate electrodes. By applying a high voltage on the spinneret, these electrodes generate a fringe electrostatic field that leads

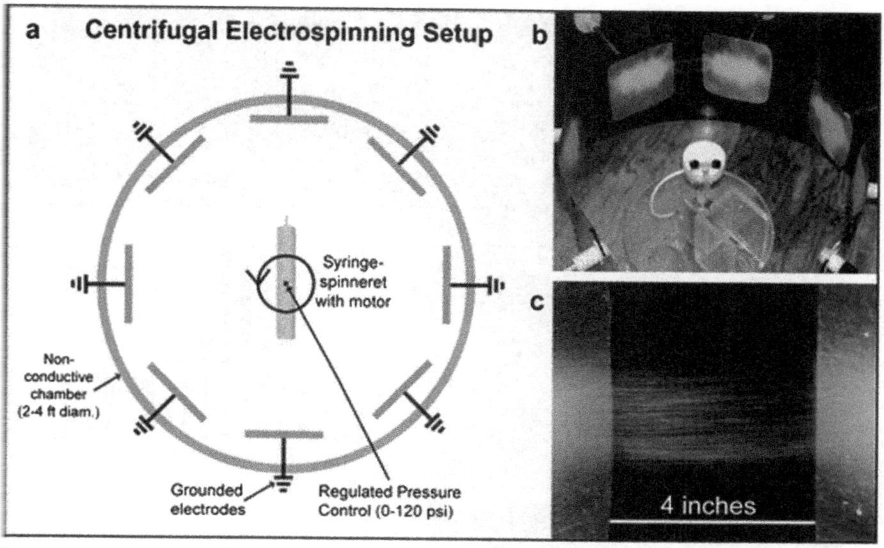

Figure 2.11 Centrifugal electrospinning (CE) system for large-area production of aligned polymer nanofibers. (a) Schematic illustration of the system configuration. (b) Photograph of the CE system with deposited PVDF nanofibers. (c) Electrospun PVDF fibers deposited across a 4-inch gap between two grounded electrodes. Reproduced from ref. 96 with permission from the Royal Society of Chemistry.

to fiber alignment across the gaps existing between the plate electrodes, as depicted in Figure 2.11.[96] Moreover, the centrifugal force generated in this configuration further contributes to fiber orientation. However, centrifugal electrospinning exhibits certain limitations, including the fact that it is incapable of producing a large number of samples comprised of highly aligned fibers.

In order to overcome these drawbacks, Erickson and coworkers designed a high-throughput centrifugal electrospinning system possessing a special configuration of wire electrodes. This modified set-up enabled the production of highly oriented fibers (production yield ~20%).[98]

Encroachment of traditional electrospinning resulted in the development of further modification processes enabling the build-up of more complex *fiber arrays and patterns*.[54,66] In 2007, D. Zhang and J. Chang described the use of electroconductive templates to fabricate electrospun mats with controllable architectures and patterns.[99] From their studies it has been concluded that the presence of protrusions and the diameters of the electroconductive wires constructing the woven collectors were key influencing parameters for the arrangement and architecture of the produced fibrous mats.

In another example, Y. H. Jeong *et al.*[100] modified the focusing and scanning functions of a HPDE set-up, by introducing: (a) cylindrical side-wall and sharp-pin ground electrodes and (b) a dielectric thin plate as a collector having X–Y motion control. These new set-up modification elements led to

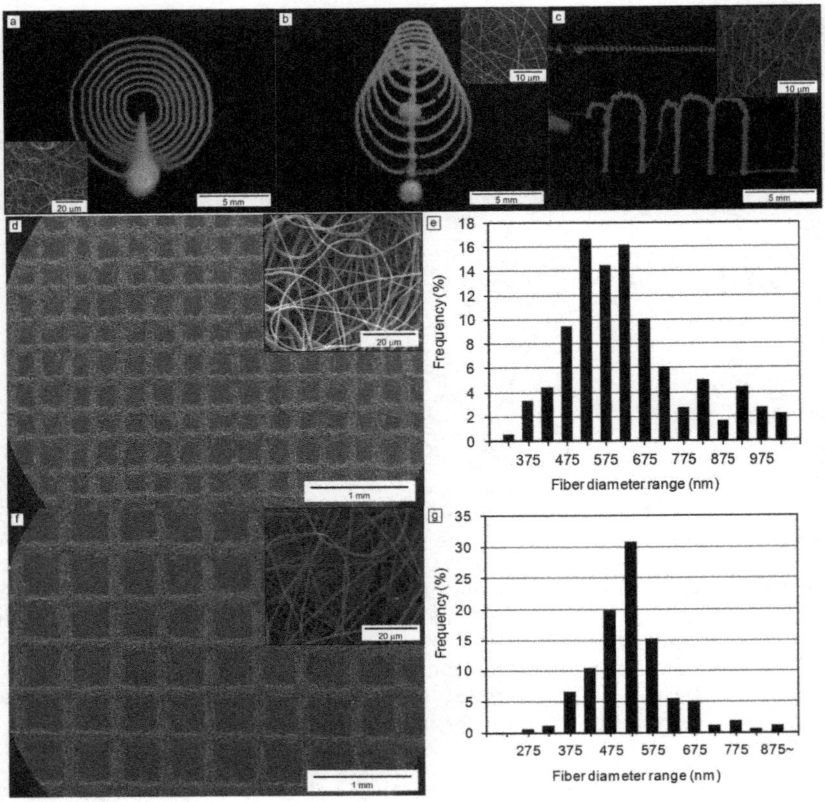

Figure 2.12 (a) Photograph of a nanofibrous pattern consisting of a set of concentric circles with a maximum radius of 5 mm. (b) Photograph of a pattern composed of stepwise reduced and shifted circles with a maximum radius of 5 mm. (c) Photograph of a pattern consisting of the alphabetic characters 'nm'. (d) Scanning electron micrograph (SEM) image and (e) fiber diameter distribution of a membrane with a single-layer lattice pattern and a grid size of 0.3×0.3 mm^2. (f) SEM images and (g) fiber diameter distribution of lattice pattern membranes with a grid size of 0.5×0.5 mm^2. The fabrication times for the circular and alphabetic patterns (a)–(c) were 13.2, 13.5, and 3.3 s, respectively. The total fabrication times for the membranes shown in (d) and (e) were 174 and 101 s, respectively. Reprinted with permission from J. Lee, S. Y. Lee, J. Jang, Y. H. Jeong and D.-W. Cho, *Langmuir*, 2012, **25**, 7267, Copyright 2012 American Chemical Society.

the successful production of electrospun nanofibrous patterns with various geometries and morphological features as seen in Figure 2.12.[100]

Highly controllable patterns of electrospun nanofibers may also be produced by means of *melt-electrospinning direct-writing*.[95] The fact that in melt electrospinning the dynamic and uncontrolled fiber deposition that is typically observed in solution electrospinning is suppressed, renders this process a bridging element between electrospinning and additive manufacturing, towards the fabrication of 3D nanofibrous structures while allowing at the same time low-resolution fabrication limits.

In a final example, Y. Zhang *et al.*[101] generated spatially-defined nanofibrous patterns by direct depositing electrospun fibers onto topographically-structured insulating substrates. The authors showed that the presence of protrusions on the surfaces of the dielectric substrates was the most important factor influencing the fiber deposition mechanism. More precisely, by employing electric field simulation studies it has been demonstrated that the presence of a strong electric field causes the polarization of the insulating protrusions, resulting in fiber accumulation onto the oppositely charged protrusions. In comparison with other methods developed for generating well-defined patterns out of electrospun fibers making use of metallic conductive collectors fabricated by more complex and costly processes,[102] in this case the use of non-conductive patterned collectors of low cost can be used instead to serve the same purpose, enabling at the same time scalability.

2.3.2 Morphology

In general, conventional electrospinning typically leads to the generation of 1D solid cylindrical fibers with smooth surfaces. However, in many research fields the development of electrospun fibers with unconventional morphologies and surface topologies is highly important because such hierarchically-structured fibers, including primary and secondary fiber structures as well as internally-structured fibers, may exhibit unusual and unique properties and functions, including optical and mechanical properties, superhydrophobicity and self-cleaning properties.[103] Such morphologies include among others helical and ribbon-like fibers, nanopillared, core–shell and hollow fibers as well as more complex morphologies such as multichannel tubular and nanowire-in-microtube fibrous structures.[103] Figure 2.13 provides examples of electrospun fibers prepared by different research groups, exhibiting unusual morphologies.

2.3.2.1 Ribbon-like, Wrinkled, Helical, and Beads-on-a-string (Necklace-like) Primary Fiber Morphologies

Many of the observed morphologies are produced by means of conventional electrospinning without the necessity for system modification, by simply varying the experimental and processing parameters. For example, morphological tuning of poly(methyl methacrylate) (PMMA) fibers from cylindrical to necklace to ribbon-like morphologies, as well as alterations in their surface texture (*i.e.* porous, smooth, wrinkled), can be realized by varying the solvent type, the polymer solution concentration and the humidity conditions.[104] Although numerous examples exist of reports on the generation of such unusual fiber morphologies by electrospinning, in this part we will be mainly focusing on literature examples where electrospinning set-up

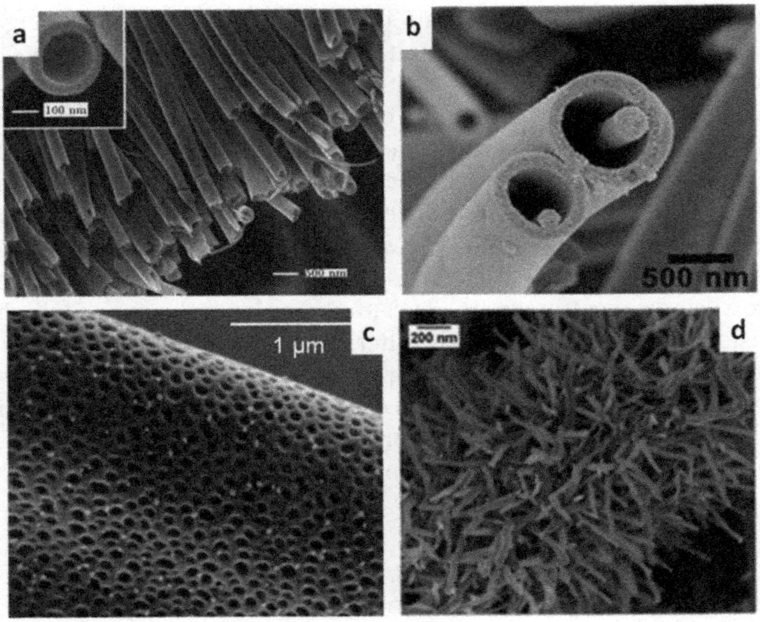

Figure 2.13 (a) SEM image of TiO_2/PVP poly hollow nanofiber. Reprinted (adapted) with permission from *Nano Lett.*, 2004, 4, 933. Copyright 2004 American Chemical Society.[105] (b) Cross-sectional SEM image of TiO_2 fibers, which shows the interesting nanowire-in-microtube structure. Reprinted with permission from C. Hongyan, N. Wang, J. Di, Y. Zhao, Y. Song and L. Jiang, *Langmuir*, 2010, **26**, 11291, Copyright 2010 American Chemical Society. (c) Scanning electron microscopy image of electrospun PS in THF at 50% relative humidity. Reprinted with permission from C. L. Casper, J. S. Stephens, N. G. Tassi, D. B. Chase, and J. F. Rabolt, *Macromolecules*, 2004, **37**, 573, Copyright 2004 American Chemical Society.[107] (d) High magnification SEM images of ZnO nanowires grown over electrospun cellulosic nanofibers. Reproduced from ref. 108 with permission from the Royal Society of Chemistry.

advancements enabled the fabrication of such unconventional electrospun fiber morphologies.

2.3.2.2 Coaxial and Multi-axial Fibers

Electrospun fibers exhibiting a core–shell morphology today attract considerable attention, mainly due to their potential applications in the biomedical field as sustained drug delivery systems.[54,109–111] By substituting the monoaxial spinneret with a coaxial (dual capillary) or a multi-axial one, core–shell electrospun fibers can be produced, consisting of different types of materials deposited on consecutive fiber layers in a concentric configuration. The use of at least two concentric capillaries enables different polymer solutions to be electrospun independently, as schematically depicted

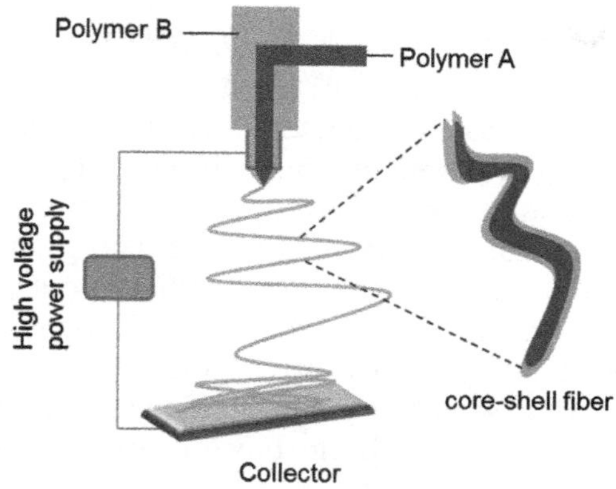

Figure 2.14 Schematic representation of a typical coaxial electrospinning set-up.

in Figure 2.14. Consequently, this set-up configuration overcomes several drawbacks appearing in the case of the single-capillary configuration, including the non-controlled distribution of the therapeutic compounds within the fiber and consequently the non-controllable drug release profile, the occurrence of burst release phenomena due to the accumulation of a certain drug amount on the fiber surfaces[112] and the use of an organic solvent that is typically employed for dissolving the polymer carrier, which can be harmful to both synthetic and biomolecular therapeutic agents, including proteins and enzymes.

The main limitation of the capillary-based coaxial electrospinning process, which restricts the commercialization of the core–shell electrospun fibers, is its low productivity. In 2013, G. C. Rutledge and coworkers described a new technology with a high potential for mass production of uniform core–shell fibers.[113] This technology is based on free-surface electrospinning, a technique that enables the production of fibers at two to three orders of magnitude higher rates compared to the spinneret-based electrospinning methods.[41] Their set-up configuration consists of a bath containing two immiscible liquids that are layered one on top of the other. By immersing metal wire electrodes (placed in a parallel orientation to the liquid surface) into the bath, a bilayer thin film is created on the surface of the wires, which in turn, due to instability phenomena, breaks up into small droplets. Upon applying an electric field, a coaxial jet is produced from each droplet, resulting in the production of solid, uniform fibers with a core–shell morphology.

Based on the same principle, multi-axial fibers have also been prepared and evaluated as sustained, multi-step drug delivery platforms.[114] Their multilayer core–shell morphology enables the incorporation of different types of therapeutic agents, including hydrophilic and hydrophobic pharmaceutical substances, proteins, antibodies, growth factors, *etc.* in each layer, and a multi-release drug profile depending on the diffusion rate of each of the above therapeutic compounds from the polymer matrix.[115]

2.3.2.3 Hollow Fibers

By selectively removing the core-forming material in fibers exhibiting a core–shell morphology, it is possible to obtain hollow fiber structures. Such materials can be interesting in various applications including biomedical, sensing and catalysis,[116,117] and energy-related applications.[118] Exemplarily, hollow magnetoactive polycaprolactone (PCL-based) fibers containing an antifungal drug were prepared by coaxial electrospinning. In this case, dimethyl silicone oil was used as the core-forming liquid whereas the outer shell consisted of PCL, Fe_3O_4 magnetic nanoparticles and the antifungal reagent. Perfusion of the silicon oil out of these core–shell structures resulted in the generation of hollow drug-loaded fibers exhibiting biocompatibility and sustained drug release properties under an auxiliary magnetic field.[119]

Multichannel hollow fibers can also be produced by one-step coaxial electrospinning, by introducing multiple capillaries in a single needle as described by L. Jiang and coworkers.[120]

2.3.2.4 Helical Fibers

Helical electrospun fibers may find use in several applications including filtration, components in micro-electromechanical systems (MEMS), photonics and electronics, sensing, *etc.*[103]

Lin and coworkers have employed a microfluidic electrospinning nozzle instead of the usual needle-like spinneret, for creating bicomponent fibers from polyacrylonitrile and elastomeric polyurethane.[121] From their study it was shown that the flow rate is a highly influencing parameter on the fiber morphology.

By using bicomponent coaxial electrospinning involving a rigid and a flexible polymer component, Chen *et al.*[122] fabricated core–shell fibers that – under certain conditions – could create very tight helical structures. In this case, the combination of the flexible and the rigid component within such core–shell bicomponent fibers is a key factor for obtaining such morphologies.

Helical polyvinylpyrrolidone (PVP) microfibers of various diameters were also fabricated on a modified electrospinning set-up, where a metal plate was welded together with the spinneret, while between the spinneret and the collecting plate a focusing aid was introduced, consisting of seven aluminium rings with seven resistors to maintain voltage gradient between adjacent rings. Furthermore, the collector is rotated and additionally, a computer-controlled translation stage is used to move the collector along the radial direction so as to avoid fiber overlapping.[123] This set-up modification causes a change in the electric field profile between the upper and lower part of the focusing aid, resulting in the formation of helical fibers with controllable lateral width. Finally, by introducing special collectors, for example tilted glass slides[124] and nail tips,[125] helical or curled fibers have also been produced.

2.3.2.5 Internally-structured Fibers

Using a multifluidic coaxial electrospinning approach, Zhao *et al.*[106] reported the fabrication of core–shell fibers having a unique nanowire-in-microtube structure. The spinneret was assembled by three coaxial stainless steel capillaries, while a compound jet was generated by feeding three different solutions into the spinneret, using an inert middle fluid as a spacer between the inner and outer solutions. By selectively removing the fluid spacer, fibers exhibiting a nanowire-in-microtube structure (Figure 2.13(b)) were produced.

2.3.3 3D Structure Development

During recent decades, there has been an increased interest in the use of electrospun fibers in tissue engineering applications owing to their high surface area, porosity, and pronounced versatility in terms of material composition and biofunctionalization. To date there have been many examples in the literature dealing with the applicability of electrospun nanofibrous mats as scaffolds in musculoskeletal and connective tissue engineering, skin, vascular and neural tissue engineering.[126–128]

However, a significant limitation of electrospun fibrous scaffolds arises from the fact that conventional electrospinning leads to the production of 3D sheet-like structures that are tightly packed, thus acting more as 2D fibrous membranes.[54] This roughened 2D surface structure restricts cell migration and infiltration.

The above-mentioned limitation prompted a number of researchers working in the field of tissue engineering and drug delivery to develop new strategies to increase porosity and decrease packing density, thus resulting in the growth of 3D electrospun fibrous structures.[54,129–133]

These include the 'fiber sacrificing' method,[134,135] where co-electrospinning of two different fibers in regards to chemical composition and physical properties is applied, followed by the removal of one of the two fibers *via* post-treatment, resulting in the generation of voids in the places where these fibers were located throughout the mixed-fiber electrospun mat. A similar process is called the 'salt leaching method'. In this case, micro-/nanosize salt particles are suspended in the polymer solution to be electrospun and they are removed afterwards through a leaching process, resulting in electrospun fibrous mats having secondary nano-/micro-scale porous structures.[136] A very interesting concept involves the use of short electrospun nanofibers to yield ultralight nanofiber-based aerogels.[137–139]

Besides the above, system modification approaches have also been adopted for fabricating 3D electrospun fibrous scaffolds. Examples include the fabrication of a 3D low-density, uncompressed ball-like scaffold by using a grounded spherical dish and an array of needle-like probes as collectors instead of the traditional flat-plate collector,[140] the use of a porous mandrel having pressurized air coming out of its pores,[141] and the introduction of a

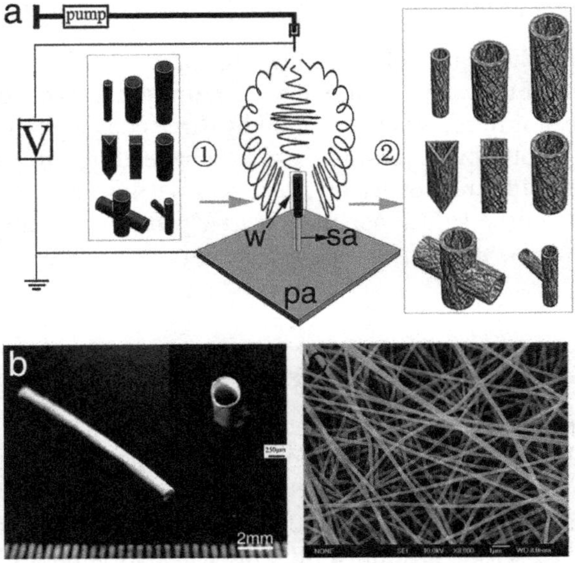

Figure 2.15 (a) Schematic illustration of the fabrication of fibrous tubes by an electrospinning technique using 3D columnar collectors (1, 3D columnar collectors; 2, relevant fibrous tubes; w, working collector; pa, plane assistant collector; sa, stick assistant collector). (b) Fibrous tube with diameter of 500 μm (inset is the cross-section image). (c) SEM image of fiber assemblies of tube shown in panel (b). Reprinted with permission from D. Chang and J. Zhang, *Nano Lett.*, 2008, **8**, 3283. Copyright 2008 American Chemical Society.

frozen rotating mandrel, a process called 'cryogenic electrospinning'.[142,143] Three-dimensional fibrous tubes of different shapes and patterns have been also developed by Zhang and Chang.[144] In this work, the authors describe a unique static collecting method by using 3D columnar collectors of different configurations, which led to the fabrication of various microscopic architectures and 3D macroscopic tubular structures as shown in Figure 2.15.

The fabrication of porous tubular structures with controlled fiber orientation using a modified electrospinning method has been also reported by S. Ramakrishna and coworkers.[145] The authors introduced a knife-edged auxiliary electrode that was charged oppositely to the spinneret, resulting in fiber alignment in a diagonal direction, forming tubular morphologies.

In another study, Kim and coworkers combined rapid prototyping and electrospinning to produce 3D hierarchical scaffolds consisting of micro-sized polymer strands and micro/nano-sized fibers.[146] There is an excellent mini-review on the combination of electrospinning and additive manufacturing aspects, targeted towards the generation of open-structured 3D scaffolds with tunable morphologies and unique functions.[95] More precisely, the authors carried out a literature survey on the combination of solution and melt electrospinning with additive manufacturing principles. Such a combination is considered to be highly powerful for the future design and manufacture of tissue engineering scaffolds. The bimodal and multiphasic tissue engineering scaffolds are also described, because this is an emerging field developed using both electrospinning and additive manufacturing. Special emphasis is given to melt electrospinning, owing to the predictable behavior of the electrified jet, thus interrelating this method with additive manufacturing strategies.

2.4 Conclusion

The tremendous industrial prospects for electrospun nanofibers has led to the encroachment of new technologies and strategies, which have brought this process out of its narrow laboratory constraints and directed it towards commercialization. There are sound academic research examples relating to the advancement of traditional electrospinning, which have already been translated to industry. Future process developments are expected to evolve through effective cooperation between the academic and industrial sectors, by suppressing the existing electrospinning weaknesses and reinforcing its strengths. Existing weaknesses that must be overcome include safety issues with the industrial-scale electrospinning process due to the use of high voltages and flammable solvents, cost-effectiveness during production and maintenance, and consistency in terms of commercialized product quality (including control over nanofiber diameter, morphology, degree of alignment, *etc.*) with the latter reflected in the performance of products.

Acknowledgements

This article is based upon work from the MPNS COST Action MP1206, 'Electrospun nano-fibres for bio inspired composite materials and innovative industrial applications', supported by COST (European Cooperation in Science & Technology, http://www.cost.eu).

References

1. BCC Research, *Nanofibers: Technology and Developing Markets*, 2007, Report Code NAN043A.
2. BCC Research, *Nanofibers: Technology and Developing Markets*, 2010, Report Code NAN043B.
3. C. J. Luo, S. D. Stoyanov, E. Stride, E. Pelanb and M. Edirisinghea, *Chem. Soc. Rev.*, 2012, **41**, 4708.
4. D. S. LeCorre-Bordes, P. Jaksons and K. Hofman, *J. Appl. Polym. Sci.*, 2017, 44836, 1–8.
5. N. Bhardwaj and S. C. Kundu, *Biotechnol. Adv.*, 2010, **28**, 325.
6. C. J. Luo, M. Nangrejo and M. Edirisinghe, *Polymer*, 2010, **51**, 1654.

7. Z. Xu, X. Huang and L. Wan, *Adv. Topics Sci. Technol.*, 2009, 306.
8. Z. M. Huang, Y. Z. Zhang, M. Kotaki and S. Ramakrishna, *Compos. Sci. Technol.*, 2003, **63**, 2223.
9. Q. P. Pham, U. Sharma and A. G. Mikos, *Tissue Eng.*, 2006, **12**, 1197.
10. M. G. Tardajos, M. Nash, Y. Rochev, H. Reinecke, C. Elvira and A. Gallardo, *Macromol. Chem. Phys.*, 2012, **213**, 529.
11. J. M. Deitzel, J. Kleinmeyer, D. Harris and N. C. Beck Tan, *Polymer*, 2001, **42**, 261.
12. J. Venugopal, Y. Z. Zhang and S. Ramakrishna, *J. Nanoeng. Nanosyst.*, 2005, **218**, 35.
13. J. Yang, S. Zhan, N. Wang, X. Wang, Y. Li, Y. Li, W. Ma and H. Yu, *J. Dispersion Sci. Technol.*, 2010, **31**, 760.
14. J. K. Hong and S. V. Madihally, *Tissue Eng., Part B*, 2011, **17**, 125.
15. C. Migliaresi, G. A. Ruffo, F. Z. Volpato and D. Zeni, Advanced electrospinning setups and special fibre and mesh morphologies, in *Electrospinning for Advanced Biomedical Applications and Therapies*, ed. N. M. Neves, Smithers Rapra, 2012, pp. 23–59.
16. N. M. Thoppey, J. R. Bochinski, L. I. Clarke and R. E. Gorga, *Polymer*, 2010, **51**, 4928.
17. M. Wojasinski, J. Golawski and T. Ciach, *J. Polym. Res.*, 2017, **76**, 1.
18. H. Niu, X. Wang and T. Lin, Needleless Electrospinning: Developments and Performances, in *Nanofibers – Production, Properties and Functional Applications*, ed. T. Lin, In Tech, 2011, pp. 17–36.
19. I. Esmaeilzadeh, V. Mottaghitalab, B. Tousifar, A. Afzali and M. Lamani, *Int. J. Ind. Chem.*, 2015, **6**, 193.
20. H. Niu, T. Lin and X. Wang, *J. Appl. Polym. Sci.*, 2009, **114**, 3524.
21. Y.-K. Fuh, L.-C. Lien and S.-Y. Chen, *J. Macromol. Sci., Part B: Phys.*, 2012, **51**, 1742.
22. X. Wang, H. Niu, X. Wang and T. Lin, *J. Nanomater.*, 2012, **2012**, 785920.
23. C. J. Angammana and S. H. Jayaram, *IEEE Trans., Ind. Appl.*, 2011, **47**, 1028–1035.
24. J. S. Varabhas, G. G. Chase and D. H. Reneker, *Polymer*, 2008, **49**, 4226.
25. O. O. Dosunmu, G. G. Chase, J. S. Varabhas, W. Kataphinan and D. H. Reneker, *Nanotechnology*, 2006, **2**, 1123.
26. G. Kim, Y.-S. Cho and W. D. Kim, *Eur. Polym. J.*, 2006, **42**, 2031.
27. A. Varesano, R. A. Carletto and G. Mazzuchetti, *J. Mater. Process. Technol.*, 2009, **209**, 5178.
28. B. Ding, E. Kimura, T. Sato, S. Fujita and S. Shiratori, *Polymer*, 2004, **45**, 1895.
29. W. Tomaszewski and M. Szadkowski, *Fibres Text. East. Eur.*, 2005, **13**, 22.
30. A. Varesano, F. Rombaldoni, G. Mazzuchetti, C. Tonin and R. Comotto, *Polym. Int.*, 2010, **59**, 1606.
31. I. G. Kim, J.-H. Lee, A. R. Unnithan, C.-H. Park and C. S. Kim, *J. Ind. Eng. Chem.*, 2015, **31**, 251.
32. S. Xie and Y. Zeng, *Ind. Eng. Chem. Res.*, 2012, **51**, 5336.
33. B. Chu, B. S. Hsiao, D. Fang and A. Okamoto, US7662332, 2010.
34. I. C. Um, D. Fang, B. S. Hsiao, A. Okamoto and B. Chu, *Biomacromolecules*, 2004, **5**, 1428.

35. Y. Liu, J. H. He, L. Xu and J. Y. Yu, *J. Polym. Eng.*, 2008, **28**, 55.
36. Y. Liu, J. H. He, J. Y. Yu, L. Xu and L. F. Liu, Chinese Patent, 200710036447.4, 2007.
37. J. H. He, H. Y. Kong, R.-R. Yang, H. Dou, N. Faraz, L. Wang and C. Feng, *Therm. Sci.*, 2012, **16**, 1263.
38. J.-H. He and Y. Liu, *Comput. Math. Appl.*, 2012, **64**, 1033.
39. R. Yang, J. He, L. Xu and J. Yu, *Polymer*, 2009, **50**, 5846.
40. H. Niu and T. Lin, *J. Nanomater.*, 2012, **2012**, 725950.
41. A. L. Yarin and E. Zussman, *Polymer*, 2004, **45**, 2977.
42. X. Wang, H. Niu, T. Lin and X. Wang, *Polym. Eng. Sci.*, 2009, **49**, 1582.
43. D. Lukas, A. Sarkar and P. Pokorny, *J. Appl. Phys.*, 2008, **103**, 084309.
44. C. J. Angammana and S. H. Jayaram, *Part. Sci. Technol.*, 2016, **34**, 72.
45. L. Wei, H. Yu, L. Jia and X. Qin, *Text. Res. J.*, 2016, 1.
46. O. Jirsak, F. Sanetrnik, D. Lukas, V. Kotek, L. Martinova and J. Chaloupek, WO 2005/024101 A1, 2005.
47. T. Lin, X. Wang, X. Wang and H. Niu, WO 2010/043002, 2010.
48. B. Lu, Y. Wang, Y. Liu, H. Duan, J. Zhou, Z. Zhang, Y. Wang, X. Li, W. Wang, W. Lan and E. Xie, *Small*, 2010, **6**, 1612.
49. X. Wang, T. Lin and X. Wang, *Fibers Polym.*, 2014, **15**, 961.
50. S. L. Liu, Y. Y. Hung, H. D. Zhang, B. Sun, J. C. Zhang and Y. Z. Long, *Mater. Res. Innovations*, 2014, **18**, 833.
51. S. Tang, Y. Zeng and X. Wang, *Polym. Eng. Sci.*, 2010, **50**, 2252.
52. Y.-Z. Long, B. Sun, H.-D. Zhang, J.-L. Duvail, C.-Z. Gu and H.-L. Yin, *Nanotechnol. Res. J.*, 2014, **7**, 155.
53. Y. S. Lee and T. Livingston Arinzeh, *Polymer*, 2011, **3**, 413.
54. H. Li, Y. Xu, H. Xu and J. Chang, *J. Mater. Chem. B*, 2014, **2**, 5492.
55. J. Zheng, X. Yan, M.-M. Li, G.-F. Yu, H.-D. Zhang, W. Pisula, X.-X. He, J.-L. Duvail and Y.-Z. Long, *Nanoscale Res. Lett.*, 2015, **10**, 1.
56. P. Sahatiya and S. Badhulika, *RSC Adv.*, 2015, **5**, 82481.
57. H.-D. Zhanga, C.-C. Tangb, Y.-Z. Longa, J.-C. Zhanga, R. Huanga, J.-J. Li and C.-Z. Gu, *Sens. Actuators, A*, 2014, **219**, 123.
58. Z. Zheng, L. Gan and T. Zhai, *Sci. China Mater.*, 2016, **59**, 200.
59. D. Todd and M. Walter, *J. Power Sources*, 2016, **311**, 182.
60. T. Tamura, R. Takemori and H. Kawakami, *J. Power Sources*, 2012, **217**, 135.
61. T. Tamura and H. Kawakami, *Nano Lett.*, 2010, **10**, 1324.
62. T. Krishnamoorthy, M. Z. Tang, A. Verma, A. S. Nair, D. Pliszka, S. G. Mhaisalkar and S. Ramakrishna, *J. Mater. Chem.*, 2012, **22**, 2166.
63. D. Joly, J.-W. Jung, I.-D. Kim and R. Demadrille, *J. Mater. Chem. C*, 2016, **4**, 10173.
64. S. Cavaliere, S. Subianto, I. Savych, D. J. Jonesa and J. Rozièrea, *Energy Environ. Sci.*, 2011, **4**, 4761.
65. J. Min, W. Min, W. Shiyong, C. Zhibao and M. Shichun, *Prog. Chem.*, 2016, **m28**, 711.
66. W. E. Teo and S. Ramakrishna, *Nanotechnology*, 2006, **17**, R89.
67. W.-E. Teo, R. Inai and S. Ramakrishna, *Sci. Technol. Adv. Mater.*, 2011, **12**, 013002, 1–19.

68. D. Li, Y. Wang and Y. Xia, *Nano Lett.*, 2003, **3**, 1167.
69. D. Li, G. Ouyang, J. T. McCann and Y. Xia, *Nano Lett.*, 2005, **5**, 913.
70. X. Mo and X.-H. Weber, *Macromol. Symp.*, 2004, **217**, 413.
71. P. Katta, M. Alessandro, R. D. Ramsier and G. G. Chase, *Nano Lett.*, 2004, **4**, 2215.
72. J. S. Kim and D. H. Reneker, *Polym. Eng. Sci.*, 1999, **39**, 849.
73. M. B. Bazbouz and G. K. Stylios, *J. Appl. Polym. Sci.*, 2008, **107**, 3023.
74. A. Theron, E. Zussman and A. L. Yarin, *Nanotechnology*, 2001, **12**, 384.
75. C. Y. Xu, R. Inai, M. Kotaki and S. Ramakrishna, *Biomaterials*, 2004, **25**, 877.
76. S. Wang, Y. Yang, Y. Zhang, X. Fei, C. Zhou, Y. Zhang, Y. Li, Q. Yang and Y. Song, *J. Appl. Polym. Sci.*, 2014, **131**, 39735.
77. A. M. dos Santos, J. Dierck, M. Troch, M. Podevijn and E. Schacht, *Macromol. Mater. Eng.*, 2011, **296**, 637.
78. I. Savva, E. Evaggelou, G. Papaparaskeva, T. Leontiou, T. Stylianopoulos, F. Mpekris, K. Stylianou and T. Krasia-Christoforou, *RSC Adv.*, 2015, **5**, 104400.
79. D. Li, Y. Wang and Y. Xia, *Adv. Mater.*, 2004, **16**, 361.
80. A. L. Szentivanyi, H. Zernetsch, H. Menzel and B. Glasmacher, *Int. J. Artif. Organs*, 2011, **34**, 986.
81. J. K. Hong and S. V. Madihally, *Tissue Eng., Part B*, 2011, **17**, 125.
82. A. Jafari, J.-H. Jeon and I.-K. Oh, *Macromol. Rapid Commun.*, 2011, **32**, 921.
83. L. S. Carnell, E. J. Siochi, N. M. Holloway, R. M. Stephens, C. Rhim, L. E. Niklason and R. L. Clark, *Macromolecules*, 2008, **41**, 5345.
84. C. Hellmann, J. Belardi, R. Dersch, A. Greiner, J. H. Wendorff and S. Bahnmueller, *Polymer*, 2009, **50**, 1197.
85. D. Yang, B. Lu, Y. Zhao and X. Jiang, *Adv. Mater.*, 2007, **19**, 3702.
86. D. Yang, J. Zhang, J. Zhang and J. Nie, *J. Appl. Polym. Sci.*, 2008, **110**, 3368.
87. J. A. Ajao, A. A. Abiona, S. Chigome, A. Y. Fasasi, G. A. Osinkolu and M. Maaza, *J. Mater. Sci.*, 2010, **45**, 2324.
88. D. Sun, C. Chang, S. Li and L. Lin, *Nano Lett.*, 2006, **6**, 839.
89. X.-X. He, J. Zheng, G.-F. Yu, M.-H. You, M. Yu, X. Ning and Y.-Z. Long, *J. Phys. Chem.*, 2017, **121**, 8663.
90. C. Chang, K. Limkrailassiri and L. Lin, *Appl. Phys. Lett.*, 2008, **93**, 123111.
91. D. D. Camillo, V. Fasano, F. Ruggieri, S. Santucci, L. Lozzi, A. Camposeo and D. Pisignano, *Nanoscale*, 2013, **5**, 11637.
92. Z. Wang, X. Chen, J. Zeng, F. Liang, P. Wu and H. Wang, *AIP Adv.*, 2017, **7**, 035310.
93. C.-T. Pan, K.-C. Tsai, S.-Y. Wang, C.-K. Yen and Y.-L. Lin, *Micromachines*, 2017, **8**, 97.
94. J. Kameoka, R. Orth, Y. Yang, D. Czaplewski, R. Mathers, G. W. Coates and H. G. Craighead, *Nanotechnology*, 2003, **14**, 1124.
95. P. D. Dalton, C. Vaquette, B. L. Farrugia, T. R. Dargaville, T. D. Brown and D. W. Hutmacher, *Biomater. Sci.*, 2013, **1**, 171.

96. D. Edmondson, A. Cooper, S. Jana, D. Wood and M. Zhang, *J. Mater. Chem.*, 2012, **22**, 18646.
97. C.-C. Liao, C.-C. Wang and C.-Y. Chen, *Polymer*, 2011, **52**, 4303.
98. A. E. Erickson, D. Edmondson, F.-C. Chang, D. Wood, A. Gong, S. L. Levengood and M. Zhang, *Polymer*, 2015, **134**, 467.
99. D. Chang and J. Zhang, *Adv. Mater.*, 2007, **19**, 3664.
100. J. Lee, S. Y. Lee, J. Jang, Y. H. Jeong and D.-W. Cho, *Langmuir*, 2012, **25**, 7267.
101. S. Zhao, Q. Zhou, Y.-Z. Long, G.-H. Suncd and Y. Zhang, *Nanoscale*, 2013, **5**, 4993.
102. B. Carlberg, T. Wang and J. Liu, *Langmuir*, 2010, **26**, 2235.
103. N. E. Zander, *Polymers*, 2013, **5**, 19.
104. L. Li, Z. Jiang, M. Li, R. Li and T. Fang, *RSC Adv.*, 2014, **4**, 52973.
105. D. Li and Y. Xia, *Nano Lett.*, 2004, **4**, 933.
106. C. Hongyan, N. Wang, J. Di, Y. Zhao, Y. Song and L. Jiang, *Langmuir*, 2010, **26**, 11291.
107. C. L. Casper, J. S. Stephens, N. G. Tassi, D. B. Chase and J. F. Rabolt, *Macromolecules*, 2004, **37**, 573.
108. T. J. Athauda, U. Buttb and R. R. Ozer, *RSC Adv.*, 2013, **3**, 21431.
109. W. Lu, J. Sun and X. Jiang, *J. Mater. Chem. B*, 2014, **2**, 2369.
110. A. L. Yarin, *Polym. Adv. Technol.*, 2011, **22**, 310.
111. Y. Hong, Z. Chen Guang, Z. Yun Hui, Z. Gong Wen and T. Xiao Yan, *Sci. China: Chem.*, 2010, **53**, 1246.
112. S.-F. Chou, D. Carson and K. A. Woodrow, *J. Controlled Release*, 2010, **220**(pt B), 584.
113. K. M. Forward, A. Flores and G. C. Rutledge, *Chem. Eng. Sci.*, 2013, **104**, 250.
114. A. Khalf and S. V. Madihally, *Mater. Sci. Eng., C*, 2017, **76**, 161.
115. A. Khalf and S. V. Sundararajan, *Eur. J. Pharm. Biopharm.*, 2017, **112**, 1.
116. P.-C. Liu, J.-H. Gong, S.-G. Yang, J.-H. Ma and J. Xu, *J. Inorg. Mater.*, 2013, **28**, 571.
117. H. Qu, S. Wei and Z. Guo, *J. Mater. Chem. A*, 2013, **1**, 11513.
118. L. Li, S. Peng, J. K. Yoong Lee, D. Ji, M. Srinivasan and S. Ramakrishna, *Nano Energy*, 2017, **39**, 111.
119. B. Wang, H. Zheng, M. W. Chang, Z. Ahmad and J. S. Li, *Colloids Surf., B*, 2016, **145**, 757.
120. Y. Zhao, X. Cao and L. Jiang, *J. Am. Chem. Soc.*, 2007, **129**, 764.
121. T. Lin, H. Wang and X. Wang, *Adv. Mater.*, 2005, **17**, 2699.
122. S. L. Chen, H. Q. Hou, P. Hu, J. H. Wendorff, A. Greiner and S. Agarwal, *Macromol. Mater. Eng.*, 2009, **294**, 265.
123. X. Cui, L. Li and F. Xu, *Appl. Phys. A*, 2011, **103**, 167.
124. J. Yu, Y. Qiu, X. Zha, M. Yu, J. Yu, J. Rafique and J. Yin, *Eur. Polym. J.*, 2008, **44**, 2838.
125. C.-C. Tang, J. C. Chen, Y. Z. Long, H. X. Yin, B. Sun and H. D. Zhang, *Chin. Phys. Lett.*, 2011, **28**, 056801.

126. S. Khorshidi, A. Solouk, H. Mirzadeh, S. Mazinani, J. M. Lagaron, S. Sharifi and S. Ramakrishna, *J. Tissue Eng. Regener. Med.*, 2016, **10**, 715.
127. K. Ghosal, S. Thomas, N. Kalarikkal and A. Gnanamani, *J. Polym. Res.*, 2014, **21**, 410, 5 pages.
128. K. Ghosal, A. Manakhov, L. Zajíčková and S. Thomas, *AAPS PharmSciTech*, 2017, **18**, 72.
129. N. H. A. Ngadiman, M. Y. Noordin, A. Idris and D. Kurniawan, *Proc. Inst. Mech. Eng., Part H*, 2017, **231**, 597.
130. W.-E. Te, R. Inai and S. Ramakrishna, *Sci. Technol. Adv. Mater.*, 2011, **12**, 013002.
131. S. E. Moulton and G. G. Wallace, *J. Controlled Release*, 2014, **193**, 27.
132. B. Sun, Y. Z. Long, H. D. Zhang, M. M. Li, J. L. Duvail, X. Y. Jiang and H. L. Yin, *Prog. Polym. Sci.*, 2014, **39**, 862.
133. S. Zhong, Y. Zhang and C. T. Lim, *Tissue Eng., Part B*, 2012, **18**, 77.
134. B. M. Baker, A. O. Gee, R. B. Metter, A. S. Nathan, R. A. Marklein, J. A. Burdick and R. L. Mauck, *Biomaterials*, 2008, **29**, 2348.
135. M. C. Phipps, W. C. Clem, J. M. Grunda, G. A. Clines and S. L. Bellis, *Biomaterials*, 2012, **33**, 524.
136. Y. Wang, B. Wang, G. Wang, T. Yin and Q. Yu, *Polym. Bull.*, 2009, **63**, 259.
137. S. Yang, Y. Jianyong, T. Xiaomin, G. Jianlong and D. Bin, *Nat. Commun.*, 2014, **5**, 5802.
138. G. Duan, S. Jiang, V. Jérôme, J. H. Wendorff, A. Fathi, J. Uhm, V. Altstädt, M. Herling, J. Breu, R. Freitag, S. Agarwal and A. Greiner, *Adv. Funct. Mater.*, 2015, **25**, 2850.
139. F. Deuber and C. Adlhart, *Chimia*, 2017, **71**, 236.
140. B. A. Blakeney, A. Tambralli, J. M. Anderson, A. Andukuri, D.-J. Lim, D. R. Dean and H.-W. Jun, *Biomaterials*, 2011, **32**, 1583.
141. M. J. McClure, P. S. Wolfe, D. G. Simpson, S. A. Sell and G. L. Bowlin, *Biomaterials*, 2012, **33**, 771.
142. M. Simonet, O. D. Schneider, P. Neuenschwander and W. J. Stark, *Polym. Eng. Sci.*, 2007, **47**, 2020.
143. O. D. Schneider, F. Weber, T. J. Brunner, S. Lohe, M. Ehrbar, P. R. Schmidlin and W. J. Stark, *Acta Biomater.*, 2009, **5**, 1775.
144. D. Chang and J. Zhang, *Nano Lett.*, 2008, **8**, 3283.
145. W. E. Teo, M. Koraki, X. M. Mo and S. Ramakrishna, *Nanotechnology*, 2005, **16**, 918.
146. G. H. Kim, J. G. Son, S. Park and W. D. Kim, *Macromol. Rapid Commun.*, 2008, **29**, 1577.

CHAPTER 3

Biomimetic Electrospun Composites: from Fundamental Insights to Commercialization

E. MELE

Materials Department, Loughborough University, Loughborough LE11 3TU, UK
*E-mail: E.Mele2@lboro.ac.uk

3.1 Introduction

Biomimetics is an interdisciplinary research field that aims to find solutions to engineering problems by studying, imitating and adapting properties, functions and processes of biological systems.[1–5] The term biomimetics was introduced for the first time by Otto Schmitt in 1957, referring to a device that mimicked the electrical action of nerves.[1,2] Since then, a variety of biomimetic and bioinspired materials and devices with multiple functionalities and significant commercial potential has been developed. Well-known examples include super-hydrophobic, self-cleaning and antifouling surfaces; membranes and bio-constructs with micro- and nano-porosity; self-healing and antireflective coatings.[6,7]

The remarkable physical and chemical properties of biological systems are the result of a complex hierarchical assembly of diverse components across multiple length scales, from molecular level to macroscale (Figure 3.1).[8] For instance, the exceptional water repellency and the self-cleaning ability of

Soft Matter Series No. 7
Electrospinning: From Basic Research to Commercialization
Edited by Erich Kny, Kajal Ghosal and Sabu Thomas

Published by the Royal Society of Chemistry, www.rsc.org

Figure 3.1 Photographs of natural objects that are formed of a hierarchical assembly of structures across multiple length scales (top row). Scanning electron microscopy (SEM) images of corresponding micro- and nanostructures (bottom row). Reprinted from *Materials Today*, **16**, X. Wang, B. Ding and B. Li, Biomimetic electrospun nanofibrous structures for tissue engineering, 229–241, Copyright 2013, with permission from Elsevier.

the surface of the lotus leaf (*Nelumbo nucifera*) is ascribed to the presence of micro-papillae, which are composed of nanoscale wax tubules organized in a three-dimensional (3D) way. When water droplets sit on these hierarchical wax asperities, the air trapped between them limits liquid penetration and results in a high water contact angle (~164°) and low hysteresis (~3°).[9] The superior mechanical properties of seashells, such as abalone shells, derive from their nacreous ceramic/organic layer (95 wt% ceramic and 5 wt% organic), which is formed by stacked platelets of aragonite polymorph of $CaCO_3$ (500 nm thickness) in between layers of biopolymers (keratin, collagen and chitin; 5–20 nm thick).[10,11] This alternation of brittle and soft materials in a 'bricks and mortar' type of structure provides nacre with its high toughness (tensile strength of 170 MPa and 140 MPa for dry and wet samples, respectively) and with a work of fracture that is 3000 times greater than that of monolithic $CaCO_3$ crystals.[10,11] By analogy to nacre, bone tissue is the result of the multiscale hierarchical arrangement of hard and soft phases. In particular, collagen fibrils and hydroxyapatite nanocrystals form mineralized collagen fibers (100 nm in size), which further assemble into lamellae (5 µm in size) and osteons (100 µm in diameter).[12] This structural heterogeneity enables bone to resist fracture and to possess high toughness.

Among the various micro- and nano-fabrication methods that have been used to produce biomimetic structures, electrospinning has emerged for its ability to generate polymeric fibers that emulate the organization and chemical composition of naturally-occurring systems.[4,13] Synthetic and natural polymers, their blends and nanocomposites have been processed by electrospinning into a wide range of morphologies, such as ribbon-like, helical, necklace-like, porous, core–shell, and hollow fibers.[13] These structures have been proposed to replicate the functions of the natural objects previously described and many others, including: the interwoven protein fibrillary networks of the native extracellular matrix that offer mechanical support and

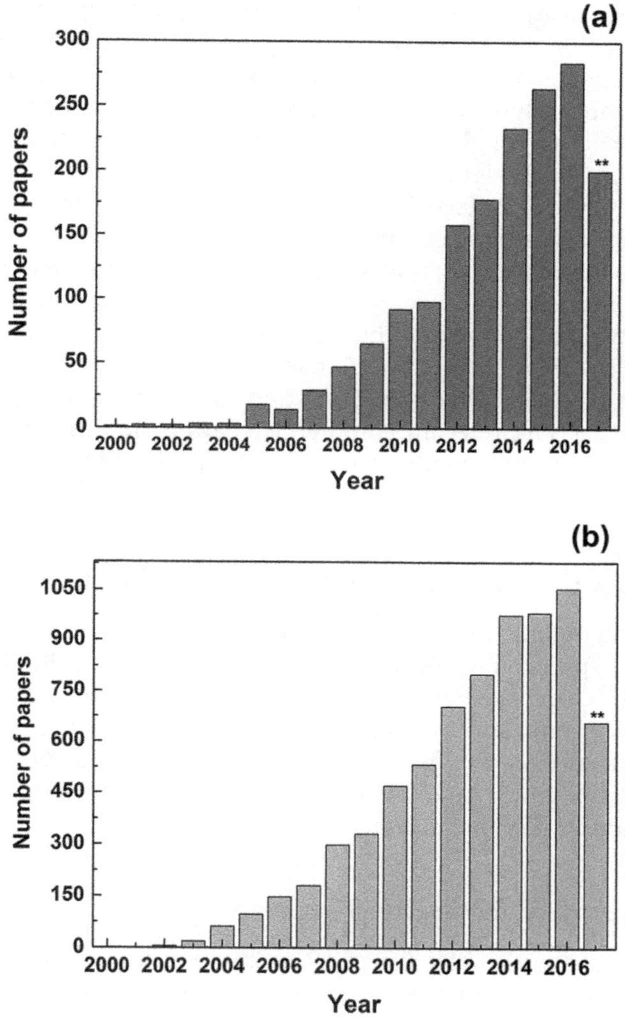

Figure 3.2 Yearly numbers of papers published from 2000 to 2017 indexed in Scopus under the topics: (a) 'electrospinning' and 'membranes' and 'filtration'; (b) 'electrospinning' and 'scaffolds' and 'tissue engineering'. **This count considers the first 8 months of 2017.

chemical cues for cell proliferation and differentiation; the curved and textured fibers that confer hydrophobicity to the leaves of the silver ragwort plant; and the hollow hairs of polar bear fur, which guide light and provide thermal insulation (Figure 3.1).

This chapter will discuss the recent scientific advances in the field of biomimetic composite fibers, with the main focus on systems that have recently attracted commercial interest, including membranes with controlled wetting properties for liquid and air filtration and electrospun scaffolds for biomedical applications. Figure 3.2 shows the trend in scientific publications related to these two topics between 2000 and 2017. The number of research outputs on electrospun membranes for filtration has progressively increased since 2005, with numbers almost doubling over a 5 year period (158 papers in 2012 *vs.* 284 papers in 2017), giving an indication of the academic attention on research area (Figure 3.2a). On the other hand, a steady rise in the number of publications on bio-constructs based on electrospun fibers has been recorded since 2002 (Figure 3.2b).

The chapter is organized in two sections. The first one deals with biomimetic hierarchical surfaces and membranes produced by electrospinning nanocomposite materials, polymer blends, block copolymers and stimuli-responsive materials. An overview of the wetting theories, including the static and dynamic behavior of liquids and super-non-wetting states, is also provided. The second part of the chapter will discuss the use of electrospinning to create scaffolds for bone and skin tissue engineering. At the end of each section, relevant commercially-available products will be discussed, such as filters for water and air purification and membranes for cell cultures.

3.2 Liquid-repellent Electrospun Fibers

Liquid-repellent and super-non-wetting materials are used in a broad range of industrial applications, from anti-corrosive coatings, antifouling, antifogging or anti-icing behavior to membranes for water purification/desalination, optoelectronic and biomedical devices.[14–16] In this section the recent scientific advances in the field of super-hydrophobic and super-oleophobic electrospun fibers will be discussed, with particular focus on self-cleaning surfaces and filtration membranes.

3.2.1 Overview of Wetting Theories

The wetting properties of a flat solid surface are described by the Young's equation (eqn (3.1)), which relates the contact angle (θ) of a water droplet deposited onto the surface to the solid–liquid interfacial tension (γ_{SL}), the solid–vapor interfacial energy (γ_{SV}) and the liquid–vapor surface tension (γ_{LV}) of the droplet–surface system (Figure 3.3a):[15–17]

$$\cos\theta = \frac{\gamma_{SV} - \gamma_{SL}}{\gamma_{LV}} \tag{3.1}$$

In order to predict how the contact angle is affected by the presence of topographical micro- and nanostructures, the Wenzel and Cassie–Baxter models are typically used. In the Wenzel state, the liquid completely wets surface cavities and asperities, leading to an increase in the solid–liquid interface (Figure 3.3b). By introducing the roughness factor r, which is the ratio of the actual surface area to its horizontal projection and is always greater than unity, the Wenzel equation (eqn (3.2)) states that surface texture increases

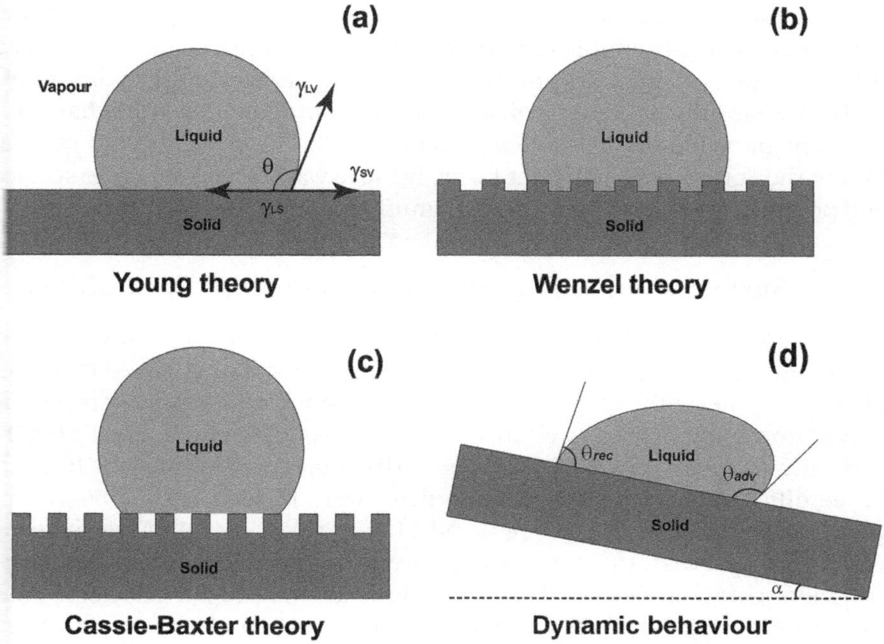

Figure 3.3 Schematic illustrations of a liquid droplet deposited onto: (a) a smooth surface following the Young theory; a rough surface following (b) the Wenzel model and (c) the Cassie–Baxter model; (d) a tilted substrate.

the wettability of intrinsically hydrophilic materials and reduces that of intrinsically hydrophobic ones:

$$\cos\theta_{\mathrm{w}} = r\cos\theta \tag{3.2}$$

By contrast, in the Cassie–Baxter state, the droplet is unable to fill the surface cavities because air remains trapped inside them (Figure 3.3c). This heterogeneous interface, where the liquid is in contact with both solid substrate and air, induces changes in the contact angle according to the Cassie–Baxter equation (eqn (3.3)):

$$\cos\theta_{\mathrm{CB}} = f\cos\theta + f - 1 \tag{3.3}$$

with f being the fraction of the surface wetted by the liquid (f varies from 0 to 1). When the surface roughness increases, more air is trapped, resulting in a decrement of f and a corresponding increment of the contact angle.

In addition to the apparent (static) contact angle, as per the Young, Wenzel or Cassie–Baxter theory, the contact angle hysteresis (CAH) and the sliding angle (α) are typically used to determine the dynamic behavior of liquids on solid surfaces, particularly in the case of super-non-wetting materials.[18] CAH is a measurement of the force required to remove a droplet from a surface and is defined as the difference between the advancing, θ_{adv}, and receding contact angle, θ_{rec} (example of a tilted substrate in Figure 3.3d); α is instead the critical angle at which the droplet rolls off the surface. Super-hydrophobic and self-cleaning surfaces exhibit an apparent contact angle higher than 150° and very low values of CAH and α (less than 10°). Therefore, a small inclination of the substrate is sufficient to easily move water drops onto the surface and potentially wash off dust and contaminants.[19]

3.2.2 Surfaces and Membranes with Engineered Wettability

Diverse groups have reported on membranes with unique wetting behavior by engineering the chemical properties and morphology of electrospun fibers.[17,20–23] Recently, Li and coworkers have fabricated super-hydrophobic nanofibrous membranes by electrospinning poly(vinylidene fluoride) (PVDF) and silica nanoparticles functionalized with octadecyltrichlorosilane (OTS).[24] Three different sizes of SiO_2 nanoparticles were mixed with PVDF: 40 nm (PVDF-S-40), 167 nm (PVDF-S-160) and 210 nm (PVDF-S-210). After the electrospinning process, the nanoparticles were stably attached onto the fibers and hierarchical surfaces with wrinkles, protrusions and particle aggregates were generated, particularly when particle sizes of 167 and 210 nm were used (Figure 3.4a–f). As expected, the increment in surface roughness resulted in a reduced contact area between the water droplets and the electrospun mats and therefore in high hydrophobicity, with water contact angle values of 152°, 155° and 163° for PVDF-S-40, PVDF-S-160 and PVDF-S-210, respectively. The electrospun composite mats were utilized for membrane distillation (MD), which is a separation process in which vapor molecules are driven through a porous membrane by temperature-induced pressure difference.[25] MD tests performed using a 3.5 wt% NaCl aqueous solution at 60 °C for 24 hours showed a permeate vapor flux of 41.1 kg m^{-2} h^{-1} for PVDF-S-40, considerably higher than that of commercial PVDF membranes (7.2 kg m^{-2} h^{-1}), due to the optimized porosity and super-hydrophobic behavior.

In another work, a super-hydrophobic electrospun layer of PVDF- and hexamethyl disilazane-modified fumed silica nanoparticles ($mSiO_2$) has been deposited onto a thermo-cohesive layer of poly(vinyl acetate) (PVAc) and PVDF fibers, in order to create a 'lotus-effect' tape.[26] Different to PVDF fibers that exhibited a smooth surface, the roughness of $mSiO_2$/PVDF mats increased as a function of the amount of SiO_2 nanoparticles. Nano-protrusions appeared on the fiber surface due to the silica nanoparticles, leading to an increment of water contact angle, from 150° for PVDF mats to 174° for $mSiO_2$/PVDF ones (2.1 $mSiO_2$/PVDF mass ratio) and to a reduction of sliding angle up to 3° (Figure 3.4g). Not only water, but also other liquids (milk, orange juice and coffee) were repelled by the nanocomposite fibers (inset Figure 3.4g). The thermoplastic PVAc/PVDF fibers instead promoted the adhesion of the 'lotus-effect' tape onto diverse substrates (glass, wood, plastic and aluminum foil), reaching a peel strength of 230 N m^{-1}, which was higher than that of a commercially-available tape (around 170 N m^{-1}).

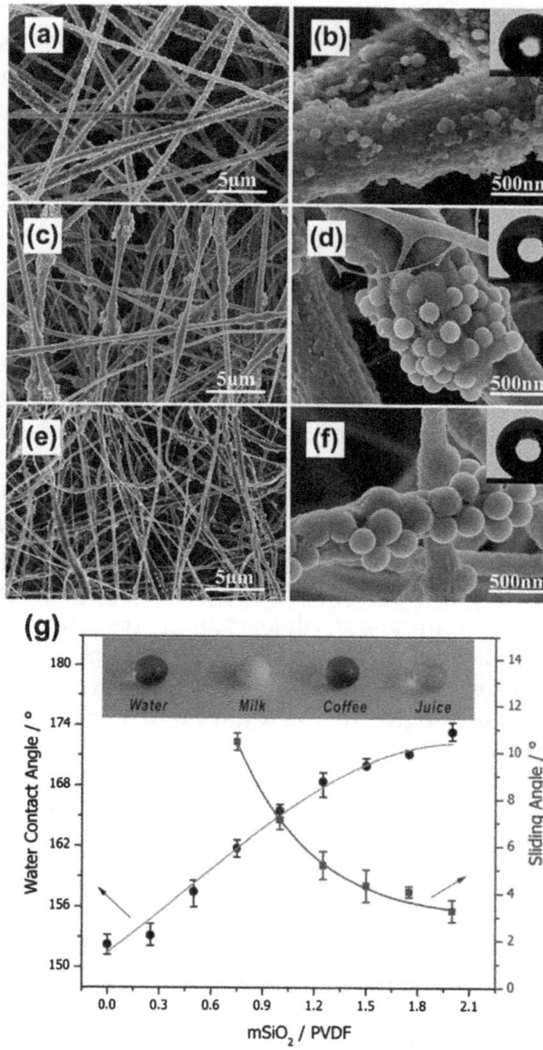

Figure 3.4 SEM images at different magnifications of PVDF electrospun fibers containing SiO_2 nanoparticles with (a, b) 40 nm, (c, d) 167 nm and (e, f) 210 nm diameter, after sonication in isopropyl alcohol for 1 hour at 80 °C. Insets: photographs of water droplets having contact angle values of (b) 152.1°, (d) 153.7° and (f) 160.2°. Adapted with permission from X. Li, X. Yu, C. Cheng, L. Deng, M. Wang and X. Wang, *ACS Appl. Mater. Interfaces*, 2015, 7, 21919, Copyright 2015 American Chemical Society. (g) Graphs of water contact angles (black circles) and sliding angles (blue squares) of electrospun composite mats for different $mSiO_2$/PVDF mass ratios. Inset: photographs of liquid droplets onto the composite mat ($mSiO_2$/PVDF mass ratio of 1.5). Adapted from ref. 26 with permission from the Royal Society of Chemistry.

Silica nanoparticles have also been combined with polyacrylonitrile (PAN) in order to produce super-hydrophobic, waterproof and breathable membranes.[27] PAN fibers were chemically modified by immersion in *n*-hexane solutions of amino-silicone oil (ASO, a low surface energy substance) and then blade-coated with acetone solutions of SiO_2 nanoparticles. ASO was introduced to induce hydrophobicity, as demonstrated by the significant increment in water contact angle, from 32° to 140°. On the other hand, SiO_2 nanoparticles created multiscale rough structures onto the nanofiber surface with a consequent increase in water contact angle up to 151° and changes in thermal conductivity (0.0028 W m^{-1} K^{-1}), water vapor transmittance rate (11.4 kg m^{-2} day^{-1}) and air permeability (20.5 mm s^{-1}). The same group has reported on the combination of electrospinning and electrospray for the production of composite super-hydrophilic and underwater super-oleophobic PAN and SiO_2 membranes.[28] The electrospun membranes were capable of separating micro-scale oil-in-water emulsions with separation fluxes higher than those of traditional membranes: Fluxes of 6456, 2585 and 2371 L m^{-2} h^{-1} were measured for petroleum ether, hexadecane and diesel oil, respectively. A similar system has recently been used as the active element in individual protection filters capable of capturing particulate matter (PM) with aerodynamic equivalent diameter lower than 2.5 µm ($PM_{2.5}$), which is dangerous for human health.[29] Membranes with a gradient structure were prepared by depositing an outer layer of PAN/SiO_2 fibers, an interlayer of PVDF/PAN-SiO_2 fibers and an inner layer of PVDF fibers. The super-hydrophilicity of PAN/SiO_2 and PVDF/PAN-SiO_2 served to transmit water vapor, whereas the hydrophobic nature of PVDF fibers prevented the formation of capillary water under high humidity. The composite filter was tested under conditions of hazardous levels of $PM_{2.5}$ (higher than 150), recording a purification efficiency of 99.99% after 600 minutes with a low drop pressure of 86 Pa. An increase in the pressure drop values was observed in service due to pore clogging, but the filter recovered the initial performance after cleaning.

The studies discussed above and others reported in the literature demonstrate that combining nanomaterials with electrospun nanofibers is an efficient way of obtaining biomimetic hierarchical structures with engineered wetting properties for a wide range of applications.[30–35] Another approach is based on the electrospinning of properly designed polymeric systems without inclusion of nanomaterials. Lee and coworkers have reported on the electrospinning of super-hydrophobic and superoleophilic polystyrene (PS) fibers deposited on stainless steel meshes for the separation of low viscosity oils from water.[36] The composite membranes exhibited a water contact angle of 155° with a sliding angle of 5°, whereas the contact angle of diesel was close to 0°. Diesel easily permeated through the membrane (100 ml in 8 minutes by gravity) and oil–water separation efficiencies of nearly 100% were achieved. In another study, super-hydrophobic fibers with controlled surface morphology (with and without nanopapillae, nanopores and microgrooves) were fabricated by electrospinning PS solutions at different concentrations (15–30 wt%).[37] A gradual decrease in contact angle values, from 155° to 127°, was measured by increasing the PS concentration due to changes in fiber morphology, from a surface with hierarchical micro/nano-scaled roughness to a flat one. The electrospun membranes showed potential for commercial MD applications due to their water distillation performance (vapor flux of 113 kg m^{-2} h^{-1} for 20 g L^{-1} NaCl salt concentration) that was superior to commercial PVDF ones (vapor flux of about 30 kg m^{-2} h^{-1} for 20 g L^{-1} NaCl salt concentration).

Together with conventional polymers and polymer blends, current research aims to create smart bioinspired fibers by electrospinning polymers that change their properties in response to external stimuli, such as gradients of pH, temperature and CO_2.[38–42] For instance, tree-like nanofiber membranes of poly(vinylidene fluoride)-*graft*-poly(acrylic acid) (PVDF-*g*-PAA) with pH responsiveness have been developed for oil–water separation applications.[39] As shown in Figure 3.5a, PVDF-*g*-PAA electrospun fibers possessed a unique morphology composed of trunk and branch fibers with different diameters. These structures were formed by splitting of the polymer jet during the electrospinning process as a consequence of high electrical conductivity of the polymer solution and phase separation phenomena within it. The tree-like structure was responsible for fibers with a specific surface area of 26.7 m^2 g^{-1}, which was significantly higher than that of common PVDF electrospun fibers (7.8 m^2 g^{-1}). Furthermore, the grafting of PAA, which is a well-known pH-responsive polymer, induced changes in the wetting properties of the fibers due to protonation and deprotonation of PAA's carboxylic acid groups. The PVDF-*g*-PAA membranes reversibly switched from super-hydrophilic when in contact with water droplets of pH 6.5 (PAA in the deprotonated state) to hydrophobic (148°) for water droplets of pH 2.0 (PAA in the protonated state), as shown in Figure 3.5b. Moreover, the fibers showed underwater oleophilicity after immersion in water of pH 2.0, and underwater oleophobicity (oil contact angle of 138°) after immersion in water of pH 6.5. Those properties were applied to achieve oil–water separation with high and stable permeation flux of about 9600 L m^{-2} h^{-1} and separation efficiency of 99.58%.

Membranes that can reversibly switch from hydrophilic to hydrophobic on demand have also been obtained by using poly(*N*-isopropylacrylamide) (PNIPAM)[41] and poly(*N*,*N*-diethylaminoethyl methacrylate) (PDEAEMA)[42] as stimuli-responsive polymers. PNIPAM is a temperature-sensitive polymer that changes from a hydrophilic to a hydrophobic state for temperature values higher than its lower critical solution temperature (LCST) of 32 °C.[43] Composite electrospun mats of cellulose acetate (CA) and PNIPAM have been developed to achieve transitions from super-hydrophilicity at room temperature (23 °C) to hydrophobicity at 40 °C (contact angle of around 130°).[41] PDEAEMA has instead been combined with poly(methyl methacrylate) (PMMA), in order to synthesize PMMA-*co*-PDEAEMA copolymers.[42] PDEAEMA is a CO_2-responsive material that undergoes extended hydrophilic chain conformation when reacting with CO_2; upon exposure to N_2 to remove CO_2, PDEAEMA recovers its hydrophobic behavior. The wettability

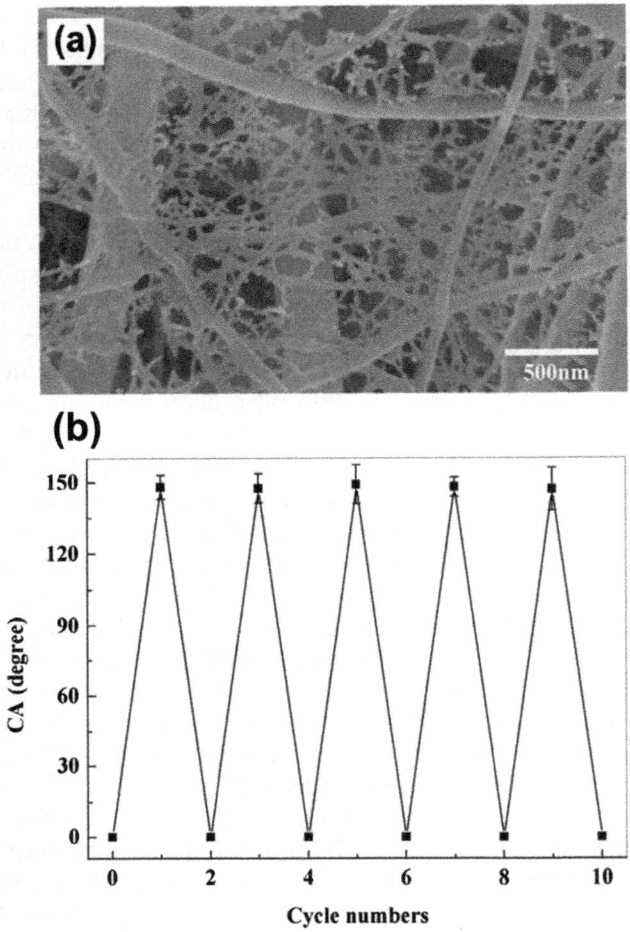

Figure 3.5 (a) SEM image of PVDF-*g*-PAA tree-like nanofiber membrane. (b) Reversible wettability behavior of PVDF-*g*-PAA membranes between superhydrophilicity and hydrophobicity. Reprinted from *Journal of Membrane Science*, **534**, B. Cheng, Z. Li, Q. Li, J. Ju, W. Kang and M. Naebe, Development of smart poly(vinylidene fluoride)-*graft*-poly(acrylic acid) tree-like nanofiber membrane for pH-responsive oil/water separation, 1–8, Copyright 2017, with permission from Elsevier.

of PMMA-*co*-PDEAEMA electrospun fibers was investigated in the presence and absence of CO_2. In a CO_2 atmosphere, the water contact angle of PMMA-*co*-PDEAEMA mats was about 140°; after purging CO_2, a significant reduction of water contact angle to 36° was measured in 5 minutes. A similar transition, which was due to protonation/deprotonation effects of PDEAEMA amine groups, was also observed for underwater oil wettability. In a neutral aqueous environment, a droplet of hexane with initial contact angle of 36° was rapidly absorbed by the fibers. On the contrary, after bubbling CO_2, the oil contact angle reached values of 150°. The switch between hydrophobicity/oleophilicity and hydrophilicity/oleophobicity was reversibly repeated under an alternating CO_2/N_2 stimulation. The smart PMMA-*co*-PDEAEMA membranes were used for oil/water separation tests, showing oil and water fluxes of 17 000 and 9554 L m^{-2} h^{-1}, respectively, in the presence of CO_2. Interestingly for commercial applications, the electrospun mats were capable of high separation efficiency, with water contents in hexane, petroleum ether and *n*-heptane of 30, 40 and 60 ppm, respectively.

The case studies discussed in this section show that the electrospinning of properly designed combinations of materials is a facile and cost-effective method for developing novel, efficient and reusable systems with the desired surface wettability. The potential of these systems has been demonstrated for purification of oil/water mixtures[44] and air filtration,[45] where high flux rates and superior separation efficiencies have been achieved, in comparison to commercially-available counterparts. In addition, the possibility of creating smart membranes that switch their oil/water affinity on and off in response to external stimuli further broadens the future industrial applicability of electrospun fibers.

3.2.3 Commercially-available Products

The academic research that has been conducted on electrospinning in the past two decades has proved that this technology is suitable for the production of innovative, robust and highly efficient surfaces for liquid repellence and membranes for oil–water separation, water desalination and air filtration. The results achieved have attracted industrial interest and companies like Coway, Liquidity Nanotech Corporation, Pardam Nanotechnology and SPUR Nanotechnologies have developed and commercialized filters made of electrospun fibers for water purification. For example, the Nanotrap filter from Coway uses a patented technology for enhancing the durability of electrospun nanofibers by coating them with alumina.[46] The filters eliminate bacteria and viruses from drinking water with high selectivity, and they reduce biofouling. Purification cartridges based on electrospun fibers have also been produced by Liquidity Nanotech Corporation, whose Naked filter mechanically purifies water from bacterial and other contaminants by exploiting the porosity of the electrospun mat.[47] The same working principle has been used by the company eSpin Technologies for the recently launched air filter Exceed, which captures submicron size particles (dust and smoke) and allergy irritants, and it experiences a low pressure drop during its lifetime.[48] The company Revolution Fibres has introduced air filters made of nanofibers infused with natural antimicrobial additives (manuka extracts).[49] Their SETA filtration devices are able to trap allergens, airborne spores and bacteria into the dense fibrous web and to inactivate the microorganisms due to the plant extracts. A wide range of electrospun nanofibers (PVDF, PCL, PAN and polyvinylbutyral fibers) that can be used as membranes for water filtration and for air purification with high efficiency are also produced by Pardam Nanotechnology.[50]

3.3 Scaffolds for Tissue Engineering

Tissue engineering (TE) aims to develop strategies for restoring and improving the biological functions of injured tissues through the combination of scaffolds (natural or synthetic supporting biomaterials), cells (autologous, heterologous, differentiated and stem cells) and bioactive molecules (growth factors, cytokines, drugs, proteins).[51] Ideally, scaffolds for TE have to mimic the 3D architecture of human tissues, which is characterized by a hierarchical assembly of structures across multiple length scales, in order to promote cell growth and consequently tissue repair.

The resemblance of electrospun fibrous mats to the structural organization of the native extracellular matrix (ECM) is one of the main motivations for applying electrospinning to the production of scaffolds for TE. ECM is a 3D network of fibrous-forming proteins, such as collagen, elastin, fibronectin, glycoproteins and glycosaminoglycans, and it offers physical support to cells but also regulates cellular processes.[52] Likewise, electrospun scaffolds provide physical and biochemical cues for tissue regeneration, thanks to their high surface-to-volume ratio, engineered morphology and chemical composition. Their porosity can be controlled in order to promote transport of oxygen and diffusion of nutrients and growth factors, and naturally-occurring bio-macromolecules can be incorporated inside the fibers in order to achieve biocompatibility and bioactivity. In this section, case studies that have recently reported the use of electrospun mats as scaffolds for bone and skin TE will be discussed.

3.3.1 Bone TE

Musculoskeletal pathologies, such as fractures, osteoporosis, tumors and rheumatic diseases, affect millions of people worldwide, and their incidence is expected to increase in the coming years due to the increase in the elderly population.[53,54] Current clinical treatments for large bone defects rely on autologous or allograft transplantation procedures that are expensive, challenging and at high risk of infection, rejection and disease transmission. In order to overcome these complications, artificial implants that mimic the native structure and biomechanics of bone have been proposed.

As illustrated in Figure 3.6, bone is a bio-nanocomposite system that is mainly formed by the hierarchical organization of type-I collagen fibrils (1.5 nm in diameter and 300 nm long) and plates of hydroxyapatite (HA, 25–50 nm in size and 1.5–4.0 nm thick) oriented along the main axis of the collagen fibrils.[55,56] These mineralized collagen fibers are arranged into lamellae and osteons in compact bone (cortical bone), which has a porosity of 20% and anisotropic mechanical properties. Instead, a network of trabeculae forms the spongy architecture of the trabecular bone (cancellous bone), which has a porosity of 90% and load-bearing capabilities.[53] The multiscale structure of bone is pivotal in regulating cellular biochemical processes (cell adhesion

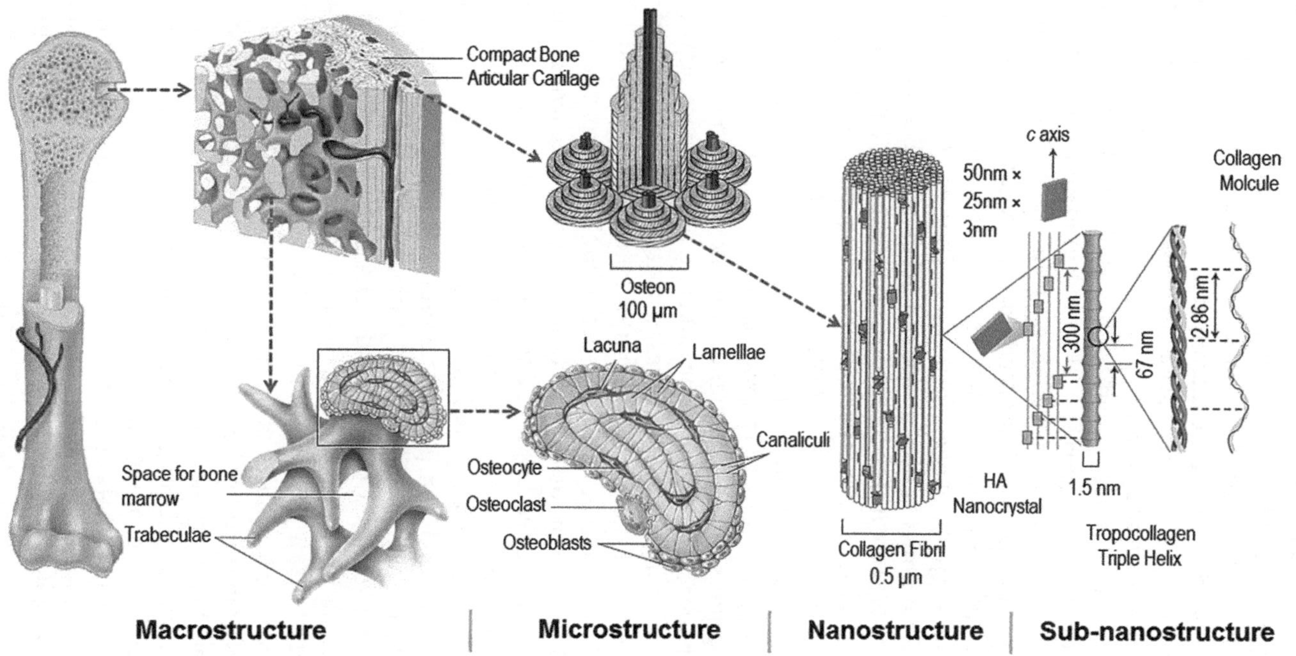

Figure 3.6 Hierarchical structure of bone, from macro- to nano-scale. Reprinted from *Biomaterials*, **83**, X. Wang, S. Xu, S. Zhou, W. Xu, M. Leary, P. Choong, M. Qian, M. Brandt and Y. M. Xie, Topological design and additive manufacturing of porous metals for bone scaffolds and orthopaedic implants: A review, 127–141, Copyright 2016, with permission from Elsevier.

and migration, vascularization, biomineralization) and tissue remodeling in response to mechanical loading.

Electrospun scaffolds for bone TE are usually based on hybrid systems: a matrix of biodegradable polymer, such as collagen, gelatine, poly(ε-caprolactone) (PCL), poly(L-lactic acid) (PLLA) and poly(lactide-*co*-glycolide) (PLGA), incorporating bioactive inorganic nanomaterials, such as HA and bioglasses, and therapeutics molecules, such as growth factors, antibiotics and drugs.[57–64] Gandhimathi and coworkers have produced composite fibers of PLLA-*co*-PCL/silk fibroin/ascorbic acid/tetracyclinehydrochloride (PLACL/SF/AA/TC) containing nano-HA (n-HA).[57] The biodegradation and tensile properties of the copolymer (PLACL) have been combined with the bioactivity of SF, AA and TC in order to obtain scaffolds that regulate cell attachment and osteogenic differentiation. In fact, AA is a supplement for osteoblast differentiation and type-I collagen formation, whereas TC is a broad-spectrum antibiotic that limits bone damage and promotes cell attachment. The resulting PLACL/SF/AA/TC/n-HA electrospun scaffolds had a porosity of 94% and a tensile strength of 4.32 MPa, which was higher than that measured for fibers without silk fibroin (1.51 MPa tensile strength of PLACL fibers), due to the exceptional mechanical properties of silk. *In vitro* tests on human mesenchymal stem cells (hMSCs) showed that the presence of HA increased the surface roughness of the fibers and stimulated osteogenesis and mineralization. Enhanced cell proliferation, alkaline phosphatase activity (ALP, primary indicator of immature osteoblast activity) and expression of osteocalcin (OCN, marker of biomineralization) were recorded for PLACL/SF/AA/TC/n-HA. This indicated that hMSCs seeded onto the electrospun nanocomposite scaffolds underwent osteogenic differentiation, reached the mineralization stage and deposited ECM. In another work, nano-HA particles were created on the surface of porous PLLA and PLLA/collagen fibers by *in vitro* mineralization through calcium and phosphorous treatments.[58] In contrast to PLLA fibers, PLLA/collagen fibers determined the formation of HA crystals due to collagen nucleation sites: HA ratios of 57.3 wt% and 23.5 wt% were measured for PLLA + Col + HA and PLLA + HA, respectively. The HA minerals made the fibers hydrophilic (water contact angles of around 5° and 117° for PLLA + Col + HA and PLLA + HA, respectively) and increased their surface area (48.82 and 21.06 $m^2\ g^{-1}$ for PLLA + Col + HA and PLLA + HA, respectively). On the other hand, collagen was required to obtain a high number of viable MSCs on the fibrous scaffolds. Cell osteogenesis was monitored by recording ALP activity, expression levels of OCN and osteopontin (OPN), and formation of mineralized ECM. ALP activity was significantly lower for mineralized scaffolds than for porous PLLA fibers, possibly due to the phosphate-rich environment that down-regulated ALP expression. On the contrary, MSCs expressed high levels of OCN, OPN and calcium deposition on PLLA + Col + HA electrospun scaffolds, proving that collagen and HA had an important effect in inducing osteoblastic differentiation of MSCs. The osteoinductive scaffolds produced, which were seeded and not seeded with MSCs, were then implanted in mice. Histological analysis revealed no bone formation for PLLA + Col, whereas new bone and neo-vasculature were observed for PLLA + Col + Ha electrospun scaffolds, due to their ability to recruit stem cells and control their differentiation. The results obtained in this study highlight how the synergy between diverse factors, like fibrous structure, nano-roughness and naturally-occurring molecules (collagen and HA), is critical for creating highly osteoinductive composite implants and promoting bone regeneration.

The group led by H.-W. Kim has encapsulated mesoporous bioactive glass nanoparticles (MBNs) loaded with osteogenic enhancer fibroblast growth factors and osteogenic drugs into electrospun fibers in order to elicit therapeutic action.[59,60] MBNs are bioactive nanocomponents that induce apatite mineralization and ionic release. Core-shell electrospun fibers have been used for the sequential release of two different growth factors:[59] fast release of fibroblast growth factor 2 (FGF2), which stimulates cellular mitosis and angiogenesis, and slow release of FGF18, which induces osteogenesis. FGF18-loaded MBNs with a size of 119 nm, a surface area of 527.2 $m^2 g^{-1}$ and 6.97 nm mesopores were added into the poly(ethylene oxide) (PEO) core of PEO/PCL core–shell fibers at different concentrations (1, 3 and 5 wt%). FGF2 was instead loaded into the core of the fibers to achieve rapid delivery. Initially, the burst release of the therapeutic molecules was limited by the PCL outer shell, which acted as diffusion barrier; then rapid release was observed due to PCL degradation over time. In addition, MBNs pre-loading significantly extended the release profile. The use of MBNs also had an effect on the mechanical properties of the scaffolds by increasing the tensile strength and the elastic modulus up to 9.1 MPa and 18.7 MPa, respectively, for 5% MBNs (1.8 MPa and 4.5 MPa, respectively, for fibers without MBNs), and decreasing the elongation at break (145% for 5% MBNs and 620% for 0% MBNs). Moreover, by increasing the MBN content the hydrolytic degradation of the fibers and their mineralization ability increased as well. Degradation analyses in phosphate-buffered saline (PBS) at 37 °C for 30 days revealed a weight loss of 5 and 15% for scaffolds with 0% and 5% MBNs, respectively. *In vitro* tests showed that, differently from fibers without MBNs, scaffolds with 5% MBNs (Figure 3.7a) were almost completely covered with calcium phosphate crystals after a 14 day immersion in simulated body fluid (SBF). Once loaded with FGF2/FGF18, the composite scaffolds promoted osteogenic differentiation of MSCs, as demonstrated by the high levels of ALP and OPN measured, and bone formation in rat calvarium defects. As shown by the microcomputed tomography (µCT) images in Figure 3.7b, the addition of the growth factors induced a substantial increase in hard tissue formation after 6 weeks of implantation, compared to the control group and to animals treated with fibers without FGF2/FGF18. Bone ingrowth was observed from the margin towards the center of the defects, with bone volume higher than 30 mm^3 (Figure 3.7c) and bone surface density higher than 4 g cm^{-2} (Figure 3.7d). This work is an example of how osteogenic stimulating platforms with controlled mechanical properties, degradation behavior and release profile can be created by incorporating mesoporous nano-carriers into electrospun scaffolds.

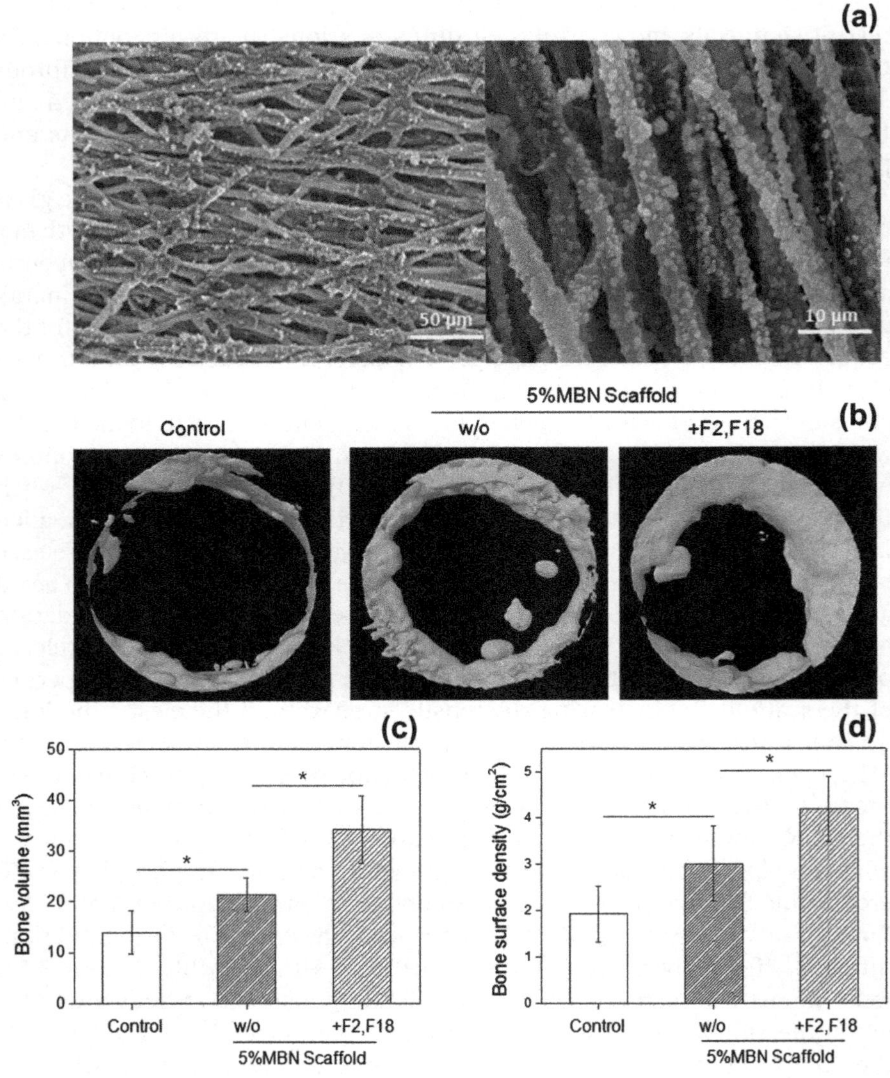

Figure 3.7 (a) SEM images of electrospun scaffolds with 5% MBNs after 14 days of immersion in SBF. (b) μCT images of rat calvarium defects for control group and mice treated with 5% MBNs scaffolds without and with FGF2/FGF18, after 6 weeks of implantation. (c) Bone volume and (d) bone surface density for control group and mice treated with 5% MBNs scaffolds without and with FGF2/FGF18. Reprinted from *Acta Biomaterialia*, **16**, M. S. Kang, J.-H. Kim, R. K. Singh, J.-H. Jang and H.-W. Kim, Therapeutic-designed electrospun bone scaffolds: mesoporous bioactive nanocarriers in hollow fiber composites to sequentially deliver dual growth factors, 103–116, Copyright 2015, with permission from Elsevier.

3.3.2 Skin TE

Skin is the largest organ of the human body (7% of the total body weight) and acts as a protective barrier against the external environment.[65,66] It also has thermoregulatory and sensory functions. Skin is formed of two layers, the epidermis and the dermis, which are mechanically supported by a layer of fat, the hypodermis.[67] The epidermis is the outermost layer and consists of keratinocytes in different stages of differentiation, Langerhans cells and melanocytes.[67,68] The dermis, instead, is composed of ECM proteins (fibers of collagen and elastin), blood vessels, glands and nerves. The main cellular components of this layer are fibroblasts, which are responsible for the synthesis of ECM molecules.

Skin injuries caused by mechanical, thermal or chemical traumas or diseases are referred to as wounds and cause a disruption of the anatomical structure of skin and its functions. The cascade of events that naturally take place after an injury in order to restore skin's properties is called wound healing. Nowadays, delayed or impaired wound healing is considered a worldwide health concern, with a negative impact on a patient's quality of life (infections, amputations and mortality) and an economic burden of over US$9.5 billion a year.[69] Chronic wounds, which are generally the result of vascular insufficiency, pressure necrosis, infections, cancer and diabetes, often persist in a pathological inflammatory state and fail to complete the healing process. The prevalence of chronic wounds is expected to rise over the coming years due to an ageing population (up to a 60% increase in the population of Europe aged over 65 by 2050) and obesity (20% of the world's adults will be obese by 2030).[69]

As illustrated in Figure 3.8, the wound healing process consists of four sequential and overlapping phases: homeostasis, inflammation, proliferation and remodeling.[69] Homeostasis happens within seconds or minutes after the injury. A fibrin cloth of polymerized fibrinogen, fibronectin, vitronectin, and thrombospondin is formed in order to stop blood loss at the wound site and protect the injured tissue from microorganisms. In the inflammatory stage (hours to days), neutrophils and macrophages are recruited by platelet-produced cytokines. They are involved in debris clearance and in the production of cytokines and growth factors that stimulate the migration of fibroblasts to the wound region. During the proliferation phase (days to weeks), fibroblasts, epithelial cells and keratinocytes are responsible for ECM formation, re-epithelialization and angiogenesis. The last phase is the remodeling, which can last from weeks to years. It results in acellular scar tissue, maturation of ECM proteins and arrangement of collagen fibers. When the healing process is completed, the skin structure is regenerated and its functions are restored.

Products for wound care, such as skin substitutes and dressings, aim to accelerate the healing process by influencing the cellular pathways of tissue regeneration, and limiting infections and disease transmission.[65] Electrospinning has been extensively used for the fabrication of such products owing to its ability to recreate the structure and chemical composition of

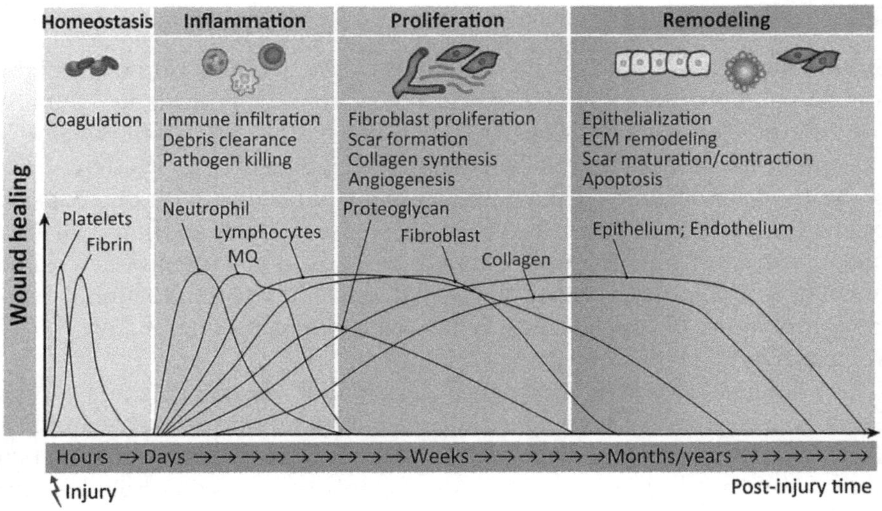

Figure 3.8 Timeline of the classical model of wound healing and relative phases: homeostasis, inflammation, proliferation and remodeling. Reproduced from ref. 69 with permission from the Royal Society of Chemistry.

ECM and deliver bioactive molecules.[70–73] In this section, current research in the field of electrospun wound dressings will be discussed, with particular attention on composite systems based on natural polymers.

Electrospun dressings are attractive materials for skin TE because their fibrous architecture emulates the native ECM and therefore stimulates cell proliferation and differentiation, while their porosity facilitates the absorption of exudates and the diffusion of oxygen and nutrients. In addition, the mechanical flexibility of electrospun mats ensures good coverage and protection of the injured tissue, even against bacteria. In particular, electrospun fibers based on natural macromolecules, such as proteins and polysaccharides, are widely used for their bioactivity, biocompatibility and biodegradability.[74–79] Recently, nanofibrous scaffolds of gelatine (a derivative of collagen) and polyglycolic acid (PGA) with a payload of monocyte chemoattractant protein-1 (MCP-1, a recruiter of macrophages and monocytes) have been developed in order to accelerate wound healing in diabetic mice.[79] *In vitro* pharmacokinetic studies have shown that the scaffolds released the MCP-1 within 3 days, allowing a rapid recruitment of macrophages and monocytes to the wound bed. Indeed, animal tests on a model of a full thickness excisional cutaneous wound in diabetic mice recorded the highest density of F4/80+ murine macrophages in the wound bed, 3 days after surgery, for animals treated with gelatine/PGA/MCP-1 scaffolds. Moreover, the composite scaffolds promoted the complete closure of the wound within 10 days, in contrast to the 14 days needed for the control group (animals without treatment). The accelerated healing process and re-epithelialization were

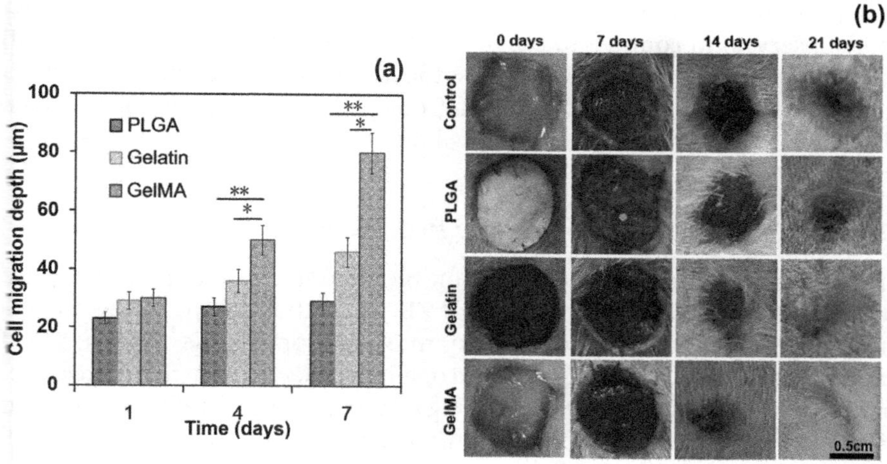

Figure 3.9 (a) Cell migration depth into GelMA, gelatine and PLGA scaffolds at different time points. *Significant difference between GelMA and gelatine scaffolds. **Significant difference between GelMA and PLGA scaffolds. (b) Photographs of wound beds for animal groups (control, PLGA, gelatine and GelMA) at different time points. Reprinted from *Acta Biomaterialia*, **49**, X. Zhao, X. Sun, L. Yildirimer, Q. Lang, Z. Y. Lin, R. Zheng, Y. Zhang, W. Cui, N. Annabi and A. Khademhosseini, Cell infiltrative hydrogel fibrous scaffolds for accelerated wound healing, 66–77, Copyright 2017, with permission from Elsevier.

attributed to the presence of macrophages that secreted vascular endothelial growth factors and pro-inflammatory cytokines.

Another example of an advanced wound dressing based on electrospun fiber has been reported by Zhao and collaborators using a photo-crosslinkable hydrogel based on gelatine methacrylamide (GelMA).[80] The crosslink density of the GelMA electrospun fibers was controlled by ultraviolet (UV) light, achieving low water permeability and high water retention capacity by increasing UV exposure. These two properties are important to maintain hydration of the wound bed. *In vitro* tests on fibroblasts demonstrated that the GelMA scaffolds promoted cell adhesion, viability and migration, due to their low stiffness. As shown in Figure 3.9a, after 7 days of cell culture, the highest penetration depth (around 80 µm) was measured for cells seeded onto GelMA fibers in comparison to electrospun control mats of gelatine (around 60 µm) and PLGA (around 40 µm). This resulted in fully cellularized 3D constructs that promoted the healing process of mouse dorsal wounds (8 mm diameter). On day 14 after surgery (Figure 3.9b), GelMA scaffolds induced 80% wound closure that was similar to that of gelatine fibers and greater than that of PLGA (60%) and the control group (no treatment). The granulation tissue was replaced by a new tissue and collagen deposition was detected. Complete healing was reached after 3 weeks for GelMA fibers, with no visible wound, in contrast to 5 and 10% wound areas for gelatine and PLGA scaffolds, respectively (Figure 3.9b). Intact epidermis (100 μm thickness) was also observed, in contrast to the other groups (PLGA scaffolds and control). The results of this work demonstrate the ability of electrospun dressings to accelerate wound healing by tailoring not only the bioactivity of the fibers but also their physical and mechanical properties.

3.3.3 Commercially-available Products

Driven by the potential of electrospun membranes for biomedical applications and particularly as scaffolds for TE, innovative companies worldwide have based their business on electrospinning of biomaterials. Mimetix® scaffolds consisting of randomly oriented or aligned electrospun fibers have been commercialized by the Electrospinning Company for clinical trials and medical devices.[81] They are supplied gamma-irradiated and in standard 6-well, 12-well, 96-well and 384-well format, ideal for cell culture studies. The company Nanofiber Solutions also produces electrospun fiber multiwell plates for cell cultures and regenerative medicine.[82] The fiber layers are 20 μm thick and are composed of either aligned (NanoAligned™) or randomly oriented (NanoECM™) PCL nanofibers. OrthoReBirth has developed ReBOSSIS, a bioresorbable bone void filler for the repair of bone defects.[83] ReBOSSIS is made of β-tricalcium phosphate (β-TCP), PLLA and siloxane-containing calcium carbonate (SiV). The interconnectedness and high porosity of this scaffold promotes the formation of new bone and the growth of blood vessels. Nicast commercializes AVflo™, a CE-certified polyurethane vascular access graft for hemodialysis patients, with self-sealing features and a multi-layered structure.[84] AVflo™ has been used in clinical studies and it has the potential to replace central venous catheters and arteriovenous fistula grafts. The same company has also developed SpinCare™, a portable bedside electrospinning system that creates dressings to fit the wound size. Finally, Arsenal Medical uses AxioCore™ core-sheath nanofiber technology to produce scaffolds that promote cell integration and tissue regeneration, and enable the delivery of bioactive compounds.[85]

3.4 Conclusions

The past 20 years have seen increasing scientific and industrial attention on composite biomimetic systems produced by electrospinning, for applications ranging from filtration to regenerative medicine. As discussed in this chapter, recent studies have demonstrated that nano-structuring functional materials (synthetic and natural polymers, proteins and inorganic compounds) in the form of fibers through electrospinning allows the creation of hierarchical structures with controlled morphology, super-non-wetting properties, stimuli responsiveness and improved bioactivity. Most of the research advances in the area of electrospun fibers for liquid filtration, air purification and TE have already been translated into commercial products, as highlighted by the examples described in this chapter.

References

1. B. Bhushan, *Phil. Trans. R. Soc., A*, 2009, **367**, 1445.
2. J. F. V. Vincent, O. A. Bogatyreva, N. R. Bogatyrev, A. Bowyer and A.-K. Pahl, *J. R. Soc., Interface*, 2006, **3**, 471.
3. P. Fratzl, *J. R. Soc., Interface*, 2007, **4**, 637.
4. J. Zhao, X. Zhao, Z. Jiang, Z. Li, X. Fan, J. Zhu, H. Wu, Y. Su, D. Yang, F. Pan and J. Shi, *Prog. Polym. Sci.*, 2014, **39**, 1668.
5. E. Mele, S. Girardo and D. Pisignano, *Langmuir*, 2012, **28**, 5312.
6. N. Zhao, Z. Wang, C. Cai, H. Shen, F. Liang, D. Wang, C. Wang, T. Zhu, J. Guo, Y. Wang, X. Liu, C. Duan, H. Wang, Y. Mao, X. Jia, H. Dong, X. Zhang and J. Xu, *Adv. Mater.*, 2014, **26**, 6994.
7. X. Wang, B. Ding and B. Li, *Mater. Today*, 2013, **16**, 229.
8. P.-Y. Chen, J. McKittrick and M. A. Meyers, *Prog. Mater. Sci.*, 2012, **57**, 1492.
9. B. Bhushan and Y. C. Jung, *Prog. Mater. Sci.*, 2011, **56**, 1.
10. M. A. Meyers, A. Y.-M. Lin, P.-Y. Chen and J. Muyco, *J. Mech. Behav. Biomed. Mater.*, 2008, **1**, 76.
11. T. Li and K. Zeng, *Nanoscale*, 2014, **6**, 2177.
12. Y. Liu, D. Luo and T. Wang, *Small*, 2016, **12**, 4611.
13. J. Lin, X. Wang, B. Ding, J. Yu, G. Sun and M. Wang, *Crit. Rev. Solid State Mater. Sci.*, 2012, **37**, 94.
14. S. Martin, P. S. Brown and B. Bhushan, *Adv. Colloid Interface Sci.*, 2017, **241**, 1.
15. J. Yong, F. Chen, Q. Yang, J. Huo and X. Hou, *Chem. Soc. Rev.*, 2017, **46**, 4168.
16. T. Darmanin and F. Guittard, *J. Mater. Chem. A*, 2014, **2**, 16319.
17. N. Nuraje, W. S. Khan, Y. Lei, M. Ceylan and R. Asmatulu, *J. Mater. Chem. A*, 2013, **1**, 1929.
18. Z. Chu and S. Seeger, *Chem. Soc. Rev.*, 2014, **43**, 2784.
19. L. Chen, Z. Guo and W. Liu, *J. Mater. Chem. A*, 2017, **5**, 14480.
20. J. Wu, N. Wang, Y. Zhao and L. Jiang, *J. Mater. Chem. A*, 2013, **1**, 7290.
21. E. Mele, I. S. Bayer, G. Nanni, J. A. Heredia-Guerrero, R. Ruffilli, F. Ayadi, L. Marini, R. Cingolani and A. Athanassiou, *Langmuir*, 2014, **30**, 2896.
22. A. Davis, E. Mele, J. A. Heredia-Guerrero, I. S. Bayer and A. Athanassiou, *J. Mater. Chem. A*, 2015, **3**, 23821.
23. C. R. Reshmi, P. S. Suja, A. Juraij and A. Sujith, *RSC Adv.*, 2017, **7**, 2092.
24. X. Li, X. Yu, C. Cheng, L. Deng, M. Wang and X. Wang, *ACS Appl. Mater. Interfaces*, 2015, 7, 21919.
25. A. Alkhudhiri, N. Darwish and N. Hilal, *Desalination*, 2012, **287**, 2.
26. B. Yue, B. Zhang, J. You, Y. Li, L. Li and J. Li, *RSC Adv.*, 2016, **6**, 17215.
27. J. Sheng, Y. Xu, J. Yu and B. Ding, *ACS Appl. Mater. Interfaces*, 2017, **9**, 15139.
28. J. Ge, J. Zhang, F. Wang, Z. Li, J. Yu and B. Ding, *J. Mater. Chem. A*, 2017, **5**, 497.
29. X. Zhao, Y. Li, T. Hua, P. Jiang, X. Yin, J. Yu and B. Ding, *Small*, 2017, **11**, 1603306.

30. V. A. Ganesh, S. S. Dinachali, A. S. Nair and S. Ramakrishna, *ACS Appl. Mater. Interfaces*, 2013, **5**, 1527.
31. Y. Liao, R. Wang and A. G. Fane, *J. Membr. Sci.*, 2013, **440**, 77.
32. Y. Liao, R. Wang and A. G. Fane, *Environ. Sci. Technol.*, 2014, **48**, 6335.
33. Y. Liao, C.-H. Loh, R. Wang and A. G. Fane, *ACS Appl. Mater. Interfaces*, 2014, **6**, 16035.
34. S. M. S. Shahabadi, H. Rabiee, S. M. Seyedi, A. Mokhtare and J. A. Brant, *J. Membr. Sci.*, 2017, **537**, 140.
35. Y. Wang, C. Lai, X. Wang, Y. Liu, H. Hu, Y. Guo, K. Ma, B. Fei and J. H. Xin, *ACS Appl. Mater. Interfaces*, 2016, **8**, 25612.
36. M. W. Lee, S. An, S. S. Latthe, C. Lee, S. Hong and S. S. Yoon, *ACS Appl. Mater. Interfaces*, 2013, **5**, 10597.
37. X. Li, C. Wang, Y. Yang, X. Wang, M. Zhu and B. S. Hsiao, *ACS Appl. Mater. Interfaces*, 2014, **6**, 2423.
38. C. Huang, S. J. Soenen, J. Rejman, B. Lucas, K. Braeckmans, J. Demeester and S. C. De Smedt, *Chem. Soc. Rev.*, 2011, **40**, 2417.
39. B. Cheng, Z. Li, Q. Li, J. Ju, W. Kang and M. Naebe, *J. Membr. Sci.*, 2017, **534**, 1.
40. J.-J. Li, Y.-N. Zhou and Z.-H. Luo, *ACS Appl. Mater. Interfaces*, 2015, **7**, 19643.
41. V. A. Ganesh, A. S. Ranganath, R. Sridhar, H. K. Raut, S. Jayaraman, R. Sahay, S. Ramakrishna and A. Baji, *Macromol. Rapid Commun.*, 2015, **36**, 1368.
42. H. Che, M. Huo, L. Peng, T. Fang, N. Liu, L. Feng, Y. Wei and J. Yuan, *Angew. Chem., Int. Ed.*, 2015, **54**, 8934.
43. C. de las Heras Alarcon, S. Pennadam and C. Alexander, *Chem. Soc. Rev.*, 2005, **34**, 276.
44. P. S. Suja, C. R. Reshmi, P. Sagitha and A. Sujith, *Polym. Rev.*, 2017, **57**, 467.
45. M. Zhu, J. Han, F. Wang, W. Shao, R. Xiong, Q. Zhang, H. Pan, Y. Yang, S. K. Samal, F. Zhang and C. Huan, *Macromol. Mater. Eng.*, 2017, **302**, 1600353.
46. https://www.coway.com.my/products/water-purifier, last accessed August 2017.
47. https://nakedfilter.com/, last accessed August 2017.
48. http://exceedfilters.com/page/about.html, last accessed August 2017.
49. https://www.revolutionfibres.com/products/, last accessed November 2017.
50. http://www.pardam.cz/products/, last accessed November 2017.
51. F. J. O'Brien, *Mater. Today*, 2011, **14**, 88.
52. A. D. Theocharis, S. S. Skandalis, C. Gialeli and N. K. Karamanos, *Adv. Drug Delivery Rev.*, 2016, **97**, 4.
53. M. A. Fernandez-Yague, S. A. Abbah, L. McNamara, D. I. Zeugolis, A. Pandit and M. J. Biggs, *Adv. Drug Delivery Rev.*, 2015, **84**, 1.
54. L. Roseti, V. Parisi, M. Petretta, C. Cavallo, G. Desando, I. Bartolotti and B. Grigolo, *Mater. Sci. Eng., C*, 2017, **78**, 1246.

55. X. Wang, S. Xu, S. Zhou, W. Xu, M. Leary, P. Choong, M. Qian, M. Brandt and Y. M. Xie, *Biomaterials*, 2016, **83**, 127.
56. Y. Liu, D. Luo and T. Wang, *Small*, 2016, **12**, 4611.
57. C. Gandhimathi, J. R. Venugopal, A. Y. Tham, S. Ramakrishna and S. D. Kumar, *Mater. Sci. Eng., C*, 2015, **49**, 776.
58. S. Liao, L. T. H. Nguyen, M. Ngiam, C. Wang, Z. Cheng, C. K. Chan and S. Ramakrishna, *Adv. Healthcare Mater.*, 2014, **3**, 737.
59. M. S. Kang, J.-H. Kim, R. K. Singh, J.-H. Jang and H.-W. Kim, *Acta Biomater.*, 2015, **16**, 103.
60. A. El-Fiqi, J.-H. Kim and H.-W. Kim, *ACS Appl. Mater. Interfaces*, 2015, **7**, 1140.
61. H. Chen, X. Huang, M. Zhang, F. Damanik, M. B. Baker, A. Leferink, H. Yuan, R. Truckenmuller, C. van Blitterswijk and L. Moroni, *Acta Biomater.*, 2017, **59**, 82.
62. S. Amjadian, E. Seyedjafari, B. Zeynali and I. Shabani, *Int. J. Pharm.*, 2016, **507**, 1.
63. T. Xu, J. M. Miszuk, Y. Zhao, H. Sun and H. Fong, *Adv. Healthcare Mater.*, 2015, **4**, 2238.
64. J. S. Fernandes, P. Gentile, M. Martins, N. M. Neves, C. Miller, A. Crawford, R. A. Pires, P. Hatton and R. L. Reis, *Acta Biomater.*, 2016, **44**, 168.
65. J. R. Dias, P. L. Granja and P. J. Bartolo, *Prog. Mater. Sci.*, 2016, **84**, 314.
66. D. Tartarini and E. Mele, *Front. Bioeng. Biotechnol.*, 2015, **3**, 206.
67. J. Wang and M. Windbergs, *Eur. J. Pharm. Biopharm.*, 2017, **119**, 283.
68. M. Norouzi, S. M. Boroujeni, N. Omidvarkordshouli and M. Soleimani, *Adv. Healthcare Mater.*, 2015, **4**, 1114.
69. E. Mele, *J. Mater. Chem. B*, 2016, **4**, 4801.
70. M. Liu, X.-P. Duan, Y.-M. Li, D.-P. Yang and Y.-Z. Long, *Mater. Sci. Eng., C*, 2017, **76**, 1413.
71. K. Ghosal, S. Thomas, N. Kalarikkal and A. Gnanamani, *J. Polym. Res.*, 2014, **21**, 410.
72. K. Ghosal, A. Manakhov, L. Zajíčková and S. Thomas, *AAPS PharmSciTech*, 2017, **18**, 72.
73. T. Uyar and E. Kny, in *Electrospun Materials for Tissue Engineering and Biomedical Applications: Research, Design and Commercialization*, Woodhead Publishing, 2017.
74. W. Zhang, S. Ronca and E. Mele, *Nanomaterials*, 2017, **7**, 42.
75. H. Hajiali, M. Summa, D. Russo, A. Armirotti, V. Brunetti, R. Bertorelli, A. Athanassiou and E. Mele, *J. Mater. Chem. B*, 2016, **4**, 1686.
76. I. Liakos, L. Rizzello, H. Hajiali, V. Brunetti, R. Carzino, P. P. Pompa, A. Athanassiou and E. Mele, *J. Mater. Chem. B*, 2015, **3**, 1583.
77. S. Bhowmick, S. Rother, H. Zimmermann, P. S. Lee, S. Moeller, M. Schnabelrauch, V. Koul, R. Jordan, V. Hintze and D. Scharnweber, *Mater. Sci. Eng., C*, 2017, **79**, 15.
78. I. Yousefi, M. Pakravan, H. Rahimi, A. Bahador, Z. Farshadzadeh and I. Haririan, *Mater. Sci. Eng., C*, 2017, **75**, 433.

79. H. Yin, G. Ding, X. Shi, W. Guo, Z. Ni, H. Fu and Z. Fu, *Colloids Surf., B*, 2016, **145**, 226.
80. X. Zhao, X. Sun, L. Yildirimer, Q. Lang, Z. Y. Lin, R. Zheng, Y. Zhang, W. Cui, N. Annabi and A. Khademhosseini, *Acta Biomater.*, 2017, **49**, 66.
81. http://www.electrospinning.co.uk/, last accessed August 2017.
82. http://www.nanofibersolutions.com, last accessed November 2017.
83. http://orthorebirth.com, last accessed August 2017.
84. http://www.nicast.com/, last accessed August 2017.
85. http://www.arsenalmedical.com, last accessed November 2017.

CHAPTER 4

Cell Electrospinning and Technology Transfer from Lab to Market Scale

A. TOWNSEND-NICHOLSON*[a] AND S. N. JAYASINGHE[b]

[a]University College London, Research Department of Structural & Molecular Biology, Darwin Building, Gower Street, London WC1E 6BT, UK; [b]BioPhysics Group, University College London, Department of Mechanical Engineering, Roberts Engineering Building, Torrington Place, London WC1E 7JE, UK
*E-mail: a.townsend-nicholson@ucl.ac.uk

4.1 Introduction

Through advances in clinical medicine, the estimated lifespan of humans has been significantly increased and the causes of morbidity and mortality have shifted substantially over the past 100 years. The leading cause of death in 2015, according to the World Health Organization, was ischemic heart disease, closely followed by stroke; these two causes of death were responsible for 15 million of the 56.4 million deaths in 2015. Chronic diseases now cause an increasing number of deaths worldwide and lead to a significantly reduced quality of life. In 2015, non-communicable diseases contributed to between 62% and 89% of deaths in all regions of the world except Africa, where only 33% of deaths were attributed to this cause. In the USA, the leading causes of death in 1900 were infectious diseases, with heart disease ranked at number 4, yet in 2014, heart disease was ranked at number 1 and

Soft Matter Series No. 7
Electrospinning: From Basic Research to Commercialization
Edited by Erich Kny, Kajal Ghosal and Sabu Thomas

Published by the Royal Society of Chemistry, www.rsc.org

the top three causes of death (heart disease, cancer and chronic lower respiratory disease) accounted for over 50% of all deaths. Over the past 5 years, the leading causes of death in the USA have remained relatively constant and it is reasonable to assume that this trend will be maintained in other developed nations.[1–3]

Tissue and organ transplantation technologies provide a means of replacing diseased and dysfunctional tissues and organs, but perusal of the national transplant activity reports from any country collecting and publishing such data reveals that the number of patients on transplant waiting lists vastly exceeds the number of available donors. Each year, significant numbers of patients die waiting for a suitable donor organ to become available. Cross-species organ transplantation has been considered a means of improving the number of organs available for transplant, with genetic modification used to overcome the difficulty of human antibody binding to animal antigens, as occurs in the case of xenotransplantation using pig organs. Under these circumstances, a number of different genetic modifications have been employed to prevent recognition of the pig's vascular endothelial galactose oligosaccharide, galactose-α1,3-galactose, by human antibodies and to provide the pig with increased resistance to human complement-mediated injury. The use of animal organs in the clinic is fast approaching the stage of clinical trials with international consensus.[4,5]

Donor availability is not the only limitation that can impact on improving patient quality of life through transplant. It can also be difficult to find a suitable donor match for patients. This can occur, for example, in the case of patients of multiracial heritage who require therapeutic treatments based on stem cells obtained from cord blood, where certain genetic markers are inherited. In such conditions, a donor with similar racial heritage may present a much-improved chance of success. Mesenchymal stem cells (MSCs) obtained from cord blood provide a very good example of this. MSCs are of great utility in the replacement of damaged and diseased tissues as a result of their ability to differentiate into a wide variety of tissues and a number of tissue engineering strategies have been developed using MSC cells.[6] The Food and Drug Administration (FDA) has approved at least five different stem cell-based medicinal products derived from cord blood.[7]

Stem cells play a central role in regenerative medicine, an emerging branch of medical sciences that enables treatment for patients whose own regenerative systems are unable to restore a lost or reduced functionality. There are six major classes of stem cells that have been identified as having clinical promise for the treatment of defects and the regeneration of tissue.[8] The World Health Organization, amongst others, provides access to a searchable international clinical trials registry platform to identify the clinical trials of stem cell-based therapies that are currently under way.[9] Stem cells coupled with tissue engineering provides a foundation for the provision of therapies for disease and for the repair of damage that cannot be achieved by conventional medicines.

The adoption of tissue engineering strategies to replace diseased or damaged tissues is particularly enticing as this approach is directly aligned with the philosophy of personalized medicine, where bespoke treatments can be created that are directly targeted at an individual's own and unique therapeutic needs. Tissue engineering has the capacity to produce a bespoke replacement tissue using cells in which pre-existing defects have been repaired, which is an extremely powerful and highly individualized approach to the treatment of disease. The generation of replacement tissues for each patient using that patient's own cells bypasses any requirement to identify suitable tissue matches from a donor pool and neatly circumvents the problems of transplant rejection as the patient in question is their own donor, thereby preventing rejection of the tissue to be transplanted. Gene therapy technologies can be included as part of the tissue engineering protocol to rectify any causal mutations encoded in the patient's own DNA prior to construction of the engineered tissues.

Tissue engineering methodologies have been undergoing a rapid evolution over the past decade and have clearly moved now from the invention phase into the development phase. Several examples have proved particularly promising and are heading for the clinic. This chapter will describe the use of one such tissue engineering technique, cell electrospinning, for the generation of 3D biological architectures for the replacement of diseased or damaged tissues. Cell electrospinning is explicitly defined here as the generation of cell-laden nanofibers produced using the electrospinning of biosuspensions containing intact cells. Comparisons will be made with competing tissue engineering techniques to establish why cell electrospinning is the tissue engineering method of choice for tissue replacement in humans and key criteria necessary for the successful introduction of this technology in the marketplace will be proposed.

4.2 Tissue Engineering and the Construction of 3D Biological Architectures

Tissue engineering provides a unique approach to personalized medicine for the repair and replacement of diseased, damaged or ageing tissues and organs. This is the primary focus of this chapter. It should be noted, however, that there is a wide variety of different biomedical and bioscience applications of the underpinning technologies used for tissue engineering. The development of prosthetics to replace bone damaged in accidents or malformed during neural crest development is an example. The chromosome 22q11.2 deletion (22qDS) under which DiGeorge and velocardiofacial syndromes fall provides a very good illustration of the benefits of tissue engineering for syndromes that have a range of congenital abnormalities affecting a number of organ systems.[10] Other applications include 3D cell culture and 3D controlled cryopreservation for clinical use.[11] Non-clinical commercial applications include the use of tissue engineering technologies for cellular agriculture to produce *in vitro* meat and biofabricated leather, oral delivery systems for macromolecular drugs and bioactive agents that have been combined with functional feeds in fish and 3D scaffolds fabricated for the large-scale expansion of anchorage-dependent cells in culture for cell enrichment and drug screening.[12,13]

In order for these potential applications of tissue engineering to come successfully to the market as products, either commercially available to the general public or used clinically, there are a number of defined attributes that the technologies must possess. Biopatterning methodologies must achieve precision placement of cells within 3D space, be scalable, keep costs to a minimum (by reducing the number of steps, reducing the time needed to fully integrate multiple layers of cells, *etc.*) and be biofriendly, to ensure conservation of the biological properties of the cells being used in the engineering process. The most promising of the techniques currently available are outlined below and the rationale for putting forward cell electrospinning as the method of choice is described.

4.2.1 3D Printing/Biofabrication

3D printing is a technology that essentially stacks layers of 2D material depositions to create a third dimensional *z*-axis. It has significant cost benefits because it adds materials when required, which prevents the material waste during the forming step that is seen with the other technologies described below. There are a large number of different ways in which these materials can be dispensed or extruded and 3D printing can generate some exquisitely complex and high-definition architectures. 3D printing is an inherently scalable technology, but it is the extrusion methods that present the principal difficulty in the adoption of this methodology as a biopatterning technique because the fine-bore needles used and the driving forces involved cause significant cellular damage.[14,15] A number of different modifications have been attempted, but none of these is currently able to avoid compromising cellular viability. To date, no material has been identified that would enable cells to be directly handled by 3D printing to create a viable self-standing, fully-cellularized bioarchitecture that requires no additional assistance during the forming stage.[11]

4.2.2 Aerodynamically-assisted Biojets and Threads

Aerodynamically-assisted jetting and threading uses a pressure field through air flow within a chamber with respect to the surrounding atmosphere to draw liquids through an exit orifice. The nature of what is formed is dependent upon the viscosity of the liquid, producing either droplets (aerodynamically-assisted jets) or continuous threads (aerodynamically-assisted threads). The AAJ/AAT process is similar to that of electrospray/electrospinning, discussed in Sections 4.2.3.2 and 4.2.3.3, and is scalable, with comparable costs, but is a non-electric field-driven competing technology. Inclusion of cells within the liquid used for AAJ/AAT creates aerodynamically-assisted biojets and biothreads (AABJ, AABT, respectively). These are biofriendly and an effective means by which to produce 3D biological architectures, in particular those that use highly conducting materials. As they are electric field-driven technologies, bio-electrosprays/cell electrospinning cannot handle highly conducting materials without discharging or damage of the conducting molecules.[11]

4.2.3 Jet-based Techniques

There are a number of jet-based methodologies that are economical, robust and capable of handling materials on the scale required for tissue engineering. These include inkjet printing, electrospraying and electrospinning.

4.2.3.1 Inkjet Printing

Inkjet printing was identified as a versatile biotechnique for printing 3D biological architectures of living cells and was the first jetting technology to develop significantly. Controlled architectures in the millimeter range have been fabricated using inkjet printing. Scaling up is feasible through direct-write assembly but has not been demonstrated yet. Inkjet printers have been used to generate cell-bearing fibers and scaffolds, however, the size of these fibers is in the hundreds of micrometers, at best, as droplets double the size of the needles can be formed and the spreading of these upon placement limits the resolution of the architectures that can be constructed. There are also limitations on the processing of high-viscosity media due to the 30–60 μM size of the needles used, which tends to promote blockage. Cost is not an issue as the resulting architectures do not require significant time in a bioreactor to achieve microintegration, but cell viability is of concern. One of the fundamental problems with inkjet printing relates to the fiber-drawing force that is imposed upon the cellular materials as part of the fabrication process. This places significant physical forces upon the cells that may result in shearing or other physical changes. To date, there has been no comprehensive study performed of the short, medium and long-term effects of inkjet printing on the key biological properties of the cells that would enable these concerns to be assuaged.

4.2.3.2 Electrospraying

Electrohydrodynamic jetting, also known as electrospraying, has the capacity to jet living cells under hostile electric field conditions without affecting their viability.[16–18] A charged polymeric liquid contained within a conducting needle is placed in an electric field with respect to a grounded or oppositely charged electrode. The potential difference between the electrodes leads to the formation of a liquid cone at the needle exit with a jet emanating from its apex. The breakdown of this jet into a 3D conical spray plume generates droplets. To achieve precision placement of these droplets, the jet must be stable and a Taylor cone formed. Electrospraying is a true 3D printing technology and requires no molds or other types of support during the forming stage. The inclusion of cells in the polymeric liquid being jetted is known as 'bio-electrospraying'.

4.2.3.3 *Electrospinning*

Electrospinning is the sister technology to electrospraying. Whilst electrospinning promotes the formation of droplets, electrospinning promotes a uniaxial elongating effect on the jet that leads to the formation of a continuously elongating thread from the jet.[19] Electrospinning can be used to generate large quantities of fibers with diameters in the nanometer range, having been spun from single, coaxial or tri-needle systems.[20–22] These fibers are able to form scaffolds and fibers. There is a tremendous capacity to control the alignment of fibers upon deposition, which allows the generation of scaffolds with directionally graded fibers as a function of scaffold depth. Scalability is readily achieved and the costs are, for the most part, determined by the polymer used, as bioreactors are not required to integrate the different cellular layers. With the inclusion of cells in the polymer, biologically active threads can be produced. This modification of the technique is known as 'cell electrospinning'. Cell electrospinning and bio-electrospraying are exceptionally biofriendly processes. The viability and other key biological properties of the cells are maintained during jetting because whilst high voltages (of the order of thousands of volts) are used, it is the current that is the effecting parameter and the currents in cell electrospinning are in the nanoampere range. Cell electrospinning is currently being developed to create scaffold architectures for a whole host of novel bioengineering and medical applications.

4.2.4 Is Cell Electrospinning Really the Answer?

Scaffolds are an incredibly important consideration in regenerative biology and medicine as they provide the much-needed support that cells require to carry out their normal functions.[11] It has been clearly demonstrated that there are significant effects on the ability of cells to undergo repair, regeneration and rejuvenation and that cells have changes in both phenotype and genotype when they lack a sufficient supporting scaffold. The post-processing coupling of cells with a scaffold fails to create a true 3D, fully microintegrated scaffold and the generation of a fully functional, cost-effective 3D biological structure enabling an integrated cell-scaffold architecture is both the key goal of tissue engineering strategies and something that is readily achieved through cell electrospinning.[23]

Is cell electrospinning the way to achieve this? Four lines of reasoning suggest that it is a reasonable means by which to do so. Firstly, conventional electrospinning using biopolymer suspensions devoid of cells creates structures with extremely small pores that cells are unable to infiltrate. Ironically, this has found applicability as a means of creating drug-eluting fibers that the smallest self-propelling human cell is unable to pass through, making it a perfect means to serve as an innovative platform technology for HIV-1 protection and sperm inhibition.[11,24] Secondly, it is difficult to produce 3D structures if there is no matrix between the cells to provide the necessary support for such an architecture. Thirdly, the post-processing seeding of scaffolds with cells is problematic. Cells are unable to remodel rigid scaffolds to generate the functional equivalent of an extracellular matrix, whereas they can remodel friendly biopolymers. Finally, with an appropriate experimental design, cell electrospinning is relatively inexpensive as it can be used to create sophisticated 3D scaffolds that do not require extensive periods of time in a bioreactor to ensure microintegration of cells with the scaffold architecture. Cell electrospinning, therefore, can produce structures that mimic native tissues whilst overcoming the limitations of producing integrated cell-scaffold architectures by other means.

4.3 Cell Electrospinning: from Concept to Lab

Whilst electrospinning has been around for over a century, the technique first emerged as an advantageous method, from a commercial standpoint, with Anton Formals' 1934 patent of a process and apparatus for preparing artificial threads and filaments by passing solubilized materials into an electric field formed between electrodes and collecting the resulting threads on a device in a form that would be of technical use.[25–31] The evolution of 'electrospinning' into 'cell electrospinning', whereby cell suspensions are used to produce cell-laden fibers and scaffolds, can be traced through a review of publications in the PubMed database, an online search engine used to query the more than 27 million citations for biomedical and bioengineering literature from MEDLINE (the US National Library of Medicine bibliographic database), life science journals and books. Figure 4.1 shows the results, by year of publication, obtained by querying the PubMed database using the search term 'cell electrospinning' (including the variant 'cell electro spinning'). This is an approach that can prove informative for the identification of points at which significant events impact upon a new technology. Such events may include technological advances, reduction of costs or a widespread adoption of the technology, which contributes significantly to the knowledge base.

From first use of the term in 2001 to 2005, 60 publications using the term 'cell electrospinning' are identified. Two-thirds of these papers describe the development of biodegradable or biocompatible polymer scaffolds suitable for use in tissue engineering as synthetic matrices to be seeded with cell types of interest or were publications focusing on the production of nanofibers containing biologically active molecules. The remaining third describes the production of nanofibers containing biologically active molecules. None of the work published during this time span describes the production of cell-laden fibers, which is the definition of cell electrospinning that was presented at the start of this chapter.

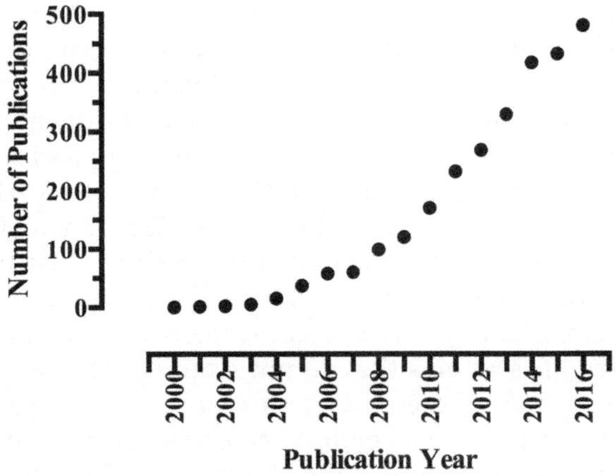

Figure 4.1 Number of publications, by year, in the PubMed database containing the term 'cell electrospinning'.

Up until this point, the ability to produce cell-laden fibers had been hampered by the fact that the conditions of high viscosity and low conductivity that were needed to achieve the spatial distribution and resolution of which the technology is capable are diametrically opposed to the requirements of low viscosity and high conductivity that are needed for viability of the cells. Initial attempts to microintegrate cells by electrospinning polymer and electrospraying cells encountered difficulties with non-uniformity of cellular integration.[32] These difficulties, however, were overcome by applying the coaxial concentric needle arrangement used to achieve precision placement of electrosprayed cell suspensions.[33] This approach uses the biopolymer to generate the properties needed for spraying in stable cone-jet mode whilst simultaneously shielding the highly conducting cell suspension from the external electric field. The first scientific paper describing the generation of active biological microthreads/scaffolds from living cells using electrospinning was published in 2006.[34] A sharp increase in the rate of publications using the term 'cell electrospinning' ensued, with almost 500 papers published in 2016 – the last complete year for which data are available.

The initial cell electrospinning work had been performed using an immortalized cell line.[34] The technique was then successfully carried out using primary cells, proving the suitability of cell electrospinning for cells of the type used in cell and gene therapy applications.[35] These findings were confirmed the following year, firmly cementing cell electrospinning as a feasible technology with which to address a wide variety of biomedical and clinical applications.[36] Cell electrospinning has been used to process a wide variety of intact eukaryotic organisms, including yeast (*Saccharomyces cerevisiae*), nematodes (*Caenorhabditis elegans*), soil-based amoeba (*Dictyostelium discoideum*) and fruit fly (*Drosophila melanogaster*), frog (*Xenopus tropicalis*) and zebrafish (*Danio rario*) embryos, and tissues generated by cell electrospinning have been successfully transplanted into mice, *in vivo*, demonstrating the viability of this technology for the production of engineered tissues that have the biocompatibility needed for successful transplantation into their intended host. These studies revealed the promise of cell electrospinning as a viable technology with significant commercial prospects.[22,37–43]

4.4 Cell Electrospinning: from Lab to Market

At present, there is neither a commercial product on the market nor a medical therapeutic in the clinic that has been derived from cell electrospinning and, to date, only two cell electrospinning patents have been filed.[44,45] Is this really surprising? As Figure 4.1 shows, it is only within the last decade that cell electrospinning has been demonstrated to generate cell-laden fibers at the scale and with the properties needed for the engineering of functional tissues. It is of interest, therefore, to appreciate how the timelines involved in the development of cell electrospinning from concept to commercialization compare with such timelines for other technologies.

Whilst there is a relative lack of information from the commercialization of other technologies with which to compare, when evaluating the elapsed time between the development of a new technology and its commercialization, studies from the energy sector have shown that innovations to replace existing products have an average time of 29 years from invention to widespread commercialization. This contrasts significantly with the 42 year average in this sector that is needed to realize commercialization for innovation directed at entirely new markets in this sector. Taken in this context, cell electrospinning is not lagging with respect to the time elapsed between invention of the technology and its widespread commercialization.[46]

Interestingly, cell electrospinning is a technology used for innovations that is directed at entirely new markets. The time elapsed since the major obstacles preventing its use for precision placement were first identified and addressed is of the order of a decade. It is not surprising, therefore, that more endeavor will be needed before the commercial advantages envisioned by Formhals can be fully realized. In fact, the timeline from innovation to commercialization in the energy sector suggests that a further decade or two remains before widespread commercialization of cell electrospinning applications can be expected, if comparison with the innovation to commercialization timelines in the energy sector prove to be informative.

Cell electrospinning is a disruptive technology, as it is a technology possessing the capacity to displace existing medical technologies; it disrupts existing pharmaceutical and medical approaches to the treatment of ageing tissues and organs and creates a new market by which this can be achieved. The stumbling blocks currently preventing cell electrospinning products from finding a successful commercial end point can be attributed to two main challenges: overcoming the remaining technical difficulties and the appropriate interrogation of cell electrospun materials prior to use.

4.4.1 Further Technical Refinements

There are at least two technical considerations that may impact upon the commercialization prospects of cell electrospinning: (i) ensuring orientation of the deposited nanofibers and (ii) the choice of polymer.

4.4.1.1 Orientating Nanofiber Deposition

One of the significant disadvantages encountered in cell electrospinning occurs when the stacked layers of nanofiber forming the 3D biological architecture under construction cause interference with the electric field, leading to the random deposition of electrospun nanofibers.[47] The use of electric field-focusing plates and rings has removed this interference-based limitation, achieving alignment of the electrospun nanofiber by deflecting the jet.[48,49] This limitation has only been overcome within the past 2–4 years and it is expected that its widespread adoption will dramatically shorten the time remaining until the first commercial application of cell electrospinning is achieved.

4.4.1.2 Choice of Polymer

A wide range of polymers has been used in cell electrospinning, including both natural and synthetic polymers.[47] Many of the cell electrospun tissues under development are intended for use in humans and there has been a great deal of interest in the use of poly(vinyl alcohol) (PVA) as a polymer for cell electrospinning, as this material has been approved by the FDA and other regulatory organizations for use in humans.[50] Sodium trimetaphosphate crosslinked PVA-based hydrogels have been shown to possess mechanical properties compatible with circulatory flow, making them of great interest for use in vascular replacements.[51] Unfortunately, PVA has been shown to deter cell attachment, leading to cell death.[52] The poor viability of cells electrospun using PVA solutions has subsequently been confirmed.[53,54]

Effective polymers for cell electrospinning must possess a number of key features, including: (i) the ability to accommodate cells in a manner that provides the functionalities essential to cell viability; (ii) the ability to undergo spinning under stable conditions; and (iii) the ability to facilitate some form of solidification or crosslinking that holds cells in place without changing their cellular properties over time. The cell electrospun structures successfully tested *in vivo* were produced using cell-friendly biopolymers, which may suggest that the successful commercial application of cell electrospinning is more likely to be achieved with the use of biopolymers, rather than with chemical polymers.[22,43]

We have explored this possibility further by using a cell electrospinning set-up immersed in a cell-friendly media bath (Figure 4.2a and b) together with a thermosensitive biopolymer, to generate a living vessel-like architecture over 6 inches in length that is comprised of an internal endothelial cell

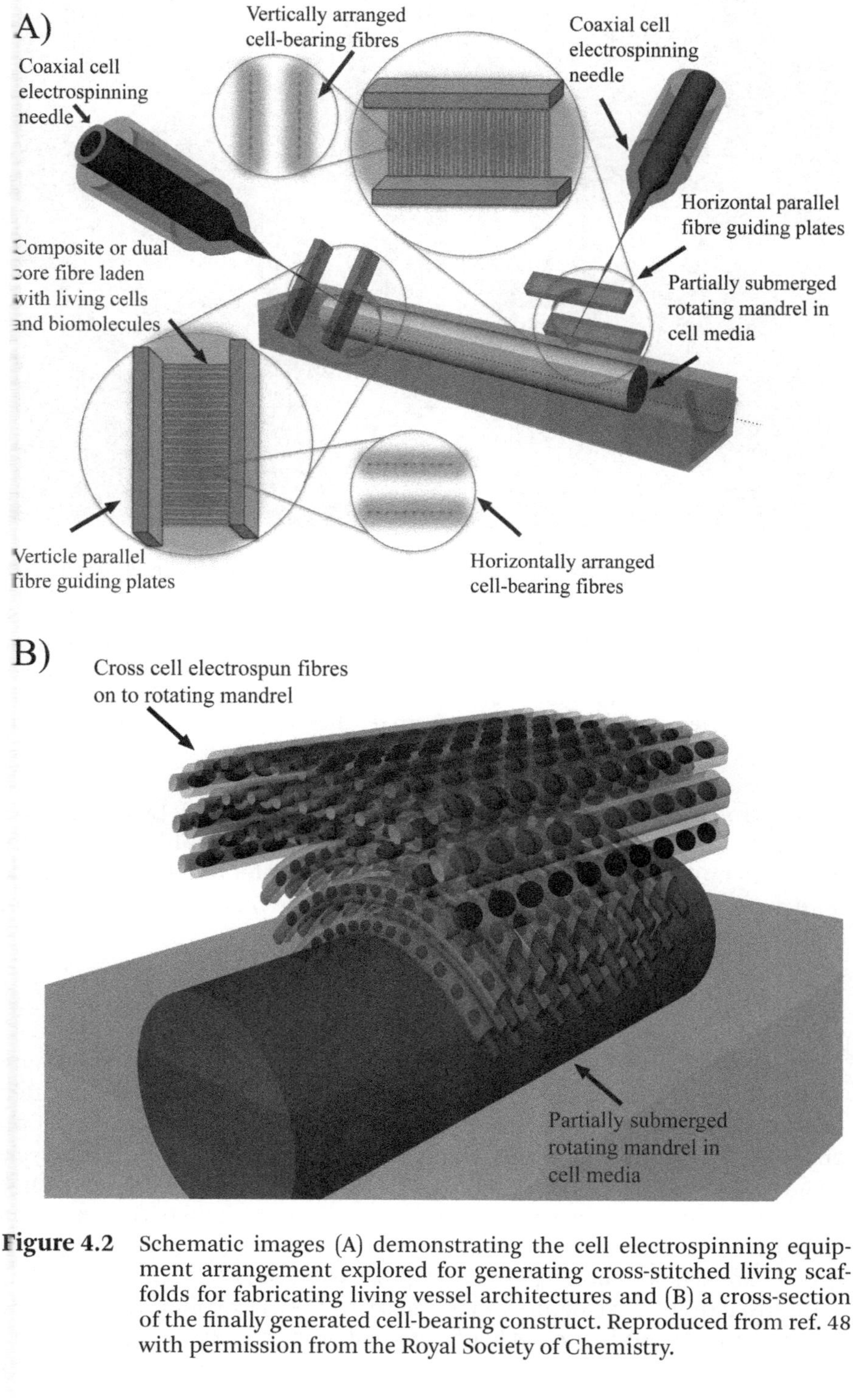

Figure 4.2 Schematic images (A) demonstrating the cell electrospinning equipment arrangement explored for generating cross-stitched living scaffolds for fabricating living vessel architectures and (B) a cross-section of the finally generated cell-bearing construct. Reproduced from ref. 48 with permission from the Royal Society of Chemistry.

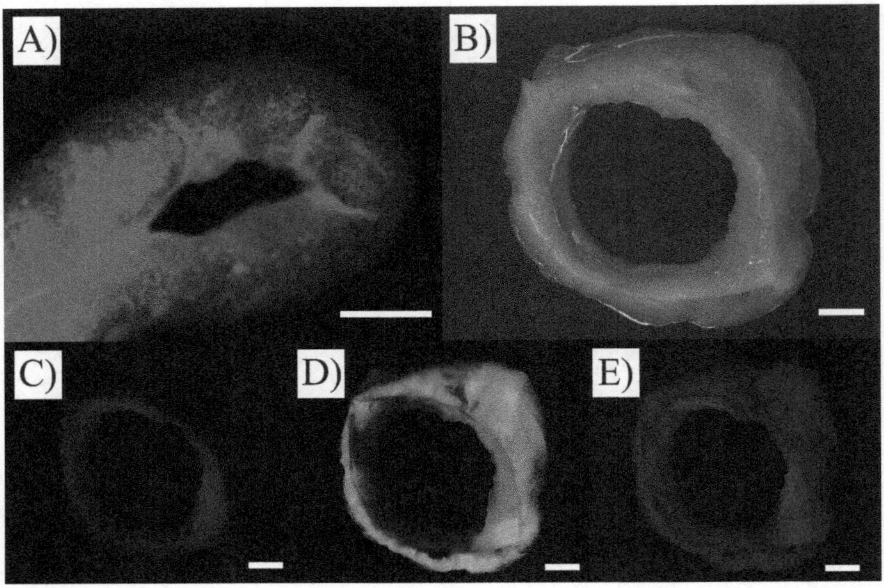

Figure 4.3 (A) Fluorescent image of a generated living vessel-like architecture. (B) An optical micrograph showing the cross-section. Scale bars in panel (A) and (B) are ~5 mm and 1 mm, respectively. Panels (C and D) depict fluorescent images of the individual cell constituents within the two-core structure. Panel (E) represents the superimposed fluorescent image of the fabricated architecture. Scale bars in panels (C) to (E) represent 1 mm. Reproduced from ref. 48 with permission from the Royal Society of Chemistry.

layer surrounded by a smooth muscle cell layer (Figure 4.3). By using a horizontal rotational mandrel, the length of the tubular vessel-like architecture is defined by the length of the mandrel and its diameter is determined by the bore diameter of the mandrel, making it possible to build bespoke vessels of any defined length and diameter.

4.4.2 Interrogation of Cell Electrospun Structures

Studies of cell electrospinning have almost always included some measure by which to assess cell viability. Other properties that have been interrogated in processed and unprocessed cells for use as a measure of cellular function post electrospinning/electrospraying include phenotypic studies, cell cycle characterization, determination of the rate of cellular proliferation, change in the rate of apoptosis/necrosis, karyotyping, measurement of the activation of cell surface receptors including G protein-coupled receptors and potassium channels, proteomic and transcriptomic analyses.[49,55]

Interrogations of structures made using cell electrospinning have been confined to a determination of key functional properties at the molecular and cellular levels. To date, there has been no testing of the mechanical properties at the tissue/structural level. For example, the muscle tone of the vessel-like structure shown in Figure 4.3 will need to be interrogated using organ bath studies to determine the pharmacological response to compounds known to elicit contraction and relaxation of vascular smooth muscle, with characterization of the time course and tension of the muscle tone responses and tests of fluidics with simulated fluids and using blood will provide a richer measure of its functional and mechanical properties. Quantification of the range of values for key properties of engineered tissues, such as the one shown in Figure 4.3, will need to be conducted before being able to conclude that what looks like a tissue actually behaves like a tissue, something that needs to be determined before one can contemplate replacing native tissue with an engineered tissue *in vivo*.

4.4.3 Getting to Market

Taken together, the identification and application of solutions for addressing the dual challenges of providing the additional necessary technical refinements and achieving an effective interrogation of the key properties of structures should rectify the issues that have prevented commercial/clinical applications of this technology from being achieved to date. Whilst the former has been accomplished, a standardized protocol for the latter remains to be implemented. The patient outcomes from the clinical application of a separate tissue engineering technology have unequivocally demonstrated that it is essential that the functional and mechanical properties of tissue-engineered architectures be shown to lie within the expected parameters of native tissues prior to any clinical use of the materials.

In 2011, the world's first bioengineered trachea were transplanted in patients. The failure to test that the mechanical and functional properties of engineered trachea were equivalent to those of normal human trachea had a catastrophic outcome for those patients into whom these tissues were grafted. In 2016, a Swedish Public Television (SVT) team was awarded the AAAS Science Journalism Gold Award for Television In-Depth Reporting for their documentary on the patient outcomes of this clinical trial.[56]

As a direct consequence, attention now needs to be placed on defining the engineering parameters needed to ensure that the mechanical and functional properties of engineered tissues are equivalent to those of their native counterparts before commencing any functional replacement *in vivo*. Definition of the required parameters and testing necessary for validation of tissues engineered for therapeutic use will enable the field to successfully move from the functional/mechanical perspective of the engineering question to the translational perspective of bespoke therapeutic treatment in the clinic. As in the case of the tissues engineered for xenotransplantation (Section 5.1), international consensus will be key in establishing protocols appropriate for defining the clinical suitability of engineered tissues.

4.5 Future Developments and Conclusion

Future developments in the field of cell electrospinning include exploitation of personalized medicine strategies, which includes consideration of the inclusion of experimental and/or medical cell and gene therapy in the tissue engineering workflow to create tissues on demand and to meet patient need. As mentioned previously, this will enable the promise of personalized medicine to be genuinely fulfilled. An area that has not yet been significantly discussed and one that holds great promise for commercialization is that of the use of engineered tissues to create templates for modeling 3D systems *in vitro*. This is an innovation that will enable the creation of 3D models of human diseases and permit drug and vaccine development and screening. Human tuberculosis is currently being modeled in this way using a bioengineered 3D model created by bio-electrospraying. Cell electrospinning could be used to do exactly the same.[57] Reduction in animal use is a key strategic priority embedded in all research funding applications in the UK and globally. In the UK, the Animals (Scientific Procedures) Act 1986 embeds the 'three Rs' (replacement, refinement and reduction), to ensure that animal research is carried out only where no practicable alternative exists.[58] Adoption of 3D models built using cell electrospinning will lead to a reduction in animal studies.

Cell electrospinning not only integrates cells and scaffolds at point of delivery, it brings together state of the art technologies in the life and medical sciences to deliver a platform technology able to contribute the key advances required for the support of human health throughout the ageing process and on a scale that has not previously been achieved. Key considerations needed for the successful application of cell electrospinning in a commercially-relevant context can be reduced to two critical elements: (i) defining the suitable biopolymers within which to suspend the cells, and (ii) ensuring that rigorous testing is carried out prior to the replacement of diseased or damaged tissues with engineered structures that are truly equivalent to the native tissues they are intended to replace.

Acknowledgements

The authors are grateful to the Royal Society, the Engineering and Physical Sciences Research Council, the National Institute for Health Research, and the British Heart Foundation, all in the UK, and the National Institutes of Health in the USA, for funding support.

References

1. http://www.who.int/mediacentre/factsheets/fs310/en/.
2. https://www.cdc.gov/nchs/data/dvs/lead1900_98.pdf.
3. https://www.medicalnewstoday.com/articles/282929.php.
4. D. K. C. Cooper, *Proc. (Bayl. Univ. Med. Cent.)*, 2012, **25**(1), 49.

5. B. Ekser, P. Li and D. K. C. Cooper, *Curr. Opin. Organ Transplant.*, 2017, **22**(6), 513.
6. C. L. Cetrula, *Stem Cell Rev.*, 2006, **2**(2), 163.
7. A. Rosemann, *J. Cell. Biochem.*, 2014, **115**(12), 2073.
8. R. S. Mahla, *Int. J. Cell Biol.*, 2016, 6940283.
9. http://apps.who.int/trialsearch/.
10. A. S. Bassett, D. M. McDonald-McGinn, K. Devriendt, M. C. Digilio, P. Goldenberg, A. Habel, B. Marino, S. Oskarsdottir, N. Philip, K. Sullivan, A. Swillen and J. Vorstman, The International 22q11.2 Deletion Syndrome Consortium, *J. Pediatr.*, 2011, **159**(2), 332.
11. S. N. Jayasinghe, *Adv. Biosyst.*, 2017, **1**, 1700067.
12. J. Vein, US Pat., 6 835 390, 2004.
13. S. Wadsworth, G. Klaric and S. N. Jayasinghe, Nor. Patent, WO2016195509, 2016.
14. K. Nair, M. Gandhi, S. Khalil, K. C. Yan, M. Marcolongo, K. Barbee and W. Sun, *Biotechnol. J.*, 2009, **4**, 1168.
15. G. M. Nishioka, A. A. Markey and C. K. Holloway, *J. Am. Chem. Soc.*, 2004, **126**, 16320.
16. S. N. Jayasinghe, A. N. Quershi and P. A. M. Eagles, *Small*, 2006, **2**, 216.
17. S. N. Jayasinghe, P. A. M. Eagles and A. N. Quershi, *Biotechnol. J.*, 2006, **1**, 86.
18. P. A. M. Eagles, A. N. Quershi and S. N. Jayasinghe, *Biochem. J.*, 2006, **394**, 375.
19. C. A. Bashur, L. A. Dahlgren and A. S. Goldstein, *Biomaterials*, 2006, **27**, 5681.
20. A. Townsend-Nicholson and S. N. Jayasinghe, *Biomacromolecules*, 2006, **7**, 3364.
21. S. N. Jayasinghe, S. Irvine and J. R. McEwan, *Nanomedicine*, 2007, **2**, 555.
22. A. N. Jayasinghe, G. Warnes and C. Scotton, *Macromol. Biosci.*, 2011, **11**, 1364.
23. M. J. Bissell and W. C. Hines, *Nat. Med.*, 2011, **17**, 320.
24. C. Ball, E. Krogstag, T. Chaowanachan and K. A. Woodrow, *PLoS One*, 2012, **7**(11), e49792.
25. J. W. G. Rayleigh, *London, Edinburgh Dublin Philos. Mag. J. Sci.*, 1882, **14**, 184.
26. J. F. Cooley, US Pat., 692 631, 1902.
27. J. Zeleny, *Phys. Rev.*, 1914, **3**, 69.
28. J. Zeleny, *Proc. Cambridge Philos. Soc.*, 1915, **18**, 71.
29. J. Zeleny, *Phys. Rev.*, 1917, **10**, 1.
30. G. I. Taylor, *Proc. R. Soc. London, Ser. A*, 1964, **280**, 383.
31. A. Formhals, US Pat., 1975504, 1934.
32. J. J. Stankus, J. Guan, K. Fujimoto and W. R. Wagner, *Biomaterials*, 2006, **27**(5), 735.
33. S. N. Jayasinghe and A. Townsend-Nicholson, *Lab Chip*, 2006, **6**, 1086.
34. A. Townsend-Nicholson and S. N. Jayasinghe, *Biomacromolecules*, 2006, **7**, 3364.
35. S. N. Jayasinghe, S. Irvine and J. R. McEwan, *Nanomedicine*, 2007, **2**(4), 555.

36. J. A. van Aalst, C. R. Reed, L. Han, T. Andrady, M. Hromadka, S. Bernacki, K. Kolappa, J. B. Collins and E. G. Loboa, *Ann. Plast. Surg.*, 2008, **60**(5), 577.
37. D. Greig and S. N. Jayasinghe, *Biomed. Mater.*, 2008, **3**(1), 34125.
38. T. Geach, N. Mongkoldhumrongkul, L. B. Zimmerman and S. N. Jayasinghe, *Analyst*, 2009, **134**, 743.
39. D. W. Clark and S. N. Jayasinghe, *Biomed. Mater.*, 2008, **3**(1), 11001.
40. P. Joly, B. Hennings and S. N. Jayasinghe, *Biomicrofluidics*, 2009, **3**(4), 44107.
41. N. Mongkoldhumrongkul, S. C. Swain, S. N. Jayasinghe and S. Sturzenbaum, *J. R. Soc., Interface*, 2010, **7**, 595.
42. N. K. Pakes, S. N. Jayasinghe and R. S. B. Williams, *J. R. Soc., Interface*, 2011, **8**, 1185.
43. S. L. Sampson, L. Saraiva, K. Gustafsson, S. N. Jaysinghe and B. D. Robertson, *Small*, 2014, **10**(1), 78.
44. L. YuanYuan, J. Chen, X. Ke, L. Shuai, Z. Ting and H. Qingxi, China Pat., CN104984405 (A), 2015.
45. L. YuanYuan, J. Chen, L. Gang, Y. Hongchen, L. Yi and H. Qingxi, China Pat., CN104862787 (A), 2015.
46. http://www.ukerc.ac.uk/programmes/technology-and-policy-assessment/innovation-timelines-from-invention-to-maturity-a-review-of-the-evidence-on-the-time-taken-for-new-technologies-to-reach-widespread-commercialisation.html.
47. R. Vasita and D. S. Katti, *Int. J. Nanomed.*, 2006, **1**(1), 15.
48. S. N. Jayasinghe, *Analyst*, 2013, **138**(8), 2201.
49. J. Walser and S. J. Ferguson, *J. Mech. Behav. Biomed. Mater.*, 2016, **58**, 188.
50. K. Saralidze, L. H. Koole and M. L. W. Knetsch, *Materials*, 2010, **3**, 3537.
51. M. Chaouat, C. Le Visage, W. E. Baille, B. Escoubet, F. Chaubet, M. A. Mateescu and D. Letourneur, *Adv. Funct. Mater.*, 2008, **18**, 2855.
52. Y. Liu, N. E. Vrana, P. A. Cahill and G. B. McGuinness, *J. Biomed. Mater. Res., Part B*, 2009, **90**(2), 492.
53. M. F. Canbolat, C. Tang, S. H. Bernacki, B. Pourdeyhimi and S. Khan, *Macromol. Biosci.*, 2011, **11**, 1346.
54. G. Zanatta, D. Steffens, D. I. Braghirolli, R. A. Fernandes, C. A. Netto and P. Pranke, *Braz. J. Med. Biol. Res.*, 2012, **45**(2), 125.
55. A. Eddaoudi, A. Townsend-Nicholson, J. F. Timms, S. Schorge and S. N. Jayasinghe, *Analyst*, 2010, **135**, 2600.
56. https://sjawards.aaas.org/awards/2016-television-depthfeature-reporting-gold.
57. L. B. Tezera, M. K. Bielecka, A. Chancellor, M. T. Reichmann, B. Al Shammari, P. Brace, A. Batty, A. Tocheva, S. Jogai, B. G. Marshall, M. Tebruegge, S. N. Jayasinghe, S. Mansour and P. T. Elkington, *eLife*, 2017, **6**, 321283.
58. Great Britain: Home Office, *Code of Practice for the Housing and Care of Animals Bred, Supplied or Used for Scientific Purposes*, The Stationery Office, 2015.

CHAPTER 5

Electrospun Fibers for Advanced Wound Care: Moving from Novel Lab-scale Curiosities to Commercial Realities

H. KRIEL*, M. P. COATES AND A. E. SMIT

The Stellenbosch Nanofiber Company (Pty) Ltd, 7 Marconi Road, Montague Gardens, Cape Town 7441, South Africa
*E-mail: haydn@sncfibers.com

5.1 Introduction to Electrospun Fibers for Advanced Wound Care

Electrospinning typically produces fibers that have diameters in the range of 100–1000 nm and are kilometers in length. They are commonly referred to as nanofibers by the electrospinning community in academia and industry, but electrospun fiber diameters may fall into a much broader fiber diameter range of 50–3000 nm. Electrospun fibers should not be compared directly with nanomaterials, which are defined as having average fiber diameters <100 nm.[1] Nanostructures such as nanotubes, nanowires and nanowhiskers are typically microns in length and present an inhalation risk during processing and handling. By comparison, electrospun fibers produced from

Soft Matter Series No. 7
Electrospinning: From Basic Research to Commercialization
Edited by Erich Kny, Kajal Ghosal and Sabu Thomas

Published by the Royal Society of Chemistry, www.rsc.org

flexible polymeric materials present no inhalation risk due to the extremely high aspect ratios.

The minuteness of electrospun fibers leads to extremely high specific surface areas, ultra-softness, drapability, microporosity and biomimicry. These characteristics result in key property benefits that are important for wound care, regenerative medicine and drug delivery. Advanced wound care has been defined as "interventions used when standard wound care has failed".[2] The term 'advanced wound care' is most commonly used in the media and literature with regards to wound dressings that are believed to be 'advanced' in that they act to control the environment of the wound bed (moisture, temperature, *etc.*), ease pain and improve the rate of healing. Advanced wound care is also used to describe other products and a more modern approach to health care; this may include advanced diagnostics, advanced therapies, products or methods, smarter approaches to care, and ultimately healing the wound as fast as possible with minimum cost and minimum patient discomfort while ensuring the functional and aesthetic properties of the rejuvenated tissue.[3]

The lines between wound care, regenerative medicine and drug delivery or nanomedicine converge when it comes to advanced wound care products as they are often required to perform functionally across all of these areas. Electrospun fiber-based products may offer competitive alternatives to the 'gold standard' products for specific indications but may also produce a category of their own and compete across multiple product categories. This chapter will cover the important considerations and state-of-the-art in electrospun fiber-based advanced wound care products.

The nature of a wound, the wound surface area and depth, the health status of the patient and the economic and environmental circumstances of the treatment facility may vary greatly and will determine what class of wound dressing and treatment regime is required/used. Wounds can be classified as acute (surgical, trauma or burn) or chronic (non-healing), or alternatively by the size and depth or whether the wound is superficial, partial-thickness or full-thickness.[4] Immediately after sustaining an injury, multiple biological pathways are activated in an attempt to stabilize the wound and start the regeneration process.[5] The five ideal stages of wound healing can be briefly described as follows.

Hemostasis, the first stage of wound healing, involves prevention of blood flow through vasoconstriction of damaged lymph and blood vessels, followed by formation of a clot by various components.[6,7] This is followed by inflammation: the arrival of inflammatory cells that produce growth factors; the removal of necrotic tissue, foreign particles and bacteria; and the stimulation of fibroblasts to initiate collagen production and re-epithelialization.[6,7] During the proliferation stage, keratinocytes migrate into the wound, re-epithelialization continues and granulation tissue is formed containing healthy extracellular matrix (ECM) components with the exception of elastin.[6] Finally, during maturation, scar tissue with disorganized ECM replaces the granulation tissue.[5,6] Scar formation, also called fibrosis, can result in poorly functional tissue because the scar tissue is mechanically weaker than undamaged tissue, may impair movement and result in lifelong disability and also have a poor cosmetic appearance.[5,6] These stages may not always be so clearly defined and in the case of chronic wounds not all stages will be achieved.[6] Figure 5.1 shows a comparison between classic wound care and advanced or regenerative wound therapy, where advanced wound therapy aims to reduce scar formation and promote wound regeneration.[5]

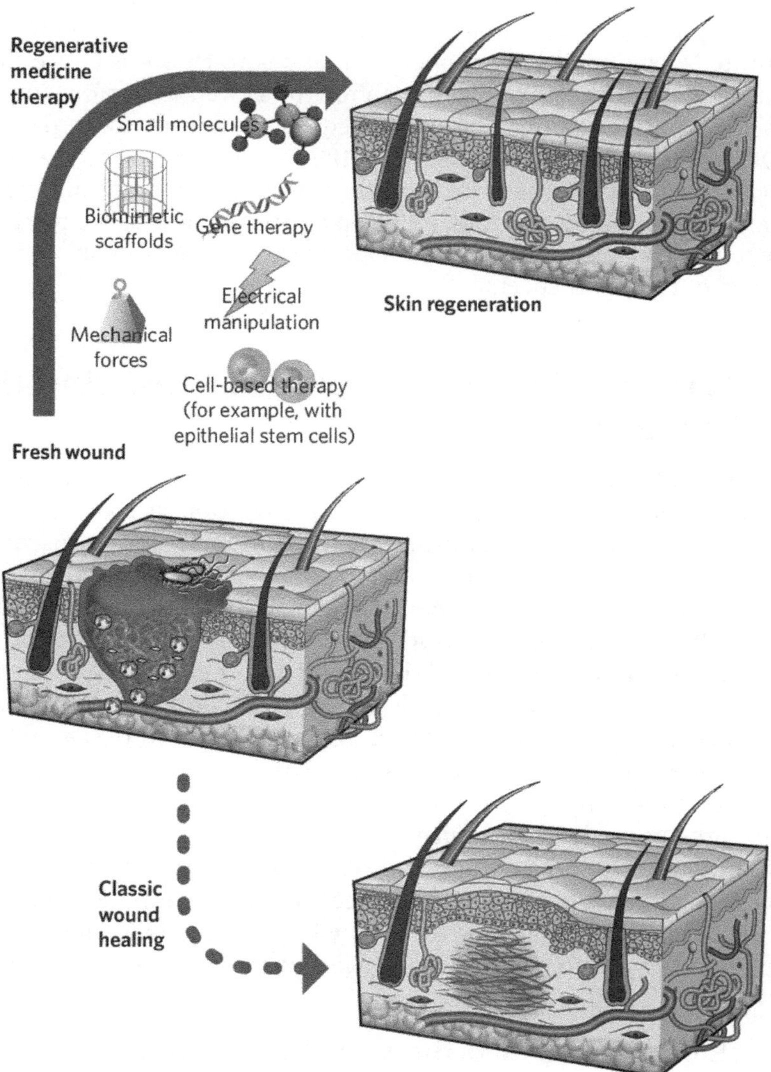

Figure 5.1 Comparison of classical wound healing *versus* advanced wound healing/regenerative therapy. Reprinted by permission from Macmillan Publishers Ltd: *Nature* (G. C. Gurtner, S. Werner, Y. Barrandon and M. T. Longaker, *Nature*, 2008, **453**, 314), copyright 2008.

Reducing the number of interventions with the wound after the initial treatment is important with regards to reducing costs but also to ensure newly formed tissue is not damaged in the dressing change and the wound is not re-exposed to infection. When it comes to costs, ter Meulen *et al.*[8] determined that the most expensive component of burn wound treatment is the cost of hospital days, followed by the surgery cost and then the cost of dressings. Easy to use, cost-effective and efficient wound care products that are readily available and which speed healing time are thus an important component of removing the delay in the wound care timeline and reducing expense.[9,10]

Electrospun fibers as primary wound dressings act as a temporary skin substitute, providing barrier properties and a simulation of native ECM to modulate and control cells for tissue growth, as well as performing a role as an excipient system for the controlled delivery of antibacterial agents and other actives for wound healing. In the following sections the research involving electrospun fibers for advanced wound care and nanomedicine will be discussed in more detail, followed by an overview of the market and the practical considerations of producing a wound care product, and a brief summary of future prospects.

5.2 Advanced Wound Care and Nanomedicine: the Research

A definition of advanced wound care is provided in Section 5.1. Nanomedicine, as defined by the European Science Foundation in 2005, is: "the science and technology of diagnosing, treating and preventing disease and traumatic injury, of relieving pain, and of preserving and improving human health, using molecular tools and molecular knowledge of the human body".[11] The sub-disciplines include analytical tools, nanoimaging, nanomaterials and nanodevices, novel therapeutics and drug systems, and clinical, regulatory and toxicological issues.[11] Other definitions include the molecular-scale treatment or repair of damaged/diseased tissue, including bone, muscle and nerves,[12] which overlaps with regenerative medicine and advanced wound care.

There are hundreds of research groups around the world identifying novel approaches to combining natural and synthetic chemistries with the aim of reducing cost, post-operative interference, and accelerating wound healing time.[6] The focus in the following sections will be on the application and use of electrospun fibers with regards to advanced wound care and nanomedicine.

5.2.1 Electrospun Fibers for Wound Care Dressings

5.2.1.1 Classification of Wounds and Dressings

As mentioned in the introduction, depending on their mechanism of healing wounds can be classified as either acute or chronic,[4,6,13] and the type of wound will dictate the treatment or dressing that is needed. Acute wounds typically heal completely with minimal scarring within 9–12 weeks, and are predominantly caused by mechanical injuries such as abrasions, tears, penetrating wounds from sharp objects or guns, and surgical wounds.[4] Burn wounds and injuries from chemicals, radiation, heat and electricity are also defined as acute wounds, with severe burn wounds requiring specialist treatment.[4] Chronic wounds either heal slowly or fail to heal and/or reoccur, and may be the result of circumstance or physiological conditions such as diabetes.[4] Treatment will also be dictated by the level of exudate production, whether the wound is infected, or whether any necrotic tissue needs to be debrided.

Wound care dressings may be categorized in various ways depending on the intended application. From a regulatory perspective, non-invasive medical devices that come into contact with injured skin, including wound dressings, are categorized as Class I–III, *e.g.* as in the guidelines in the European Commission Medical Devices Guidance Document (MEDDEV 2.4/1 Rev. 9).[14]

Another way of classifying wound dressings is as traditional or advanced, where advanced dressings are expected to have greater environmental control over the wound, and purposely affect other factors to improve healing. Important properties for a wound dressing to allow optimal healing include:[13,15]

- maintain a moisture balance
- remove excess wound exudate
- allow gas diffusion
- control temperature and pH
- act as a barrier to infection
- non-toxic/non-shedding/non-allergenic
- provide mechanical protection
- non-adherent/easy to remove
- comfortable
- conform to the wound
- minimize pain
- stimulate healing
- low cost
- ease of use
- compatibility with other treatment regimes

5.2.1.2 Natural and Synthetic Electrospun Fibers

In the application of electrospun fibers for wound dressings, they are typically separated into two types depending on whether they are made from natural or synthetic polymers. Selected examples of these, with some formulation details, are shown in Table 5.1. Synthetic polymers have the benefit of controlled and reproducible physical and mechanical properties, are typically soluble in a wider range of solvents and in some instances are of lower cost compared to the natural polymers.[6,16] Some natural polymers are

Table 5.1 Selected synthetic and natural polymers used in the production of electrospun fibers for wound care applications.[a]

Source	Polymer	Solvent system	Additives	Reference
Synthetic	Nylon 6	Acetic acid/formic acid	Silver nitrate	18
		Formic acid	5,5-Dimethyl hydantoin	19
		HFIP	MWCNTs	20
		Formic acid	–	21
	Poly(ε-caprolactone) (PCL)	DCM	–	22
		Chloroform/methanol	–	23
		Chloroform/methanol	Halloysite nanoclay	24
		Chloroform/DMF	–	25
		Chloroform/methanol	Curcumin	26
		Chloroform/DMF	Wool protein coating	27
		DMF/THF	–	28
	Poly(ethylene oxide) (PEO)	Chloroform	Chlorhexidine	29
	Poly(lactic acid) (PLA)	DMF/DCM	Silver nitrate	30
		Chloroform/DMAc	Curcumin	31
		HFIP	Mupurocin; lidocaine hydrochloride	32
		Chloroform/acetone	–	25
		DMF	Biotin, PLA-b-PEG block copolymer	33
		Chloroform	–	34
	Poly(lactide-*co*-glycolide) (PLGA)	DCM/HFIP	Ginsenoside Rg3	35
		HFIP	Basic fibroblast growth factor	36
		DMF/THF	Silver nitrate	37
		DCM	Fusidic acid	38
		DMF/THF	–	39
		DCM/DMF	–	40
		DCM	Ibuprofen	41

	Polystyrene (PS)	Chloroform	–	42
		DMF	–	43,44
	Polyurethane (PU)	DMAc/toluene	–	45
		DMF/THF	–	46
		Chloroform/methanol	–	47
		DMF	Murivenna oil	48
		DMF/THF	ZnO NPs, MWCNTs	49
	Polyvinyl alcohol (PVOH)	Water	–	50
		Water/phosphoric acid	Gentamicin	45
		Water with Triton X-100	Silver nitrate, wool protein	27
	Polyvinylpyrrolidone (PVP)	Water	Chlorhexidine	29
		Ethanol	Curcumin-loaded mesoporous silica nanoparticles	51
		Water	–	52
		Water/ethanol	Silver nitrate	53
		Ethanol	–	54
Natural	Chitosan	Acetic acid/water	–	55
	Cellulose acetate	Acetone	–	56
		Acetone/isopropanol	–	56
		Acetone/DMAc	–	56
	Collagen	HFIP	–	23,57–59
		Acetic acid/water	–	57,60
		TFE	–	59
		Ethanol/PBS	–	59
		Acetic acid/sodium acetate buffer	–	61
	Dextran	Water	BSA, lysozyme	62
		DMSO/DMF	–	62
	Gelatin	Acetic acid/water	–	28,63
		Acetic acid/ethyl acetate/water	–	64
		Ethanol/formic acid/water	–	65
	Hyaluronic acid (HA)	DMF/water	–	66

(continued)

Table 5.1 *(continued)*

Source	Polymer	Solvent system	Additives	Reference
Blends (natural/synthetic)	Cellulose acetate/pullulan	DMAc/DMSO	–	67
	Chitin butyrate/nylon 6	Acetic acid/formic acid	–	68
	Chitosan/collagen/PEO	Acetic acid	–	69
	Chitosan/PEO	Acetic acid/water	–	69,70
	Chitosan/PVOH	Acetic acid/water	Silver nitrate	29
	Chitosan/PVOH	Water	Silver nitrate	29
	Chitosan/PVOH/PVP	Water	Iodine	71
	Chondroitin sulfate/gelatin	TFE/water	–	72
	Chondroitin sulfate/PVOH	Water	Combretastatin A-4 phosphate	73
	Collagen/hyaluronic acid	DMF/NaOH	–	74
	Collagen/PCL	HFIP	–	75
	Collagen/PEO	Acetic acid/ethanol	–	59
	Collagen/PLA	HFIP	–	75
	Dextran/PU	DMF/THF	Ciprofloxacin HCl	76
	Dextran/PVP	Water	Ibuprofen sodium salt, acetylsalicylic acid	77
	Hyaluronic acid/PCL	Formic acid/HFIP	–	78
	Hyaluronic acid/PEO	DMEM	–	79
	Pullulan/PVOH	Water	Silver nanoparticles	80
	Sodium alginate/PCL	DCM/DMF	–	81
	Sodium alginate/PEO	Water	–	82
	Sodium alginate/PEO	Water	Heparin	83
	Sodium alginate/PEO	Water/Triton X-100/DMSO	–	84

[a]DMF, *N,N*-dimethylformamide; DCM, dichloromethane; DMAc, *N,N*-dimethylacetamide; HFIP, hexafluoroisopropanol; THF, tetrahydrofuran; DMEM, Dulbecco's modified eagle's medium; ZnO NPs, zinc oxide nanoparticles; MWCNTs, multiwalled carbon nanotubes; TFE, 2,2,2-trifluoroethanol; PBS, phosphate-buffered saline; NaOH, sodium hydroxide; HEC, hydroxyethyl cellulose; DMSO, dimethyl sulfoxide; BSA, bovine serum albumin.

considered to be more biocompatible, with good cellular interactions and they come from renewable sources.[6,17] However, natural polymers that come from animal sources may be prohibited in certain cases due to cultural and religious beliefs, and may also have some concerns relating to infection and transfer of disease.

5.2.1.3 *Beneficial Properties of Electrospun Fibers for Wound Dressings*

The benefits of electrospun fibers for wound dressings will be discussed in more detail in the following sections.

5.2.1.3.1 Pain Relief and Patient Comfort. Pain receptors in the epidermis and dermis, called nociceptors, can become more sensitive following inflammation of the tissue, and partial damage to nerve endings and abnormalities in healing nerve endings can add to the difficulty of pain control.[85] Painful dressing changes lead to patient non-compliance and increased costs in terms of medical practitioner time, as well as additional antiseptics and analgesics.[86] The benefit of electrospun fiber dressings is that they do not need to be changed frequently because their small pore size excludes bacteria and viruses, keeping the wound free of infection while allowing the wound to breathe and preventing build-up of exudate.[46,87,88] Analgesic-loaded electrospun fibrous membranes can be administered to address inflammation and reduce wound adhesion.[89] Novel zwitterionic poly(carboxybetaine-*co*-methyl methacrylate) fibers prepared by electrospinning were shown to be inert to cell adhesion and blood clotting and also had a bactericidal effect on *S. aureus* and *E. coli.*[90]

Electrospun fiber dressings are more comfortable than conventional dressings because of their nano-scale smoothness, because at fiber diameters of less than 1 μm the texture of the fiber surface is too fine to feel rough to the touch. There is also less irritation caused by protruding fibers as they are too small in diameter to stimulate the nociceptors,[91] and the flexibility and softness (depending on the polymer used and resultant crystallinity) of the fine electrospun fibers can improve patient comfort. Generally, the polymer chains in as-spun fibers are arranged in an amorphous manner, contributing to softness and drapability.

5.2.1.3.2 Exudate Control. In the case of wounds with a high level of exudate, an absorbent primary or secondary dressing should be used to remove excess exudate and prevent infection. Electrospun fibers made of hydrophilic polymers have higher absorption than regular films and so are more efficient at exudate control due to their high surface area to volume ratio, and for more hydrophobic polymers the exudate can pass through the highly porous fiber layer by wicking or capillary flow mechanisms to be absorbed by an outer/secondary dressing.[46,92] Surface wettability may be varied through the use of hydrophilic or hydrophobic polymers, or the incorporation of additives such as surfactants in the spin solution.[93,94] It must also be considered that additives may have multiple effects on the fibers, as in the case of addition of surfactant, where the surface wettability of the electrospun fiber material was enhanced, the cell–material interactions were improved with increased cell growth, and the degradation rate of the polymer was also increased.[93]

5.2.1.3.3 Moisture Balance. When the skin is damaged, the loss of fluid through blood loss, exudate production and water vapor loss can be very detrimental to the patient, leading to a drop in body temperature from evaporative cooling and potentially to hypovolemic shock as the body struggles to maintain adequate circulation.[85,95,96] It has also been shown that wounds heal better in a moist environment, resulting in faster healing and a reduction in scarring, pain, infection and costs.[97–99] The macro-porosity of the electrospun fiber webs allows a controlled loss of moisture as they are highly permeable to water vapor, and this can be tailored by modifying the fiber diameter, density and polymer type (natural, synthetic, hydrophilic or hydrophobic) to approximate intact skin.[88,100,101]

5.2.1.3.4 Gas Diffusion and Barrier Properties. The porosity of electrospun fibers allows for 'breathability' or the passage of gases, in particular oxygen, carbon dioxide and water vapor, which are essential for wound health and healing, without resulting in dehydration of the wound.[27,102] The fine pores of the fiber web also work to exclude bacteria and viruses from the wound bed, preventing infection.[103,104] Polyvinyl alcohol electrospun fibers containing honey and polyhexamethylene biguanide have been shown to have adequate breathability and antimicrobial activity suitable for wound dressing applications.[102]

5.2.1.3.5 Stimulation of Healing. The structural similarity of electrospun fibers to the ECM and the use of biocompatible components can stimulate healing, as the topography cues trigger fibroblast, keratinocyte, and progenitor cell adhesion, proliferation, their expression of ECM components (*e.g.* proteins such as collagen, elastin, fibronectin) and migration into the wound.[105] In addition to the structural benefits, electrospun fibers can be made from polymers or with additives that are bioactive, which is discussed in more detail in Section 5.2.2. A combination of natural and synthetic polymers is also beneficial, as synthetic polymers typically have better mechanical properties, while natural polymers tend to have better cell compatibility and enhance cell attachment and viability, as discussed in Section 5.2.4.[76] A considerable amount of research has been undertaken to reduce the degree of scarring – a major issue with normal healing and chronic wounds resulting in psychological (social and emotional) and physical problems (restricted movement as well as pain in keloid-type scarring).[106] Reparative therapies result in imperfect tissue remodeling. Electrospun fibers have shown regenerative, rather than reparative, properties, which has resulted in reduced scarring and wound healing time, reduced scar stiffness and reduced scar contraction.[107,108]

5.2.1.4 Characterization of Physical Properties

Wound dressings require characterization of their physical characteristics to compare properties such as the water vapor transmission rate (WVTR), absorption capacity, porosity, wettability, tensile strength and shelf life between different products. These will be discussed briefly below.

5.2.1.4.1 Water Vapor Transmission Rate. As mentioned above, the moisture-retention capacity of dressings is important in preventing excessive water vapor loss, and the WVTR is a useful indicator of this that can be used to select the correct dressing.[109] The WVTR can be measured using an evaporimeter to determine the grams of water transferred through the test sample, preferably using a standard protocol such as ASTM method E96-90,[110] ASTM E398-03,[70] or BS EN 13726-2:2002.[111] The WVTR for electrospun fiber dressings tends to be high due to the very high permeability of these materials, for example a WVTR of around 1296 g m^{-2} 24 h^{-1} was reported for a chitosan/PEO blended fiber mat.[70]

5.2.1.4.2 Absorption Capacity. Absorption capacity can be an important measure for dressings that are designed to have this as a positive feature, for example PVP/poly(acrylic acid) (PAA) hydrogel electrospun fibers were found to have swelling ratios between 500 and 3700%, which was double that obtained with PVP/PAA films,[112] and the incorporation of alginate in PCL electrospun fibers increased the water absorption to over 200%, compared to 25% for the pure PCL fibers.[81]

5.2.1.4.3 Pore Size. Porosity and pore size of the dressings can be measured using techniques and/or equipment such as physical properties (*e.g.* density/gravimetric methods),[113] scanning electron microscopy (SEM) analysis,[114] microcomputed tomography,[113] CAD models,[113] gas adsorption,[115] porosity calculations,[27] pycnometer,[116] mercury porosimeter,[117] liquid intrusion[118] or capillary flow porometer.[119] Regarding computation of porosity from 2D SEM images this can be achieved with open source software ImageJ using a custom plug-in called DiameterJ.[114] It is important to note that there are a number of trade-offs for each method, *e.g.* 2D imaging of pores does not elucidate the 3D pore volume and total porosity and intrusion methods use pressure that may distort samples during the measurement. Also data cannot be compared between the various methods as they are generally calculated by correlation of some physical property measurement.

5.2.1.4.4 Surface Wettability. Surface wettability is another feature that may be measured, typically using contact angle measurements.[27,93] Contact angle tests may be done using a contact angle analyzer, which measures the angle between a water droplet and the surface of the material using photographs.[94,120]

5.2.1.4.5 Mechanical Testing. Tensile strength, elongation at break, and Young's modulus or modulus of elasticity are all parameters that can be measured for electrospun fiber dressings. Parameters such as tensile strength may be determined using standard protocols such as ASTM D3379,[121] or ASTM D882-97.[58] Mechanical tests can be performed to determine adhesion of fiber layers to substrates and to determine the flexibility of electrospun fiber-based wound dressings after repeated strain. These are a peel test according to ASTM D2261 and Gelbo flex according to ASTM F 392-93, respectively.[122]

5.2.2 Electrospun Fibers for Nanomedicine

The high surface-to-volume ratio and interconnected pores make electrospun fiber webs an ideal platform for delivering therapeutics to a wound site.[123] The following section details the problems associated with wound infections, the benefits that electrospun fibers offer in nanomedicine and drug delivery, methods of incorporating these additives in fibers and, briefly, some methods of characterizing drug loading and release.

5.2.2.1 Applications for Nanomedicine and Drug Delivery

Nosocomial-acquired secondary infections affect hundreds of millions of patients and result in hundreds of thousands of deaths globally per year, placing a significant financial burden on the economy.[124] These infections are typically the result of antimicrobial resistance and are even more severe in cases of comorbidity, *e.g.* cancer treatment, organ transplant and diabetes. Due to the benefit and inherent simplicity of combining active pharmaceutical ingredients (APIs) into electrospun fibers *via* solution-based electrospinning, there has been a strong academic focus on this for wound treatment.[6] Some examples of additives incorporated into electrospun fibers are found in Table 5.1.

The high surface area offered by electrospun fiber materials offers site-specific nanomedicine strategies directly at the biointerface.[125] Controlled localized delivery is necessary to ensure safe dosing in the target site while reducing side effects.[126–128] Therapeutic agents such as antimicrobials, antioxidants, anti-inflammatories, analgesics and bioactive molecules (*e.g.* proteins, peptides, enzymes and growth factors) can be combined with electrospun fibers in many ways. Solution-based electrospinning allows for therapeutic agents, particularly those that are heat-labile, to be incorporated into electrospun fiber carriers or excipients. When added to the spin solution, therapeutic agents can change solution parameters such as viscosity, surface tension and conductivity.[129] These in turn can affect the resultant fiber morphology and production rates. Inclusion of therapeutics into electrospun fibers enhances the bioavailability of poorly soluble synthetic and plant-derived therapeutics (*e.g.* hydrophobic actives) that otherwise would need to be taken at much higher doses and have low bioavailability when administered orally.[130]

5.2.2.2 *Methods for Incorporation of Additives in Electrospun Fibers*

Various strategies exist for combining active pharmaceutical ingredients (APIs) with electrospun fibers (see Figure 5.2). The most common methods for incorporating actives include dissolution and dispersion. Although solution electrospinning is generally conducted at room temperature, APIs may still be prone to denaturation and degradation. For this reason, and because some active and excipient combinations are not electrospinnable on their own due to lower molecular masses, methods such as emulsion and core–shell, also known as coaxial electrospinning when produced from concentric needle-in-needle spinnerets, have been developed.[131] Emulsion electrospinning can be used for example for water-in-oil emulsions, where the water-soluble drug may be sensitive to the organic solvents,[132] and to produce core–shell fibers.[133] Inclusion of APIs *via* blending and emulsion electrospinning is generally compatible with existing high-throughput solution-based electrospinning technologies such as free-surface electrospinning.[134–141] Yan *et al.*[142] demonstrated the slit surface electrospinning method and Jiang and Qin[143] showed a modified free-surface coaxial electrospinning set-up to overcome the

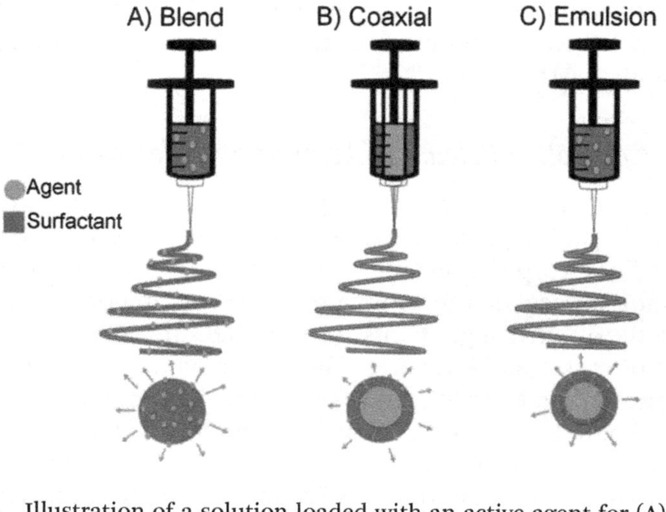

Figure 5.2 Illustration of a solution loaded with an active agent for (A) blend, (B) coaxial and (C) emulsion electrospinning showing the cross-section of the resulting fibers. Reproduced from ref. 6 with permission from the Royal Society of Chemistry.

challenges of coaxial needle electrospinning and provide commercially relevant alternatives.

In addition to these procedures, functionalization of polymers prior to electrospinning or post-processing of electrospun fibers offers alternatives to the aforementioned entrapment approaches.[144] Post-functionalization by surface modification[145] has the advantage of physically or chemically attaching active compounds to the surface of the fibers to ensure that the total surface area of functional groups is exposed, compared to blending where the active in the core of the fibers is not directly exposed to the wound environment. Additionally, post-processing electrospun fibers with actives may be required in some cases to prevent severe damage of the actives that are not electrospinnable or are prone to degradation during blending procedures, *e.g.* sonication.[146] The major disadvantage with post-processing electrospun fibers is the probability of changing the fiber morphology, for example, wetting, or vapor treatment with chemicals or heat treating above the glass transition temperature of the polymer can relax and change the shape of the fiber material.[147] Another trade-off is that some reactive chemistries may be cytotoxic and may be difficult to remove from the fiber materials and complicate the regulatory pathway of the device, *e.g.* glutaraldehyde when crosslinking gelatin and collagen.[148]

The rapid evaporation of the solvent during electrospinning of solutions containing polymer–API formulations results in the rapid but gentle drying and precipitation of the electrospun fiber-active. Electrospun fibers therefore act as a carrier to protect heat- and chemical-labile actives in a dry state until they are released into a wound. Zussman has shown that electrospun fibers encapsulating living cells can be prepared for regenerative medicinal applications.[149] Electrospinning in this context thus provides a cost-effective alternative to complex freeze-drying procedures. This offers a great possibility for the inclusion of many other natural biopharmaceuticals, *e.g.* proteins, peptides and growth factors.[150]

5.2.2.3 Controlled Release of Actives from Electrospun Fibers

Custom release profiles from electrospun fibers are dependent on multiple parameters including polymer composition, type of additive, additive loading, method of incorporation and fiber diameter.[32,151–153] Fiber diameter and the resultant surface area to volume ratio have an important effect on the release profile of active ingredients. The specific surface area or surface area to volume or mass can be theoretically calculated as described by Eichorn and Sampson[154] or experimentally determined by methods such as gas adsorption analysis using BET analysis.[155] The high surface area to volume ratio of electrospun fibers makes burst release an unavoidable reality with blend electrospinning of actives. This favors wound applications that require rapid release of non-steroidal anti-inflammatory drugs (NSAIDs) for pain management, or antimicrobial release to disrupt biofilm formation, or to treat an infected wound where burst release is required.[41] Tseng and Liu have comprehensively reviewed the topic of analgesics loaded into electrospun fibers, where they specifically discuss applications in wound dressing and burn care.[89] Compatibility of the active with the polymer also has an effect, as shown in a study comparing the encapsulation of lipophilic *versus* hydrophilic drugs in PLA electrospun fibers, which found that the hydrophilic drug was found on or near the surface of the fibers and had burst release, compared to zero-order release kinetics observed with the lipophilic drug through polymer degradation.[130]

Blending results in more superficial aggregation of the drug and therefore more prominent initial burst release, whereas core–shell electrospinning entraps the drug in the core and has a long sustained release over time. Hu *et al.*[156] showed that it is possible to obtain varying drug release kinetics and mechanical properties through the combination of braiding technology with cefotaxime sodium loaded PLLA electrospun fibers that were prepared by either blend or core–shell electrospinning.

Combination therapies can be achieved *via* concurrently electrospinning from multiple sources carrying different formulations. This is particularly useful when drugs are not compatible with each other or are not compatible to be electrospun together from the same fiber-forming polymer/solvent, *e.g.* in the case of two drugs having different lipophilicity.[32] This approach is also beneficial when dual-release profiles are required to address various stages of wound healing. Choi *et al.*[157] surface modified fibroblast growth factor (bFGF)-loaded core–shell electrospun fibers by conjugating epidermal growth factor (EGF) to the surfaces of the fibers. bFGF showed a burst release profile whereas the EGF showed sustained release in a biphasic profile to enhance wound healing. Delaying the release of the EGF reduces the probability of attack of hydrolytic enzymes and increases efficacy of the therapy.[157] Combination antimicrobial therapeutics have also been suggested to help prevent bacterial resistance due to only single antibiotic administration.[158]

5.2.2.4 Antimicrobial Electrospun Fibers

Silver formulations that have been used for their antimicrobial efficiency in wound care include silver nitrate ($AgNO_3$), silver sulfadiazine (AgSD) and silver nanoparticles, otherwise known as nanocrystalline silver (NCS). Dressings with different preparations of silver have been used to allow easier handling and better control of dosage as well as sustained release of silver, which limits dressing changes and correspondingly reduces patient discomfort, and cost in terms of time and resources.[159] Silver has been incorporated in electrospun fibers through addition of silver salts such as $AgNO_3$ in the spin solution,[18] electrospinning with silver nanoparticles through *in situ* reduction of $AgNO_3$ in the spin solution,[18,160–164] or post-treatment with $AgNO_3$ followed by reduction to give a silver nanoparticle coating,[110,165,166] as well as electrospinning with AgSD,[167–169] or with a silver nanoparticle dispersion.[80]

There are conflicting reports on cytotoxicity of silver towards human cells,[163,166,170–175] and bacterial resistance to silver has also been a concern.[170,176–180] Because of this, there has been a lot of interest in finding natural alternatives to silver. Natural options for antimicrobial agents include honey, essential and edible oils, and antimicrobial peptides.[101] Some oils such as tea tree oil have been reported to have anti-inflammatory properties, and other natural extracts such as aloe vera gel and certain plant and fruit extracts may have regenerative properties.[101] Curcumin reportedly has antioxidant and regenerative effects.[31] Many of these natural products still require further evidence to prove their effectiveness. Those that have been used in electrospun fibers include curcumin,[26,31,51] honey,[181,182] tea tree oil,[183] fruit extract[184] and aloe vera,[185] among others.[101,186]

The use of inherently antimicrobial polymers in electrospun fiber-based wound dressings offers significant advantages of high concentration of biocide groups at the large surface area of the fiber and less concern over cytotoxicity in the wound. Products that do not release APIs into the wound will also follow a less stringent regulatory path. There are also trade-offs that do need to be considered in product design, such as reduced efficacy compared to free antimicrobial molecules due to steric hindrance from the polymer backbone. Other design considerations such as toxicology, shelf life and stability in the case of a polymer with biocide covalently attached to the polymer backbone, spectrum of activity against a broad range of microbes, complexity and expense in manufacture should be taken into account.[144] Lindner *et al.*[187] showed that poly-*N*-acetyl-glucosamine electrospun fibers can be used to enhance wound closure and antimicrobial activity *via* triggering immune response signaling of defensins in animal model wounds infected with *S. aureus*. Chitosan and its derivatives have been shown to have antimicrobial, antimycotic and hemostatic properties due to the presence of free amino groups in the polymer structure.[110,188,189] Bshena[190] synthesized styrene-maleimide copolymer derivatives by modification of poly(styrene-*co*-maleic anhydride) with primary amine compounds and the resultant electrospun fibers inhibited the growth of methicillin-resistant *S. aureus* (MRSA), *Y. pestis* and *V. cholerae*.

Metal chelators such as ethylenediaminetetraacetic acid (EDTA),[191] 2,3-dihydroxybenzoic acid (DHBA)[192] and lactoferrin[193] have been incorporated into electrospun fibers and were shown to prevent biofilm formation and formation of pathogenic bacteria, provide bactericidal activity and a synergistic antimicrobial activity when combined with other antimicrobial actives.

Oxygen-generating actives have also been used to provide antimicrobial activity. Calcium peroxide generates radicals of hydrogen peroxides and calcium hydroxides when in contact with water and delivers a bactericidal effect.[194] Wang *et al.*[194] prepared calcium peroxide-loaded PCL electrospun fibers that on exposure inhibited the growth of *E. coli* and *S. epidemidis*.

5.2.2.5 *Characterization of Drug Release from Electrospun Fibers*

Other than testing the effect of antimicrobial drugs on bacterial growth through inhibition assays, there are various methods for characterizing drug loading in and release from electrospun fibers. In the case of silver, multiple characterization methods in addition to the commonly used SEM are possible, including transmission electron microscopy (TEM),[30,80,162,195] energy-dispersive X-ray (EDX) analysis,[18,30,162,166] X-ray diffraction (XRD),[30,80] X-ray photoelectron spectroscopy (XPS) and UV-vis spectroscopy.[164,196] In terms of characterizing release, inductively coupled plasma atomic emission spectrometry (ICP-OES)[162,164,195] and atomic absorption spectroscopy (AAS)[30,166] can be used to look at the release of silver from the electrospun fibers in solution. For release of drugs or other additives, methods such as UV-vis spectroscopy[197–199] and high-performance liquid chromatography (HPLC) coupled to mass spectrometry (HPLC/MS)[45] or UV-vis (HPLC/UV-vis)[32] are used to give a release profile and to study drug release kinetics.

5.2.3 Electrospun Fibers for Healthcare Diagnostics and Point-of-care

With regards to healthcare diagnostics, point-of-care devices that allow rapid detection of diseases or deficiencies are very valuable for speeding up treatment and reducing both monetary and other resource costs. The market for point-of-care diagnostic testing is large, in the billion dollar range for the US market alone.[44]

Electrospun fibers show particular promise in diagnostics due to the much larger surface-to-volume area, their flexibility in terms of choice of components, chemistry, and options for functionalization. Immunoassay methodology, based on the standard enzyme-linked immunosorbent assay (ELISA), has been used to create microfluidic lab-on-a-chip as well as paper-based lateral flow type devices with electrospun fibers showing improved sensitivity compared to conventional substrates.[39,44,200–204] Certain polymers are chosen due to their ability to adsorb proteins through non-covalent bonding, such as polycarbonate for immunoglobulin G,[200] and polystyrene (PS) for monoclonal antibody to C-reactive protein.[44] In other cases, biomolecules can be attached covalently *via* functional groups, such as in the case of streptavidin bonding through the carboxylic acid groups of PLGA electrospun fibers,[201] or bonding of a protein cancer biomarker to plasma-treated zinc oxide fibers[205] through *N*-hydroxysuccinimide (NHS) activation. Additional polymers and additives can be included to improve the specificity of the electrospun fibers.[33,202] Incorporating a coating such as polypyrrole has been used to improve the conductivity in a sensor utilizing PS fibers.[44] Electrospun fibers have in specific cases shown improved sensitivity for detection of HIV,[200] circulating tumor cells,[39,201] *E. coli*,[202] glucose,[206] breast cancer biomarker[205] and malaria biomarker,[207] as brief examples of the diversity of diagnostic applications.

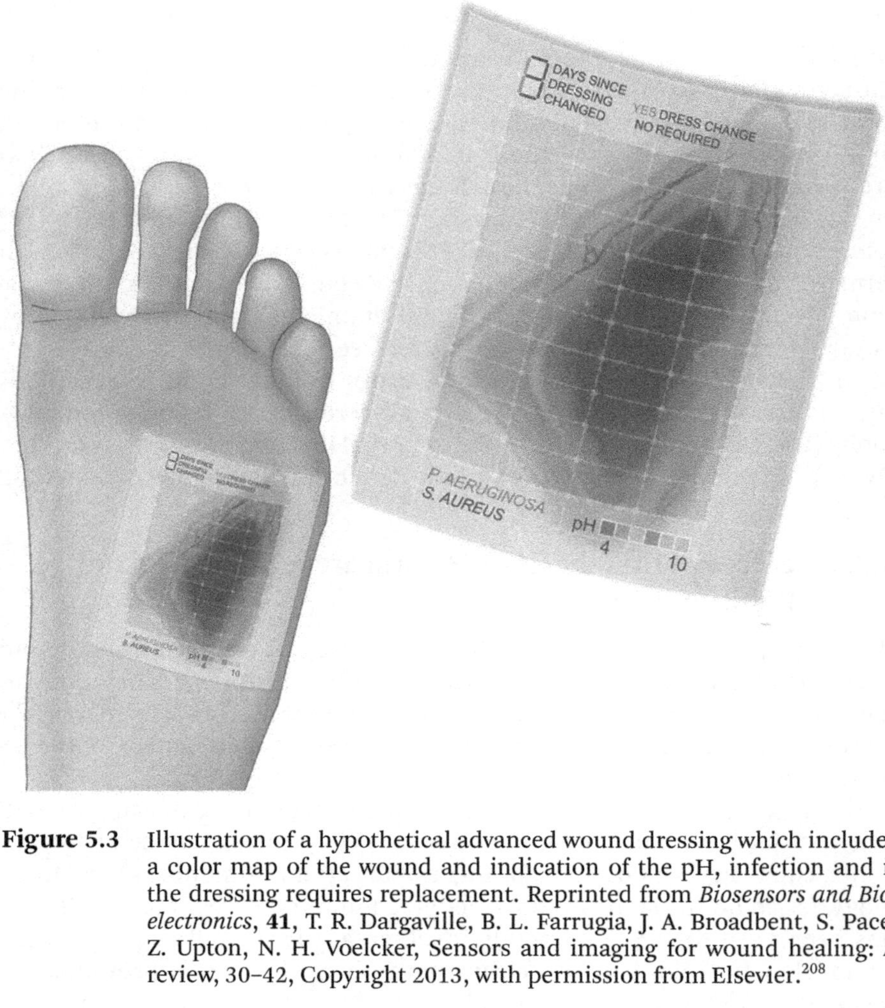

Figure 5.3 Illustration of a hypothetical advanced wound dressing which includes a color map of the wound and indication of the pH, infection and if the dressing requires replacement. Reprinted from *Biosensors and Bioelectronics*, **41**, T. R. Dargaville, B. L. Farrugia, J. A. Broadbent, S. Pace, Z. Upton, N. H. Voelcker, Sensors and imaging for wound healing: A review, 30–42, Copyright 2013, with permission from Elsevier.[208]

The intersection of diagnostics and wound care can lead to novel wound care indicators that are very attractive for giving immediate information on the state of the wound. The rapid assessment of the state of the wound means faster and easier decisions for the medical practitioner, shorter treatment time and more effective wound management.[208] Two interesting and comprehensive reviews by Dargaville *et al.*[208] and Tegl *et al.*[209] cover the types of wound markers that can be used, such as pH, bacterial load, temperature, and biomolecules such as growth factors and inflammatory mediators, and examples of sensors for detecting these markers, as shown in the illustration by Dargaville *et al.*[208] in Figure 5.3. Additives can be included in the wound dressing construction that indicate, for example by a color change or electrochemical signal, changes in the wound pH,[210,211] or the presence of pathogenic bacteria.[212,213] This is particularly important for the treatment of chronic and extensive wounds, where infection may gravely affect the prognosis for the patient's recovery. The work by Zhou *et al.*[212,213] describes the use of synthetic phospholipid vesicles as false 'cells' that are lysed by toxins from *S. aureus* and *P. aeruginosa* to release an indicating dye for visualizing the infection,[213] and/or antimicrobials/antibiotics to treat the infection.[212]

Electrospun fibers could be incorporated and may improve the sensitivity of these diagnostic dressings by allowing attachment of more active agents compared to regular dressing materials, and remove/reduce the necessity for dressing changes, typically done to inspect the wound.

5.2.4 Modulation and Characterization of Cell–Fiber Interactions

Depending on the intended market and the product or device class, certain tests for medical devices may be required by the regulatory body. An example of standard testing protocols is the biological evaluation of medical devices International Standards Organization (ISO) standard 10993. The ultimate goal is to prove that the product is safe for use, and these tests should always be considered and carried out appropriately as required from a commercial perspective. However, a product needs to be effective as well as safe, and modifications that can be made to the electrospun fiber to modulate the interactions between the cells and fibers and then characterization of these interactions on a cellular level are described here.

5.2.4.1 Cytotoxicity/Cell Viability

All medical devices and products that come into contact with human tissue should be tested for cytotoxicity/cell viability, or any adverse effects.[214] Test results from *in vitro*, *ex vivo* and animal models should be considered before moving on to tests in humans, with the understanding that these results may differ when moving to human testing and may be dependent on the location of the device and the specific tissue interactions at the site of contact. Ethical testing compels comprehensive *in vitro* and *ex vivo* testing before using animal models. *In vitro* tests for cytotoxicity rely on culturing the cells and then exposing them directly to the device, or indirectly to an extract fluid.[215] There are standard ways of preparing extracts, in terms of extraction time, temperature, and ratio.[216] However, only a few articles in the literature relating to electrospun fibers for wound care mention that standards such as ISO 10994-5 were followed.[56,69,111,181]

The direct contact method is commonly used with cytotoxicity tests involving electrospun fibers in literature, and cells are cultured directly on the fibers and tested through the use of various standard assays in comparison to a control.[63,217] For example, the MTS assay is used to evaluate cell viability/cytocompatibility/cytotoxicity[82,83,110] and measure cell proliferation of human dermal fibroblasts.[82,218,219] Cell nuclei and components can be visualized using staining and fluorescence imaging, phase contrast, electron microscopy and/or confocal microscopy after fixing the cells.[27,63,220,221] These different techniques can be used to look at effectiveness of the products in terms of retaining confluent cell morphology, as well as straightforward cytotoxicity.

Careful choice of additives and polymers and the use of non-toxic solvents (as compared to the commonly used toxic solvents HFIP and trifluoroacetic acid (TFA)) during the electrospinning can greatly reduce the toxicity risk. If cytotoxicity testing is done through contracted services, sufficient materials information should be supplied for the laboratory to determine the correct protocol.

5.2.4.2 Adhesion and Migration

Cell adhesion is measured because cells derived from many tissues require an anchoring surface to grow and it is essential for the cell to achieve the correct morphology, migration, proliferation and function.[63,222] Cellular adhesion can be measured by removing non-adherent cells and quantifying the number of adherent cells through an assay.[223] Cell migration, as can be seen particularly when the actin cytoskeleton has been stained and visualized with fluorescence microscopy, is an indication of good cellular adhesion.[63,224]

Materials choice can greatly affect the cellular adhesion to and migration over electrospun fibers. Properties can be tuned either through the choice of the polymer, through the inclusion of additives and/or blending of polymers,[78,82,225–227] or through post-processing such as functionalization,[224] coating[58] and plasma treatment.[42,228–231] Surface wettability and hydrophilicity, for example, have been found to have an effect on cell adhesion in some reports.[78,93,224,226,228] Venugopal *et al.*[23] found that using a blend of natural and synthetic polymers gave a good balance of results for cell adhesion and mechanical handling in a PCL/collagen blend.

Cell migration can be optimized by the electrospun fiber structure, as a number of articles have shown directed spreading and migration along aligned scaffolds.[221,232–235] Alignment of the scaffold can affect the morphology of the cells which can be visualized by staining.[42,236] Fu and Wang demonstrated that aligned PCL/collagen electrospun fibers supported the proliferation and migration of human adipose-derived stromal cells to a greater degree and demonstrated a higher synthesis of new ECM compared with random electrospun fiber arrangements.[237] Random electrospun fibers may, however, promote cell growth into the fiber mat, which is important for greater tissue production in deep wounds, whereas aligned fibers may promote lateral cell growth on the mat surface, which is important for rapid tissue repair in shallow wounds.[238]

5.2.4.3 Infiltration

The importance of porosity and surface area for cell interaction has been comprehensively reviewed by Loh and Choong.[113] The larger the fiber diameter, the larger the pore size for the electrospun scaffolds, and consequently the better the cellular infiltration into the matrix.[239] However, there is a critical fiber diameter size beyond which the cells cannot adhere to multiple fibers, do not spread, and elongate along a single fiber axis.[239] When the pore size of the electrospun fiber matrix is too small compared to the cell size, infiltration is only into the surface layers of the fibers.[218] This can be circumvented by cell seeding on both sides of a thin scaffold, but does need to be taken into account during scaffold design and construction.[218] Some wound applications require that a product be lofty, *e.g.* thick with larger interconnected pores but low in density. To overcome the densely packed electrospun fibers produced by electrospinning there are a number of proposed strategies that are concisely summarized by Teo.[240] Baker *et al.*[220] did a study of selectively removing sacrificial polymer fibers and looking at the infiltration of mesenchymal stem cells (MSCs) into the scaffolds. This work showed the slow infiltration of cells into the electrospun fiber scaffolds, and the improvement when the sacrificial polymer fibers were removed and larger pores were present.

5.2.4.4 Interaction of Bacteria with Electrospun Fibers

Mortimer *et al.*[241] and Said *et al.*[38,242] comprehensively discussed microbial interaction with nanostructures covering most of the important aspects that influence this interaction and that should be considered in the development of electrospun fiber advanced wound care products. Said *et al.*[242] showed that for fusidic acid (FA) (a bacterial protein synthesis inhibitor that is used as a bacteriostatic active) loaded PLGA electrospun fibers, drug delivery could be triggered by bioburden or biofilm formation. This was attributed to bacterial enzyme hydrolysis of the PLGA polymer, which resulted in acidic degradation products that decreased the pH of the wound environment but also released FA in the process to inhibit biofilm formation.[38,242] The accelerated surface erosion of the electrospun fibers comes with the price of reducing structural and mechanical properties of the electrospun fibers, which should be taken into account before designing the product.

5.3 The Market: From Lab-scale Curiosities to Commercial Realities

5.3.1 Overview

The first patents describing electrospinning were filed by Cooley[243] and Morton.[244] The first significant endeavor to improve on the process for upscaling and the commercialization of electrospinning of electrospun fibers was described in a series of 22 patents by Formals between 1931 and 1944.[245] Nearly a century later electrospinning was revived by Doshi and Reneker,[246] and since then literally hundreds of patents[247] have been issued and thousands of publications[248] have been published describing electrospun fiber concepts that have potential in a plethora of fields from electronics to biomedical applications. Due to the low production rates required, lower-hanging fruit such as filtration has seen commercial application of electrospun fibers for filtration products for nearly three decades, *e.g.* Donaldson and its Ultra-web® electrospun fiber filter.[249]

5.3.2 Electrospun Fiber Product Development for Commercial Advanced Wound Care

The advanced wound care global market was estimated to reach US$3.6 billion in 2011 with a compound annual growth rate (CAGR) of 4.2%.[250] To benefit economically sensitive areas and to reduce the load on the health system, advanced wound care has been targeted as an area of focus over traditional methods to promote faster healing and reduce patient follow-ups.[250]

Over the past 20 years there has been significant advancement of high-throughput electrospun fiber production technologies.[136,251,252] Commercial contract development and manufacturing companies that use electrospinning to develop and manufacture electrospun fibers and their relevant certification for production of medical devices are summarized in Table 5.2.

The electrospun fiber-based advanced wound care products identified in this literature review are summarized in Table 5.3. Although some of the products are 'advertised', a distinction should be made between those that are commercially sold and those that are not, *e.g.* have pre-market approval but are not commercially available.

5.3.2.1 Improvement of Mechanical Properties of Electrospun Webs for Wound Dressings

Although electrospun fibers can be used as standalone dressings, in some cases they are combined with other substrates (*e.g.* gauze), because of insufficient strength or limitations on forming thick fiber webs at scale, low crystallinity and random alignment of electrospun fibers. Electrospinning onto substrates results in poor delamination strength of the coated fibers, which has led researchers to seek other methods for improving the mechanical properties of electrospun fibers. Chong *et al.*[218] showed that fibers could be electrospun onto an existing skin substitute product, Tegaderm™ (3M Medical), for dermal wound healing, to overcome the mechanical deficiencies of a thin electrospun fiber layer and the higher costs of artificial dermal analogues. Further research is required for enhancing the mechanical properties of electrospun wound dressings, borrowing concepts from non-woven textile manufacture such as laminating[282] and calendering[283]

Table 5.2 Commercial electrospun fiber contract development and manufacturing companies and their certifications for medical device production.

Company name	Country	Relevant certification for medical device manufacture	Reference
Arsenal Medical, Inc. (AxioCore™)	USA	None found	253
Bioinicia (Fluidnatek®)	Spain	GMP and ISO 13485	254
Espin	USA	GMP and ISO 13485	255
Fuence	Japan	None found	256
IME technologies	The Netherlands	ISO 13485	257
Leonardino S.r.l.	Italy	ISO 13485 and MDD 93/42	258
Nanofiber Solutions, Inc.	USA	None found	259
Nanospun Technologies	Israel/USA	None found	260
NEI Corporation	USA	None found	261
Pardam	Czech Republic	None found	262
Revolution Fibres	New Zealand	None found	263
SNC (Stellenbosch Nanofiber Company)	South Africa	GMP, ISO 9001 and ISO 13485	264
SNS Nano Fiber Technology	USA	GMP, ISO 9001 and ISO 13485	265
Technical Fine Co., Ltd	Japan	None found	266
The Electrospinning Company Ltd	UK	GMP, ISO 9001 and ISO 13485	267
Yflow	Spain	None found	268

to allow layers of fibers and substrates to be combined to yield a stronger product where required.

5.3.2.2 Sterilization

An important consideration on the path to regulatory approval for advanced wound care products is to ensure that the bioburden is kept low and that sterilization is performed to achieve a sterility assurance level (SAL) of 10^{-6}. Various sterilization processes may alter the electrospun fiber morphology, physical properties or activity of the actives included in the product, which should be considered in the design phase of the product to avoid unnecessary reformulation. Commonly used processes for sterilization in the medical industry include steam sterilization, dry-heat sterilization, gamma irradiation, E-beam irradiation, ethylene oxide sterilization (ETO), plasma sterilization and UV sterilization.[284,285] Gamma irradiation and E-beam irradiation are the most viable strategies as products can be irradiated post-packaging. The important concept with these methods is to not over-irradiate and to assess the bioburden prior to selecting the irradiation dose, otherwise this may result in a significant decrease in molecular weight and mechanical strength.[286] It has been shown in some studies that the dose should not be above 10 kGy when using gamma or E-beam irradiation on

Table 5.3 Commercially marketed electrospun fiber-based wound care products.

Product name	Supplier	Market status	Electrospinning equipment or contract manufacturer used	Product type: primary mode of action	Reference
FastClot/SurgiClot®: hemostatic dressing	St Teresa Medical, Inc.	Available on the market in New Zealand, CE mark pending, dossier submitted February 2017	Unknown	Non-woven nanofiber dextran with human thrombin and fibrinogen (clotting proteins) hemostasis dressing for cancellous bone tissue bleeding	269 and 270
TuneCoat™: bandages for localized pain relief	Luna Innovations, Inc.	Technology readiness level 4. Phase I US Army funded SBIR program completed. Started a Phase II program	Elmarco NanoSpider® NSLab system	Localized delivery of analgesics, particularly for burn pain and peripheral neuropathies	271
BIOcular™: ocular repair dressing	Luna Innovations, Inc.	Technology readiness level 4. Phase I US Army funded SBIR program completed. Started a Phase II program	Elmarco NanoSpider® NSLab; custom multi-chemistry needle-free electrospinning system	Synthetic bandages for the treatment of the ocular surface following traumatic injury to stabilize open globe injuries or stimulate corneal wound healing	272
Nanotrix®: temporary skin substitute	Southern Medical, under Medika SA (South Africa)	Available on the market in South Africa, CE mark pending	Stellenbosch Nanofiber Company (using SNC's ball electrospinning technology or SNC BEST™)	Synthetic, self-adhesive membrane comprising an apo-lactoferrin hyaluronic acid polymer. Partial thickness wounds, split-thickness skin graft donor sites and abrasions	273

Cerafix® Dura Substitute (FDA clearance)	Acera Surgical, Inc. (USA)	Commercially available/post-market	Acera Surgical, Inc.	Cerafix® Dura Substitute is indicated as a dura substitute for the repair of dura mater	274 and 275
Restrata® Advanced Wound Matrix (FDA clearance)	Acera Surgical, Inc. (USA)	Commercially available/post-market	Acera Surgical, Inc.	Restrata® Wound Matrix is indicated for use in the management of wounds, including partial and full thickness wounds, pressure sores/ulcers, venous ulcers, diabetic ulcers, chronic vascular ulcers, tunneled/undermined wounds, surgical wounds, trauma wounds, and draining wounds	276
ReDura™: dural patch	MEDPRIN (China)	CE mark certified	MEDPRIN Regenerative Medical Technologies Co., Ltd (unknown)	Poly-L-lactic acid electrospun fiber; biomimetic-synthetic-absorbable dural substitute	277 and 278
Rivelin® patch: for treatment of oral lichen planus (mucosal diseases)	Dermtreat ApS (Denmark)	Phase 2b study planned to launch in 2018	Bioinicia (Fluidnatek®)	Treatment of autoimmune inflammatory disease oral lichen planus by incorporating the drug clobetasol propionate in the Rivelin® patch	279–281

synthetic biodegradable polymers and that with suitable handling, packaging under argon atmosphere and maintaining the materials below 8 °C it is possible to sterilize without significantly affecting product morphology and shape.[287]

5.3.2.3 *Packaging*

Advanced wound care product packaging should be suitably insulated, have adequate bacterial barrier properties, be mechanically and chemically protective and not affect the product in any way during storage, transportation, sterilization or handling throughout the expected shelf life of the product. Compliance with standards such as ISO 11607 and EN 868 are required for CE technical files and Food and Drug Administration (FDA) submissions.[288]

5.3.2.4 *Shelf Life*

Products should ideally have a long shelf life and should be able to be stored at room temperature, to ensure lower costs of storage and transport. It is understood that synthetic polymers have the advantage of a longer shelf life over natural polymers.[68,101,188] For shelf life determination, standard protocols such as ASTM F1980 and BS EN 868-8 may be used.

5.3.3 Regulatory Pathway for Electrospun Fiber-based Advanced Wound Care Products

In order for advanced wound care product manufacturers to reduce the time to market, a thorough understanding of the regulatory pathway options is required and a regulatory strategy (including clinical and reimbursement strategies) suitable for the class of medical device should be formulated early on in the design phase of the product. Delays due to an unsuccessful application for market approval or to repeat clinical trials in the regulatory process can be costly in direct financial terms and in foregone opportunities of time in the market and to leverage a competitive advantage.

5.3.3.1 *Requirements for Market Approval in the USA and Europe*

The two largest markets for advanced wound care products are the USA and the European Union.[250] Medical devices are regulated by the Food and Drug Administration (FDA) in the USA and various National Competent Authorities in each member state of the European Union.[289] These authorities ensure compliance with the Code of Federal Regulations Title 21 for Food and Drug Administration (CFR 21) and the Medical Device Directive

93/42/EEC, respectively. FDA approval and CE marking is required in order to sell medical devices in these respective markets. In both regions, devices are classed according to the risks associated with the device, which is determined by the complexity, sterility requirements, length of use (short term, long term, permanent) and invasiveness of the product, and require quality systems and risk management. The higher the risk involved, the more complex and stringent the requirements are and therefore the longer the regulatory pathway. Typically, advanced wound care devices are classified as either Class II or Class III devices.

5.3.3.2 Comparison of FDA Approval and CE Marking

It is generally understood that FDA approval is slightly more comprehensive, with a focus on product efficacy and the benefit to the health system as a whole. This focus, the strict regulations on good clinical practice (GCP) and the conduct of clinical trials, generally results in a 1- to 3-year delay in market approval compared to the CE process.[290,291] It is argued that there is more onus and trust placed on European manufacturers and surgeons and that there is a significant benefit in getting products to market rapidly without incurring high incidences of product recalls. The situation is slightly different for Class III devices in the USA Medicaid/Medicare reimbursement scheme. This may be a simpler route to follow than acquiring the approval of each European country's healthcare administration.[292]

In the USA, subcontractors and product owners require compliance with Current Good Manufacturing Practices (cGMP) as detailed in the CFR 21. Medical devices are regulated either through a 510(k) clearance or pre-market approval (PMA) scheme.[293] An overview of the FDA process can be found on the FDA website.[294] In the European Union (28 countries), subcontractors and product owners require compliance with the Medical Devices Directive (93/42/EEC). An overview of the steps to CE marking for manufacturers of medical devices is provided by the European Commission[295] and the process is discussed by French-Mowat and Burnett.[296]

5.3.3.3 Other Factors to Consider for Market Approval

Obtaining either FDA clearance or the CE mark registration will allow a medical device to be marketed but ‘getting to market’[297] requires additional consideration of a range of other factors including health economics (insurance or reimbursement strategy), unique selling points, key opinion leaders, and independent studies. A reimbursement strategy for payment through Medicare and Medicaid Services (CMS) and health insurance providers in the regions for which it will be sold should be part of a medical device manufacturer's regulatory strategy. In the USA the reimbursement code (necessary for claims) for a product is determined by the Center for

Devices and Radiological Health (CDRH) and is based on medical device product classifications designated under 21 CFR Parts 862–892.[298] CMS require continued clinical and economic data collection and reporting to determine safety, efficacy and cost-to-benefit analyses. Medical devices that do not receive medical reimbursement cover or receive less than competing products will be at a significant disadvantage compared to those that receive a higher level of cover.

5.4 Conclusion: the Future

In general, after describing the promise of electrospun fibers for tissue engineering, drug delivery and wound dressings, a significant number of review papers end up concluding that electrospinning has not reached commercial levels due to slow production rates. In this chapter it was shown that there are in fact a number of contract manufacturers of electrospun fiber products that can produce biomedically relevant electrospun fibers at various commercial volumes. Future research and development of electrospun fiber-based advanced wound care dressings should include collaboration and consultation between all stakeholders (*e.g.* doctors, nurses, regulatory experts, engineers, scientists and business developers) at the design phase of the product development cycle. This will ensure the pathway to market is mapped out sufficiently and that there is no scope creep or redesign of the product in the later stages of development. There is a generally held idea that the ultimate goal is FDA approval or CE marking but the reimbursement strategy of the product is also a major inhibitor for the product to reach the market.

A frequently encountered tendency is to overdesign a product and therefore not have an equivalent device to draw comparisons with. It also complicates the regulatory pathway. A fundamental aspect of the pathway should begin with product development on a scalable electrospinning platform using formulations that will scale adequately, *e.g.* cGMP grade polymers, class 3 solvents that are non-toxic and non-volatile, and cost-effective solvents (moving away from HFIP and TFA). The emerging research in diagnostics, regenerative medicine, advanced wound care and nanomedicine could lead to the development of smart products that allow real-time understanding of the state of the wound as well as localized release of APIs. To support electrospun fiber diagnostics, manufacturers of wound care products may look toward developing a database of different wound classes linked to known treatment regimens and dressing products to assist healthcare workers in distinguishing the correct wound product to use.[299,300]

Electrospun fiber-based wound care products may not initially compete directly with gold standard product and may be classed in a category of their own. It will take the combined effort of 'electrospinners' around the globe to get these products to market, and to consistently prove the remarkable property benefits in order to gain wide acceptance.

References

1. A. Zhang, G. Zheng and C. M. Lieber, *Nanowires*, Springer International Publishing, Cham, 2016.
2. N. Greer, N. A. Foman, R. MacDonald, J. Dorrian, P. Fitzgerald, I. Rutks and T. J. Wilt, *Ann. Intern. Med.*, 2013, **159**, 532.
3. *Defining Advanced Wound Care in 2016: A Panel Discussion*, http://www.todayswoundclinic.com/articles/defining-advanced-wound-care-2016-panel-discussion, accessed August 2017.
4. J. S. Boateng, K. H. Matthews, H. N. E. Stevens and G. M. Eccleston, *J. Pharm. Sci.*, 2008, **97**, 2892.
5. G. C. Gurtner, S. Werner, Y. Barrandon and M. T. Longaker, *Nature*, 2008, **453**, 314.
6. K. A. Rieger, N. P. Birch and J. D. Schiffman, *J. Mater. Chem. B*, 2013, **1**, 4531.
7. C. L. Baum and C. J. Arpey, *Dermatol. Surg.*, 2005, **31**, 674.
8. E. W. ter Meulen, M. J. Poley, M. van Dijk, A. D. Rogers and H. Rode, *S. Afr. Med. J.*, 2016, **106**, 1120.
9. H. Rode, A. Rogers, S. Adams, W. Kleintjes, L. Whitelock-Jones, A. Muganza and N. Allorto, *S. Afr. Med. J.*, 2013, **103**, 608.
10. A. D. Rogers, S. Adams and H. Rode, *Plast. Surg. Int.*, 2011, **2011**, 1.
11. E. S. Foundation, *Nanomedicine - an ESF – European Medical Research Councils (EMRC) Forward Look Report*, 2005.
12. T. J. Webster, *Int. J. Nanomed.*, 2006, **1**, 115.
13. K. Vowden and P. Vowden, *Surgery*, 2017, **35**, 489.
14. European Commission, *Medical Devices: Guidance Document - Classification of Medical Devices, MEDDEV 2. 4/1 Rev. 9*, 2010.
15. B. Gupta, R. Agarwal and M. Alam, *Indian J. Fibre Text. Res.*, 2010, **35**, 174.
16. A. Asti and L. Gioglio, *Int. J. Artif. Organs*, 2014, **37**, 187.
17. E. Mele, *J. Mater. Chem. B*, 2016, **4**, 4801.
18. P. Kleyi, C. L. Frost, Z. R. Tshentu and N. Torto, *J. Appl. Polym. Sci.*, 2014, **131**, 1.
19. M. H. El-Newehy, S. S. Al-Deyab, E.-R. Kenawy and A. Abdel-Megeed, *J. Nanomater.*, 2011, **2011**, 1.
20. A. L. S. Borges, E. A. Münchow, A. C. de Oliveira Souza, T. Yoshida, P. K. Vallittu and M. C. Bottino, *J. Mech. Behav. Biomed. Mater.*, 2015, **48**, 134.
21. E. Fatarella, D. Spinelli, M. Ruzzante and R. Pogni, *J. Mol. Catal. B: Enzym.*, 2014, **102**, 41.
22. M. Boguń, I. Krucińska, A. Kommisarczyk, T. Mikołajczyk, M. Błażewicz, E. Stodolak-Zych, E. Menaszek and A. Ścisłowska-Czarnecka, *Molecules*, 2013, **18**, 3118.
23. J. R. Venugopal, Y. Zhang and S. Ramakrishna, *Artif. Organs*, 2006, **30**, 440.
24. G. Nitya, G. T. Nair, U. Mony, K. P. Chennazhi and S. V. Nair, *J. Mater. Sci.: Mater. Med.*, 2012, **23**, 1749.

25. Z. Khatri, A. W. Jatoi, F. Ahmed and I.-S. Kim, *Mater. Lett.*, 2016, **171**, 178.
26. J. G. Merrell, S. W. McLaughlin, L. Tie, C. T. Laurencin, A. F. Chen and L. S. Nair, *Clin. Exp. Pharmacol. Physiol.*, 2009, **36**, 1149.
27. X. Liu, T. Lin, J. Fang, G. Yao, H. Zhao, M. Dodson and X. Wang, *J. Biomed. Mater. Res., Part A*, 2010, **94**, 499.
28. M. Dubský, Š. Kubinová, J. Širc, L. Voska, R. Zajíček, A. Zajícová, P. Lesný, A. Jirkovská, J. Michálek, M. Munzarová, V. Holáň and E. Syková, *J. Mater. Sci.: Mater. Med.*, 2012, **23**, 931.
29. A. Hassiba, M. El Zowalaty, T. Webster, A. Abdullah, G. Nasrallah, K. Khalil, A. Luyt and A. Elzatahry, *Int. J. Nanomed.*, 2017, **12**, 2205.
30. X. Xu, Q. Yang, Y. Wang, H. Yu, X. Chen and X. Jing, *Eur. Polym. J.*, 2006, **42**, 2081.
31. T. T. T. Nguyen, C. Ghosh, S. G. Hwang, L. D. Tran and J. S. Park, *J. Mater. Sci.*, 2013, **48**, 7125.
32. R. A. Thakur, C. A. Florek, J. Kohn and B. B. Michniak, *Int. J. Pharm.*, 2008, **364**, 87.
33. E. González, L. Shepherd, L. Saunders and M. Frey, *Materials*, 2016, **9**, 47.
34. S. Chen, Y. Chang, K. Lee and J. Lai, *J. Membr. Sci.*, 2014, **450**, 224.
35. L. Cheng, X. Sun, B. Li, C. Hu, H. Yang, Y. Zhang and W. Cui, *J. Mater. Chem. B*, 2013, **1**, 4428.
36. S. Sahoo, S. L. Toh and J. C. H. Goh, *Biomaterials*, 2010, **31**, 2990.
37. K. A. Khalil, H. Fouad, T. Elsarnagawy and F. N. Almajhdi, *Int. J. Electrochem. Sci.*, 2013, **8**, 3483.
38. S. S. Said, A. K. Aloufy, O. M. El-Halfawy, N. A. Boraei and L. K. El-Khordagui, *Eur. J. Pharm. Biopharm.*, 2011, **79**, 108.
39. G. Xu, Y. Tan, T. Xu, D. Yin, M. Wang, M. Shen, X. Chen, X. Shi and X. Zhu, *Biomater. Sci.*, 2017, **5**, 752.
40. H. M. Al-Youssef, M. Amina, S. Hassan, T. Amna, J. W. Jeong, K.-T. T. Nam and H. Y. Kim, *Macromol. Res.*, 2013, **21**, 589.
41. I. Cantón, R. Mckean, M. Charnley, K. A. Blackwood, C. Fiorica, A. J. Ryan and S. MacNeil, *Biotechnol. Bioeng.*, 2010, **105**, 396.
42. S. C. Baker, N. Atkin, P. A. Gunning, N. Granville, K. Wilson, D. Wilson and J. Southgate, *Biomaterials*, 2006, **27**, 3136.
43. M. Abrigo, P. Kingshott and S. L. McArthur, *ACS Appl. Mater. Interfaces*, 2015, 7, 7644.
44. V. Kunduru, M. Bothara, J. Grosch, S. Sengupta, P. K. Patra and S. Prasad, *Nanomed.: Nanotechnol., Biol. Med.*, 2010, **6**, 642.
45. J. Sirc, S. Kubinova, R. Hobzova, D. Stranska, P. Kozlik, Z. Bosakova, D. Marekova, V. Holan, E. Sykova and J. Michalek, *Int. J. Nanomed.*, 2012, 7, 5315.
46. M. Khil, D. Cha, H.-Y. Kim, I. Kim and N. Bhattarai, *J. Biomed. Mater. Res., Part B Appl. Biomater.*, 2003, **67**, 675.
47. P. Davoudi, S. Assadpour, M. A. Derakhshan, J. Ai, A. Solouk and H. Ghanbari, *Mater. Sci. Eng., C*, 2017, **80**, 213.
48. A. Manikandan, M. P. Mani, S. K. Jaganathan, R. Rajasekar and M. Jagannath, *Polym. Test.*, 2017, **61**, 106.

49. B. K. Shrestha, S. Shrestha, A. P. Tiwari, J.-I. Kim, S. W. Ko, H.-J. Kim, C. H. Park and C. S. Kim, *Mater. Des.*, 2017, **133**, 69.
50. N. Shankhwar, M. Kumar, B. B. Mandal, P. S. Robi and A. Srinivasan, *J. Biomater. Sci., Polym. Ed.*, 2016, **27**, 247.
51. D. Li, W. Nie, L. Chen, Y. Miao, X. Zhang, F. Chen, B. Yu, R. Ao, B. Yu and C. He, *RSC Adv.*, 2017, **7**, 7973.
52. M. Ignatova, N. Markova, N. Manolova and I. Rashkov, *J. Biomater. Sci., Polym. Ed.*, 2008, **19**, 373.
53. R. M. Rosa, J. C. Silva, I. S. Sanches and C. Henriques, *Mater. Lett.*, 2017, **207**, 145.
54. U. M. Subramanian, S. V. Kumar, N. Nagiah and U. T. Sivagnanam, *Int. J. Polym. Mater.*, 2014, **63**, 476.
55. H. Homayoni, S. A. H. Ravandi and M. Valizadeh, *Carbohydr. Polym.*, 2009, **77**, 656.
56. K. Rodríguez, P. Gatenholm and S. Renneckar, *Cellulose*, 2012, **19**, 1583.
57. T. Liu, W. K. Teng, B. P. Chan and S. Y. Chew, *J. Biomed. Mater. Res., Part A*, 2010, **95**, 276.
58. K. S. Rho, L. Jeong, G. Lee, B.-M. Seo, Y. J. Park, S.-D. Hong, S. Roh, J. J. Cho, W. H. Park and B.-M. Min, *Biomaterials*, 2006, **27**, 1452.
59. J. Bürck, S. Heissler, U. Geckle, M. F. Ardakani, R. Schneider, A. S. Ulrich and M. Kazanci, *Langmuir*, 2013, **29**, 1562.
60. J. Lin, C. Li, Y. Zhao, J. Hu and L. M. Zhang, *ACS Appl. Mater. Interfaces*, 2012, **4**, 1050.
61. P. Qi, Y. Zhou, D. Wang, Z. He and Z. Li, *RSC Adv.*, 2015, **5**, 87180.
62. H. Jiang, D. Fang, B. S. Hsiao, B. Chu and W. Chen, *Biomacromolecules*, 2004, **5**, 326.
63. K. Jalaja, P. R. A. Kumar, T. Dey, S. C. Kundu and N. R. James, *Carbohydr. Polym.*, 2014, **114**, 467.
64. K. Tonsomboon and M. L. Oyen, *J. Mech. Behav. Biomed. Mater.*, 2013, **21**, 185.
65. H.-C. Chen, W.-C. Jao and M.-C. Yang, *Polym. Adv. Technol.*, 2009, **20**, 98.
66. J. Li, A. He, C. C. Han, D. Fang, B. S. Hsiao and B. Chu, *Macromol. Rapid Commun.*, 2006, **27**, 114.
67. D. Atila, D. Keskin and A. Tezcaner, *Mater. Sci. Eng., C*, 2016, **69**, 1103.
68. H. R. Pant, H. J. Kim, L. R. Bhatt, M. K. Joshi, E. K. Kim, J. I. Kim, A. Abdal-hay, K. S. Hui and C. S. Kim, *Appl. Surf. Sci.*, 2013, **285**, 538.
69. J.-P. Chen, G.-Y. Chang and J.-K. Chen, *Colloids Surf., A*, 2008, **313–314**, 183.
70. N. Naseri, C. Algan, V. Jacobs, M. John, K. Oksman and A. P. Mathew, *Carbohydr. Polym.*, 2014, **109**, 7.
71. F. Gökmeşe, İ Uslu and A. Aytimur, *Polym.-Plast. Technol. Eng.*, 2013, **52**, 1259.
72. M. Pezeshki-Modaress, H. Mirzadeh, M. Zandi, S. Rajabi-Zeleti, N. Sodeifi, N. Aghdami and M. R. K. Mofrad, *J. Biomed. Mater. Res., Part A*, 2017, **105**, 2020.
73. J. Guo, H. Zhou, M. Y. Akram, X. Mu, J. Nie and G. Ma, *Carbohydr. Polym.*, 2016, **143**, 239.

74. R. L. Fischer, M. G. McCoy and S. A. Grant, *J. Mater. Sci.: Mater. Med.*, 2012, **23**, 1645.
75. M. R. Ladd, S. J. Lee, J. D. Stitzel, A. Atala and J. J. Yoo, *Biomaterials*, 2011, **32**, 1549.
76. A. R. Unnithan, N. A. M. Barakat, P. B. Tirupathi Pichiah, G. Gnanasekaran, R. Nirmala, Y. S. Cha, C. H. Jung, M. El-Newehy and H. Y. Kim, *Carbohydr. Polym.*, 2012, **90**, 1786.
77. N. N. Maslakci, S. Ulusoy, E. Uygun, H. Çevikbaş, L. Oksuz, H. K. Can and A. Uygun Oksuz, *Polym. Bull.*, 2017, **74**, 3283.
78. Y. Qian, L. Li, C. Jiang, W. Xu, Y. Lv, L. Zhong, K. Cai and L. Yang, *Int. J. Biol. Macromol.*, 2015, **79**, 133.
79. Y. Ji, K. Ghosh, X. Z. Shu, B. Li, J. C. Sokolov, G. D. Prestwich, R. A. F. Clark and M. H. Rafailovich, *Biomaterials*, 2006, **27**, 3782.
80. M. S. Islam and J. H. Yeum, *Colloids Surf., A*, 2013, **436**, 279.
81. M. S. Kim and G. Kim, *Carbohydr. Polym.*, 2014, **114**, 213.
82. S. I. Jeong, M. D. Krebs, C. A. Bonino, S. A. Khan and E. Alsberg, *Macromol. Biosci.*, 2010, **10**, 934.
83. S. I. Jeong, O. Jeon, M. D. Krebs, M. C. Hill and E. Alsberg, *Eur. Cells Mater.*, 2012, **24**, 331.
84. V. Leung, R. Hartwell, S. S. Elizei, H. Yang, A. Ghahary and F. Ko, *J. Biomed. Mater. Res.: Part B Appl. Biomater.*, 2014, **102**, 508.
85. D. P. Tarantino, *Tech. Reg. Anesth. Pain Manag.*, 2002, **6**, 33.
86. H. Schwarze, M. Küntscher, C. Uhlig, H. Hierlemann, L. Prantl, N. Noack and B. Hartmann, *Burns*, 2007, **33**, 850.
87. M. Hromadka, J. B. Collins, C. Reed, L. Han, K. K. Kolappa, B. A. Cairns, T. Andrady and J. A. van Aalst, *J. Burn Care Res.*, 2008, **29**, 695.
88. M. Abrigo, S. L. McArthur and P. Kingshott, *Macromol. Biosci.*, 2014, **14**, 772.
89. Y.-Y. Tseng and S.-J. Liu, *Nanomedicine*, 2015, **10**, 1785.
90. A. R. Unnithan, A. Ghavami Nejad, A. R. K. Sasikala, R. G. Thomas, Y. Y. Jeong, P. Murugesan, S. Nasseri, D. Wu, C. H. Park and C. S. Kim, *Chem. Eng. J.*, 2016, **287**, 640.
91. G. Ö Kayseri, N. Özdil and G. S. Mengüç, in *Woven Fabrics*, ed. H.-Y. Jeon, InTech, 2012, p. 235.
92. P. Zahedi, I. Rezaeian, S. O. Ranaei-Siadat, S. H. Jafari and P. Supaphol, *Polym. Adv. Technol.*, 2010, **21**, 77.
93. M. A. Yassin, K. N. Leknes, Y. Sun, S. A. Lie, A. Finne-Wistrand and K. Mustafa, *J. Biomed. Mater. Res., Part A*, 2016, **104**, 2049.
94. Y. Sun, Z. Xing, Y. Xue, K. Mustafa, A. Finne-Wistrand and A. C. Albertsson, *Biomacromolecules*, 2014, **15**, 1259.
95. D. Church, S. Elsayed, O. Reid, B. Winston and R. Lindsay, *Clin. Microbiol. Rev.*, 2006, **19**, 403.
96. H.-F. Liu, F. Zhang and W. C. Lineaweaver, *Ann. Plast. Surg.*, 2017, **78**, S2.
97. G. D. Winter, *Nature*, 1962, **193**, 293.
98. C. D. Hinman and H. Maibach, *Nature*, 1963, **200**, 377.
99. L. G. Ovington, *Clin. Dermatol.*, 2007, **25**, 33.

100. M. Hromadka, C. Reed, L. Han, T. Andrady, K. K. Kolappa and J. A. van Aalst, *J Burn Care Res.*, 2008, 695.
101. V. Andreu, G. Mendoza, M. Arruebo and S. Irusta, *Materials*, 2015, **8**, 5154.
102. U. Turaga, V. Singh, A. Gibson, S. Maharubin, C. Korzeniewski, S. Presley, E. Smith, R. J. Kendall and S. Ramkumar, *Tappi J.*, 2016, **15**, 655.
103. J. Quirós, K. Boltes and R. Rosal, *Polym. Rev.*, 2016, **56**, 631.
104. A. J. Hassiba, M. E. El Zowalaty, G. K. Nasrallah, T. J. Webster, A. S. Luyt, A. M. Abdullah and A. A. Elzatahry, *Nanomedicine*, 2016, **11**, 715.
105. X. Jiang, M. H. Nai, C. T. Lim, C. Le Visage, J. K. Y. Chan and S. Y. Chew, *J. Biomed. Mater. Res., Part A*, 2015, **103**, 959.
106. M. Rahimnejad, S. Derakhshanfar and W. Zhong, *Burns Trauma*, 2017, **5**, 4.
107. E. R. Lorden, K. J. Miller, L. Bashirov, M. M. Ibrahim, E. Hammett, Y. Jung, M. A. Medina, A. Rastegarpour, M. A. Selim, K. W. Leong and H. Levinson, *Biomaterials*, 2015, **43**, 61.
108. E. R. Lorden, K. J. Miller, M. M. Ibrahim, L. Bashirov, E. Hammett, S. Chakraborty, C. Quiles-Torres, M. A. Selim, K. W. Leong and H. Levinson, *Acta Biomater.*, 2016, **32**, 100.
109. D. W. Brett, *J. Wound, Ostomy Cont. Nurs.*, 2006, **33**, S3.
110. L. Tan, J. Hu, H. Huang, J. Han and H. Hu, *Int. J. Biol. Macromol.*, 2015, **79**, 469.
111. K. P. Chellamani, P. Sundaramoorthy and T. Sureshram, *Int. J. Emerg. Technol. Adv. Eng.*, 2012, **2**, 176.
112. D. Lubasova, H. Niu, X. Zhao and T. Lin, *RSC Adv.*, 2015, **5**, 54481.
113. Q. L. Loh and C. Choong, *Tissue Eng., Part B*, 2013, **19**, 485.
114. N. A. Hotaling, K. Bharti, H. Kriel and C. G. Simon, *Biomaterials*, 2015, **61**, 327.
115. R. S. Barhate, S. Koeppl and S. Ramakrishna, *Chem. Eng. Res. Des.*, 2011, **89**, 621.
116. S. Sreedhara and N. Tata, *J. Eng. Fabr. Fibers*, 2013, **8**, 132.
117. W. J. Li, C. T. Laurencin, E. J. Caterson, R. S. Tuan and F. K. Ko, *J. Biomed. Mater. Res.*, 2002, **60**, 613.
118. E. Llorens, L. J. del Valle, R. Ferrán, A. Rodríguez-Galán and J. Puiggalí, *J. Polym. Res.*, 2014, **21**, 360.
119. M. Frey and L. Li, *J. Eng. Fibers Fabr.*, 2007, **2**, 31.
120. B.-Y. Wang, S.-Z. Fu, P.-Y. Ni, J.-R. Peng, L. Zheng, F. Luo, H. Liu and Z.-Y. Qian, *J. Biomed. Mater. Res., Part A*, 2012, **100A**, 441.
121. J. J. Elsner and M. Zilberman, *Acta Biomater.*, 2009, **5**, 2872.
122. R. Nawalakhe, Q. Shi, N. Vitchuli, M. A. Bourham, X. Zhang and M. G. McCord, *Int. J. Polym. Mater. Polym. Biomater.*, 2015, **64**, 709.
123. K. Ulubayram, S. Calamak, R. Shahbazi and I. Eroglu, *Curr. Pharm. Des.*, 2015, **21**, 1930.
124. WHO online, *Health Care-associated Infections Fact Sheet*, http://www.who.int/gpsc/country_work/gpsc_ccisc_fact_sheet_en.pdf, accessed August 2017.

125. S. K. Bhullar and H. S. Buttar, *AIMS Mater. Sci.*, 2017, **4**, 370.
126. N. Alhusein, I. S. Blagbrough and P. A. De Bank, *Drug Delivery Transl. Res.*, 2013, **3**, 542.
127. C. Dwivedi, H. Pandey, A. C. Pandey and P. W. Ramteke, *Curr. Pharm. Des.*, 2016, **22**, 1460.
128. T. D. J. Heunis and L. M. T. Dicks, *J. Biomed. Biotechnol.*, 2010, **2010**, 1.
129. D.-G. Yu, L.-M. Zhu, K. White and C. Branford-White, *Health*, 2009, **1**, 67.
130. J. Zeng, L. Yang, Q. Liang, X. Zhang, H. Guan, X. Xu, X. Chen and X. Jing, *J. Controlled Release*, 2005, **105**, 43.
131. Z. Sun, E. Zussman, A. L. Yarin, J. H. Wendorff and A. Greiner, *Adv. Mater.*, 2003, **15**, 1929.
132. X. Xu, L. Yang, X. Xu, X. Wang, X. Chen, Q. Liang, J. Zeng and X. Jing, *J. Controlled Release*, 2005, **108**, 33.
133. C. Wang, L. Wang and M. Wang, *Mater. Lett.*, 2014, **124**, 192.
134. K. M. Forward, A. Flores and G. C. Rutledge, *Chem. Eng. Sci.*, 2013, **104**, 250.
135. D. Lukas, A. Sarkar and P. Pokorny, *J. Appl. Phys.*, 2008, **103**, 1.
136. H. Niu, X. Wang and T. Lin, *Nanofibers – Prod. Prop. Funct. Appl.*, 2011, 17.
137. D. Nurwaha and X. Wang, *Fibers Polym.*, 2015, **16**, 850.
138. C. Pringle, MSc Thesis, Stellenbosch University, 2011.
139. A. C. Willemse, MSc Thesis, Stellenbosch University, 2013.
140. A. E. Smit and R. D. Sanderson, US Pat., US 9205453 B22015.
141. A. E. Smit and R. D. Sanderson, EU Pat., EP 2142687 B1, 2014.
142. X. Yan, J. Marini, R. Mulligan, A. Deleault, U. Sharma, M. P. Brenner, G. C. Rutledge, T. Freyman and Q. P. Pham, *PLoS One*, 2015, **10**, 1.
143. G. Jiang and X. Qin, *Mater. Lett.*, 2014, **128**, 259.
144. M. Santos, A. Fonseca, P. Mendonça, R. Branco, A. Serra, P. Morais and J. Coelho, *Materials*, 2016, **9**, 599.
145. J. S. Choi, H. S. Kim and H. S. Yoo, *Drug Delivery Transl. Res.*, 2015, **5**, 137.
146. Y. Yang, X. Li, M. Qi, S. Zhou and J. Weng, *Eur. J. Pharm. Biopharm.*, 2008, **69**, 106.
147. P. Siridamrong, P. Phrotjanatharee and N. Thamronganaskul, *Key Eng. Mater.*, 2016, **695**, 273.
148. K. Sisson, C. Zhang, M. C. Farach-Carson, D. B. Chase and J. F. Rabolt, *Biomacromolecules*, 2009, **10**, 1675.
149. E. Zussman, *Polym. Adv. Technol.*, 2011, **22**, 366.
150. T. D. J. Heunis, C. Smith and L. M. T. Dicks, *Antimicrob. Agents Chemother.*, 2013, **57**, 3928.
151. F. Yao and J. K. Weiyuan, *Expert Opin. Drug Delivery*, 2010, **7**, 429.
152. M. Zamani, M. P. Prabhakaran and S. Ramakrishna, *Int. J. Nanomed.*, 2013, **8**, 2997.
153. K. Kataria, A. Gupta, G. Rath, R. B. Mathur and S. R. Dhakate, *Int. J. Pharm.*, 2014, **469**, 102.

154. S. J. Eichhorn and W. W. Sampson, *J. R. Soc., Interface*, 2010, **7**, 641.
155. J. Širc, R. Hobzová, N. Kostina, M. Munzarová, M. Juklíčková, M. Lhotka, Š. Kubinová, A. Zajícová and J. Michálek, *J. Nanomater.*, 2012, **2012**, 1.
156. W. Hu, Z.-M. Huang and X.-Y. Liu, *Nanotechnology*, 2010, **21**, 315104.
157. J. S. Choi, S. H. Choi and H. S. Yoo, *J. Mater. Chem.*, 2011, **21**, 5258.
158. A. G. Ashbaugh, X. Jiang, J. Zheng, A. S. Tsai, W.-S. Kim, J. M. Thompson, R. J. Miller, J. H. Shahbazian, Y. Wang, C. A. Dillen, A. A. Ordonez, Y. S. Chang, S. K. Jain, L. C. Jones, R. S. Sterling, H.-Q. Mao and L. S. Miller, *Proc. Natl. Acad. Sci.*, 2016, **113**, E6919.
159. D. J. Leaper, *Int. Wound J.*, 2006, **3**, 282.
160. W.-J. Jin, H. K. Lee, E. H. Jeong, W. H. Park and J. H. Youk, *Macromol. Rapid Commun.*, 2005, **26**, 1903.
161. H. S. Abdo, K. A. Khalil, S. S. Al-Deyab, H. Altaleb and E. S. M. Sherif, *Fibers Polym.*, 2013, **14**, 1985.
162. P. Dubey, B. Bhushan, A. Sachdev, I. Matai, S. Uday Kumar and P. Gopinath, *J. Appl. Polym. Sci.*, 2015, **132**, 1.
163. L. Lu, R. Fu, C. Li, C. Yu, Z. Li, H. Guan, D. Hu and D. Zhao, *Int. J. Nanomed.*, 2013, **8**, 4131.
164. L. Jeong, D. Cho, O. H. Kwon, B.-M. Min and W. H. Park, *Macromol. Res.*, 2014, **22**, 796.
165. M. Gopiraman, A. W. Jatoi, S. Hiromichi, K. Yamaguchi, H.-Y. Jeon, I.-M. Chung and K. Ick Soo, *Carbohydr. Polym.*, 2016, **149**, 51.
166. M. M. Lim and N. Sultana, *3 Biotech*, 2016, **6**, 211.
167. S. J. Lee, D. N. Heo, J.-H. Moon, H. N. Park, W.-K. Ko, M. S. Bae, J. B. Lee, S. W. Park, E.-C. Kim, C. H. Lee, B.-Y. Jung and I. K. Kwon, *J. Nanosci. Nanotechnol.*, 2014, **14**, 7488.
168. S. J. Lee, S. A. Park, D. N. Heo, D. Lee, H.-J. Jang, K. S. Kim, J.-H. Moon and I. K. Kwon, *J. Nanosci. Nanotechnol.*, 2016, **16**, 8554.
169. L. Jeong, M. H. Kim, J. Y. Jung, B. M. Min and W. H. Park, *Int. J. Nanomed.*, 2014, **9**, 5277.
170. S. L. Percival, P. G. Bowler and D. Russell, *J. Hosp. Infect.*, 2005, **60**, 1.
171. X. Cheng, W. Zhang, Y. Ji, J. Meng, H. Guo, J. Liu, X. Wu and H. Xu, *RSC Adv.*, 2013, **3**, 2296.
172. V. K. M. Poon and A. Burd, *Burns*, 2004, **30**, 140.
173. P. V. AshaRani, G. Low Kah Mun, M. P. Hande and S. Valiyaveettil, *ACS Nano*, 2009, **3**, 279.
174. A. Burd, C. H. Kwok, S. C. Hung, H. S. Chan, H. Gu, W. K. Lam and L. Huang, *Wound Repair Regener.*, 2007, **15**, 94.
175. C. Greulich, D. Braun, A. Peetsch, J. Diendorf, B. Siebers, M. Epple and M. Köller, *RSC Adv.*, 2012, **2**, 6981.
176. J. B. Wright, K. Lam and R. E. Burrell, *Am. J. Infect. Control*, 1998, **26**, 572.
177. K. Bridges, A. Kidson, E. J. Lowbury and M. D. Wilkins, *BMJ*, 1979, **1**, 446.
178. M. Rai, A. Yadav and A. Gade, *Biotechnol. Adv.*, 2009, **27**, 76.

179. S. Silver, *FEMS Microbiol. Rev.*, 2003, **27**, 341.
180. S. Silver, L. T. Phung and G. Silver, *J. Ind. Microbiol. Biotechnol.*, 2006, **33**, 627.
181. A. Arslan, M. Şimşek, S. D. Aldemir, N. M. Kazaroğlu and M. Gümüşderelioğlu, *J. Biomater. Sci., Polym. Ed.*, 2014, **25**, 999.
182. H. Maleki, A. A. Gharehaghaji and P. J. Dijkstra, *J. Appl. Polym. Sci.*, 2013, **127**, 4086.
183. M. Y. Bai, T. C. Chou, J. C. Tsai and W. C. Yu, *J. Biomed. Mater. Res., Part A*, 2014, **102**, 2324.
184. O. Suwantong, P. Pankongadisak, S. Deachathai and P. Supaphol, *Chiang Mai J. Sci.*, 2013, **40**, 517.
185. S. Suganya, J. Venugopal, S. Ramakrishna, B. S. Lakshmi and V. R. G. Dev, *Int. J. Biol. Macromol.*, 2014, **68**, 135.
186. M. Norouzi, S. M. Boroujeni, N. Omidvarkordshouli and M. Soleimani, *Adv. Healthcare Mater.*, 2015, **4**, 1114.
187. H. B. Lindner, A. Zhang, J. Eldridge, M. Demcheva, P. Tsichilis, A. Seth, J. Vournakis and R. C. Muise-Helmericks, *PLoS One*, 2011, **6**, e18996.
188. R. Jayakumar, M. Prabaharan, P. T. S. Kumar, S. V Nair and H. Tamura, *Biotechnol. Adv.*, 2011, **29**, 322.
189. M. Ignatova, N. Manolova and I. Rashkov, *Macromol. Biosci.*, 2013, **13**, 860.
190. O. E. S. Bshena, PhD thesis, Stellenbosch University, 2012.
191. N. Charernsriwilaiwat, P. Opanasopit, T. Rojanarata and T. Ngawhirunpat, *Int. J. Pharm.*, 2012, **427**, 379.
192. J. J. Ahire and L. M. T. Dicks, *Antimicrob. Agents Chemother.*, 2014, **58**, 2098.
193. J. Padrão, R. Machado, M. Casal, S. Lanceros-Méndez, L. R. Rodrigues, F. Dourado and V. Sencadas, *Int. J. Biol. Macromol.*, 2015, **81**, 608.
194. J. Wang, Y. Zhu, H. K. Bawa, G. Ng, Y. Wu, M. Libera, H. C. van der Mei, H. J. Busscher and X. Yu, *ACS Appl. Mater. Interfaces*, 2011, **3**, 67.
195. H. Kong and J. Jang, *Langmuir*, 2008, **24**, 2051.
196. H. Kong and J. Jang, *Biomacromolecules*, 2008, **9**, 2677.
197. A. Y. A. Kaassis, N. Young, N. Sano, H. A. Merchant, D.-G. Yu, N. P. Chatterton and G. R. Williams, *J. Mater. Chem. B*, 2014, **2**, 1400.
198. E. R. Kenawy, F. I. Abdel-Hay, M. H. El-Newehy and G. E. Wnek, *Mater. Sci. Eng. A*, 2007, **459**, 390.
199. X. Y. Dai, W. Nie, Y. C. Wang, Y. Shen, Y. Li and S. J. Gan, *J. Mater. Sci.: Mater. Med.*, 2012, **23**, 2709.
200. D. Yang, X. Niu, Y. Liu, Y. Wang, X. Gu, L. Song, R. Zhao, L. Ma, Y. Shao and X. Jiang, *Adv. Mater.*, 2008, **20**, 4770.
201. S. Hou, L. Zhao, Q. Shen, J. Yu, C. Ng, X. Kong, D. Wu, M. Song, X. Shi, X. Xu, W. H. Ouyang, R. He, X. Z. Zhao, T. Lee, F. C. Brunicardi, M. A. Garcia, A. Ribas, R. S. Lo and H. R. Tseng, *Angew. Chem., Int. Ed.*, 2013, **52**, 3379.

202. S. J. Reinholt, A. Sonnenfeldt, A. Naik, M. W. Frey and A. J. Baeumner, *Anal. Bioanal. Chem.*, 2014, **406**, 3297.
203. W. S. Lee, V. Sunkara, J.-R. Han, Y.-S. Park and Y.-K. Cho, *Lab Chip*, 2015, **15**, 478.
204. L. Matlock-Colangelo and A. J. Baeumner, *Lab Chip*, 2012, **12**, 2612.
205. M. A. Ali, K. Mondal, C. Singh, B. D. Malhotra and A. Sharma, *Nanoscale*, 2015, 7, 7234.
206. M. Taguchi, A. Ptitsyn, E. S. McLamore and J. C. Claussen, *J. Diabetes Sci. Technol.*, 2014, **8**, 403.
207. B. K. Paul, A. K. Panigrahi, V. Singh and S. G. Singh, *Analyst*, 2017, **142**, 2128.
208. T. R. Dargaville, B. L. Farrugia, J. A. Broadbent, S. Pace, Z. Upton and N. H. Voelcker, *Biosens. Bioelectron.*, 2013, **41**, 30.
209. G. Tegl, D. Schiffer, E. Sigl, A. Heinzle and G. M. Guebitz, *Appl. Microbiol. Biotechnol.*, 2015, **99**, 4595.
210. M. Omidi, A. Yadegari and L. Tayebi, *RSC Adv.*, 2017, **7**, 10638.
211. S. Trupp, M. Alberti, T. Carofiglio, E. Lubian, H. Lehmann, R. Heuermann, E. Yacoub-George, K. Bock and G. J. Mohr, *Sens. Actuators, B*, 2010, **150**, 206.
212. J. Zhou, A. L. Loftus, G. Mulley and A. T. A. Jenkins, *J. Am. Chem. Soc.*, 2010, **132**, 6566.
213. J. Zhou, T. N. Tun, S. Hong, J. D. Mercer-Chalmers, M. Laabei, A. E. R. Young and A. T. A. Jenkins, *Biosens. Bioelectron.*, 2011, **30**, 67.
214. ISO 10993–1, *Biological Evaluation of Medical Devices – Part 1: Evaluation and Testing within a Risk Management Process. International Standard*, 2009.
215. ISO 10993–5, *Biological Evaluation of Medical Devices – Part 5: Tests for In vitro Cytotoxicity. International Standard*, 2009.
216. ISO 10993–10, *Biological Evaluation of Medical Devices – Part 10: Tests for Irritation and Skin Sensitization*, 2010.
217. S. Shahzad, M. Yar, S. A. Siddiqi, N. Mahmood, A. Rauf, Z. ul A. Qureshi, M. S. Anwar and S. Afzaal, *J. Mater. Sci.: Mater. Med.*, 2015, **26**, 1.
218. E. J. Chong, T. T. Phan, I. J. Lim, Y. Z. Zhang, B. H. Bay, S. Ramakrishna and C. T. Lim, *Acta Biomater.*, 2007, **3**, 321.
219. S. G. Kumbar, S. P. Nukavarapu, R. James, L. S. Nair and C. T. Laurencin, *Biomaterials*, 2008, **29**, 4100.
220. B. M. Baker, A. O. Gee, R. B. Metter, A. S. Nathan, R. A. Marklein, J. A. Burdick and R. L. Mauck, *Biomaterials*, 2008, **29**, 2348.
221. M. Ottosson, A. Jakobsson and F. Johansson, *PLoS One*, 2017, **12**, 1.
222. A. A. Khalili and M. R. Ahmad, *Int. J. Mol. Sci.*, 2015, **16**, 18149.
223. A. Ruiz, C. E. Flanagan and K. S. Masters, *J. Biomed. Mater. Res., Part A*, 2013, **101**, 2870.
224. M. S. Peach, R. James, U. S. Toti, M. Deng, N. L. Morozowich, H. R. Allcock, C. T. Laurencin and S. G. Kumbar, *Biomed. Mater.*, 2012, **7**, 45016.

225. X. J. Loh, P. Peh, S. Liao, C. Sng and J. Li, *J. Controlled Release*, 2010, **143**, 175.
226. W.-W. Hu and H.-N. Yu, *Carbohydr. Polym.*, 2013, **95**, 716.
227. J. Song, H. Gao, G. Zhu, X. Cao, X. Shi and Y. Wang, *Carbon N. Y.*, 2015, **95**, 1039.
228. D. Yan, J. Jones, X. Y. Yuan, X. H. Xu, J. Sheng, J. C. M. Lee, G. Q. Ma and Q. S. Yu, *J. Biomed. Mater. Res., Part A*, 2013, **101A**, 963.
229. W. Liu, J. Zhan, Y. Su, T. Wu, C. Wu, S. Ramakrishna, X. Mo, S. S. Al-Deyab and M. El-Newehy, *Colloids Surf., B*, 2014, **113**, 101.
230. J. P. Chen and C. H. Su, *Acta Biomater.*, 2011, **7**, 234.
231. A. Martins, E. D. Pinho, S. Faria, I. Pashkuleva, A. P. Marques, R. L. Reis and N. M. Neves, *Small*, 2009, **5**, 1195.
232. Y. Kim, V. K. Haftel, S. Kumar and R. V. Bellamkonda, *Biomaterials*, 2008, **29**, 3117.
233. A. Jain, M. Betancur, G. D. Patel, C. M. Valmikinathan, V. J. Mukhatyar, A. Vakharia, S. B. Pai, B. Brahma, T. J. MacDonald and R. V. Bellamkonda, *Nat. Mater.*, 2014, **13**, 308.
234. Y. Xi, H. Dong, K. Sun, H. Liu, R. Liu, Y. Qin, Z. Hu, Y. Zhao, F. Nie and S. Wang, *ACS Appl. Mater. Interfaces*, 2013, **5**, 4821.
235. J. Xie, M. R. MacEwan, X. Li, S. E. Sakiyama-Elbert and Y. Xia, *ACS Nano*, 2009, **3**, 1151.
236. C. Yang, G. Deng, W. Chen, X. Ye and X. Mo, *Colloids Surf., B*, 2014, **122**, 270.
237. X. Fu and H. Wang, *Tissue Eng., Part A*, 2012, **18**, 631.
238. U. Stachewicz, T. Qiao, S. C. F. Rawlinson, F. V. Almeida, W.-Q. Li, M. Cattell and A. H. Barber, *Acta Biomater.*, 2015, **27**, 88.
239. Y. M. Ju, J. S. Choi, A. Atala, J. J. Yoo and S. J. Lee, *Biomaterials*, 2010, **31**, 4313.
240. *Comparison of 3D Structure and Fabrication Technique*, http://electrospintech.com/3dcompare.html#.Wcz3iFuCypq, accessed September 2017.
241. C. J. Mortimer and L. Burke, *J. Microb. Biochem. Technol.*, 2016, **8**, 195.
242. S. S. Said, O. M. El-Halfawy, H. M. El-Gowelli, A. K. Aloufy, N. A. Boraei and L. K. El-Khordagui, *Eur. J. Pharm. Biopharm.*, 2012, **80**, 85.
243. J. F. Cooley, US Pat., US 692631, 1900.
244. W. J. Morton, US Pat., US 705691, 1900.
245. N. Tucker, J. Stanger, M. Staiger, H. Razzaq and K. Hofman, *J. Eng. Fiber. Fabr.*, 2012, **7**, 63.
246. J. Doshi and D. H. Reneker, *Ind. Appl. Soc. Annu. Meet. 1993., Conf. Rec. 1993 IEEE*, 1993, **3**, 1698.
247. M. Gagliardi, *BCC Market Research Report: Global Markets and Technologies for Nanofibers*, Wellesley, 2016, vol. NAN043D.
248. C. Laurencin, S. Kumbar, S. Nukavarapu, R. James and M. Hogan, *Recent Pat. Biomed. Eng.*, 2008, **1**, 68.
249. T. Fang, J. Wang and X. Lin, in *Nanofibers – Production, Properties and Functional Applications*, ed. L. Tong, InTech, 1st edn, 2017, p. 135.

250. E. Melissa and T. Cotthoff, *BCC Market Research Report: Markets for Advanced Wound Care Technologies*, Wellesley, 2011, vol. PHM011G.
251. L. Persano, A. Camposeo, C. Tekmen and D. Pisignano, *Macromol. Mater. Eng.*, 2013, **298**, 504.
252. N. A. Hotaling, V. Khristov, Q. Wan, R. Sharma, B. S. Jha, M. Lotfi, A. Maminishkis, C. G. Simon and K. Bharti, *J. Ocul. Pharmacol. Ther.*, 2016, **32**, 272.
253. AxioCore™ Fiber Technology, http://www.arsenalmedical.com/technology/axiocore-drug-delivery-platform, accessed September 2017.
254. Bioinicia, http://bioinicia.com/, accessed September 2017.
255. eSpin, http://www.espintechnologies.com/, accessed September 2017.
256. Fuence, http://fuence.co.jp/en/, accessed September 2017.
257. IME Technologies, https://www.imetechnologies.com/, accessed September 2017.
258. Leonardino, http://leonardino.eu/, accessed September 2017.
259. Nanofiber Solutions, http://www.nanofibersolutions.com/, accessed September 2017.
260. NanoSpun, http://www.nanospuntech.com/, accessed September 2017.
261. NEI Corporation, http://www.neicorporation.com/services/electrospinning/, accessed September 2017.
262. Pardam, http://pardam.cz/services/, accessed September 2017.
263. Revolution Fibres, https://www.revolutionfibres.com/, accessed September 2017.
264. SNC Fibers, https://sncfibers.com/, accessed September 2017.
265. SNS NanoFiber, http://www.snsnano.com/, accessed September 2017.
266. Technical Fine Co., Ltd, http://www.technicalfine.com/, accessed September 2017.
267. The Electrospinning Company, http://www.electrospinning.co.uk/, accessed September 2017.
268. Yflow SD, http://www.yflow.com/, accessed September 2017.
269. C. E. Olson and P. A. Messina, US Pat., US 20130096479 A1, 2013.
270. S.T. Medical Press Release, http://www.stteresamedical.com/press-releases-st-teresa-medical, accessed September 2017.
271. C. Tison, *TuneCoat™ Bandages for Localized Pain Relief*, http://lunainc.com/wp-content/uploads/2017/03/20170313_TuneCoat-1-pager.pdf, accessed September 2017.
272. *Nanofiber-reinforced Hydrogels for Ocular Repair Technology Description*, http://lunainc.com/wp-content/uploads/2017/03/20170307-Ocular-One-Pager.pdf, accessed September 2017.
273. Nanotrix®, http://www.medikasa.co.za, accessed September 2017.
274. Cerafix®, http://acera-surgical.com/cerafix/, accessed September 2017.
275. *Comparative Analysis of the Mechanical Strength of Cerafix® Dura Substitute and DuraGen Plus™ Dural Regeneration Matrix*, http://acera-surgical.com/wp-content/uploads/2017/02/Comparitive-Mechanical-Assessment-of-Cerafix-and-DuraGen-Plus.pdf, accessed September 2017.

276. Restrata™, http://acera-surgical.com/restrata/, accessed September 2017.
277. Medprin R&D Team, *A Multi-center, Randomized, Single-blind Clinical Trial of a Novel Regenerative Synthetic Absorbable Dural Repair Patch*, http://www.dastavardesina.com/images/uploadfiles/000 ReDura Clinical Paper White (1).pdf, accessed September 2017.
278. ReDura™, http://www.medprin.cn/vancheerfile/files/2017/2/20170220 11042808.pdf, accessed September 2017.
279. Dermtreat secures GMP manufacturing - Lundbeckfonden, https://www.lundbeckfonden.com/en/8255/, accessed September 2017.
280. Dermtreat ApS | Patch technology targetting mucosal diseases, http://rivelin.eu/, accessed September 2017.
281. M. E. Santocildes-Romero, L. Hadley, K. H. Clitherow, J. Hansen, C. Murdoch, H. E. Colley, M. H. Thornhill and P. V. Hatton, *ACS Appl. Mater. Interfaces*, 2017, **9**, 11557.
282. K. Graham, H. Schreuder-gibson and M. Gogins, in *INTC 2003*, Baltimore, 2003, p. 15.
283. R. Pedoja, US Pat., US 20130330516 A1, 2013.
284. Z. Dai, J. Ronholm, Y. Tian, B. Sethi and X. Cao, *J. Tissue Eng.*, 2016, **7**, 1.
285. T. Annala and M. Kellomäki, in *Degradable Polymers for Skeletal Implants*, ed. P. I. J. M. Wuisman and T. H. Smit, Nova Science Publishers, Amsterdam, 2010, ch. 6, p. 123.
286. L. A. Bosworth, A. Gibb and S. Downes, *J. Polym. Sci., Part B: Polym. Phys.*, 2012, **50**, 870.
287. P. Rychter, N. Śmigiel-Gac, E. Pamuła, A. Smola-Dmochowska, H. Janeczek, W. Prochwicz and P. Dobrzyński, *Materials*, 2016, **9**, 64.
288. ISO 11607–1, *Packaging for Terminally Sterilized Medical Devices — Part 1: Requirements for Materials, Sterile Barrier Systems and Packaging Systems, International Standard*, 2006.
289. European Commission, https://ec.europa.eu/growth/single-market/goods/building-blocks/notified-bodies_en, accessed September 2017.
290. *Which Way to Go: CE Mark or FDA Approval?*, https://www.mdtmag.com/article/2012/02/which-way-go-ce-mark-or-fda-approval, accessed September 2017.
291. CRSTEurope, *CE Mark Versus FDA Approval: Which System Has it Right?*, https://crstodayeurope.com/articles/2015-feb/ce-mark-versus-fda-approval-which-system-has-it-right/, accessed September 2017.
292. *CE Mark or FDA Approval? Medical Device Professionals Weigh in - MedCity News*, https://medcitynews.com/2012/02/ce-mark-or-fda-approval-medical-device-professionals-weigh-in/?rf=1, accessed September 2017.
293. About FDA, https://www.fda.gov/AboutFDA/Transparency/Basics/ucm194460.htm, accessed September 2017.
294. FDA online, https://www.fda.gov/MedicalDevices/DeviceRegulationandGuidance/Overview/default.htm, accessed September 2017.
295. European Commission, https://ec.europa.eu/growth/single-market/ce-marking/manufacturers, accessed September 2017.

296. E. French-Mowat and J. Burnett, *J. R. Soc. Med.*, 2012, **105**, S22.
297. P. J. Ogrodnik, *Med. Device Des.*, 2013, 309.
298. FDA online, https://www.fda.gov/MedicalDevices/DeviceRegulationandGuidance/Overview/ClassifyYourDevice/ucm051530.htm, accessed September 2017.
299. Dressing Selection on the App Store, https://itunes.apple.com/us/app/dressing-selection/id587368964, accessed September 2017.
300. Dermicus Wound, http://www.dermicus.com/products/dermicus-wound/, accessed September 2017.

CHAPTER 6

Electrospinning and 3D Printing: Prospects for Market Opportunity

S. SAGHATI[a], ABOLFAZL AKBARZADEH*[c,d], A. R. DEL BAKHSHAYESH[a], R. SHEERVALILOU[b] AND EBRAHIM MOSTAFAVI[d]

[a]Department of Tissue Engineering, Tabriz University of Medical Sciences, University Street, Tabriz, 5165665931, Iran; [b]Department of Molecular Medicine, Tabriz University of Medical Sciences, University Street, Tabriz, 5165665931, Iran; [c]Department of Medical Nanotechnology, Faculty of Advanced Medical Sciences, Tabriz University of Medical Sciences, Tabriz, 5165665931, Iran; [d]Department of Chemical Engineering, Northeastern University, Boston, MA, 02115
*E-mail: akbarzadehab@tbzmed.ac.ir

6.1 Introduction

Bioprinting is a fast developing and wide-reaching innovation that is having a major impact in medical and pharmaceutical sciences.[1,2] Bioprinting can be defined as a computer-aided transfer procedure, using the simultaneous writing of living cells and biomaterials with a layer-by-layer design for the production of bioengineered constructs.[3] This technology developed in robotics, optics, and chemical research more than 30 years ago to simplify

Soft Matter Series No. 7
Electrospinning: From Basic Research to Commercialization
Edited by Erich Kny, Kajal Ghosal and Sabu Thomas

Published by the Royal Society of Chemistry, www.rsc.org

the creation of prototypes.[4] It quickly became a regular tool in the automotive, consumer goods and aerospace industries.[5–9]

Electrospinning was broadly used to construct tissue engineering scaffolds. However, it is very difficult to introduce the cells into the nanoporous structure of a normal electrospun scaffold by electrospinning.[10] Furthermore, electrospinning is not capable of fabricating exceedingly intricate and complex biomimetic scaffolds. 3D bioprinting is a method that can be used to solve these problems.[11–14]

The technique can be used to make engineered 3D tissues or organs in a programmed, systematized manner, because it is capable of simulating the concrete cellular arrangement of a tissue or organ.[15] Such made-to-order 3D printed constructs can be used for the regeneration of multipart tissues such as nerves, vessels and cartilage.[16] Several cell types can be positioned next to one another at targeted locations, with high accuracy during the construct.[17]

Mimicking the precise *in vivo* behavior of human organs can help to reduce the cost of drug discovery as well as enabling more accurate estimates of therapeutic/toxic responses. In addition, producing constructs with dysfunctional or diseased human cells can lead to more *in vitro* disease/cancer models for studying tissue pathology or testing new drugs.[18] Nevertheless, due to the limited choice of printable inks, 3D printing methods often offer partial control over the resolution and negotiated mechanical properties. To address these limitations, new approaches combine 3D printing and electrospinning techniques to produce a unique 3D biomimetic scaffold with a harmonious porous structure surrounded by aligned fibers.[16]

This chapter focuses on the general principles, concepts, and approaches relating to the application of 3D bioprinting technologies for generating representative *in vitro* 3D tissue models. We present the latest examples, chosen to show key ideas. We also discuss the amazing opportunities for 3D bioprinting technologies for further ultimate and translational studies on tissue engineering and cancer models, toxicological tests, and biological models. Some indications of the potential markets are also described.

6.2 Materials for 3D Bioprinting

There are three steps involved in bioprinting: (i) preparing data to print, (ii) choosing proper bioprinting methods, and (iii) formulating the components of bioink (biomaterials and cells) (Figure 6.1). These material and technology selections reflect planning considerations at several steps of bioprinting.[19]

6.2.1 Bioinks

Bioinks normally contain a hydrogel pre-polymer solution and cells.[20] There are two main types of hydrogels used for bioink production: natural polymer based and synthetic polymer based hydrogels.[21–23] Bioinks have been

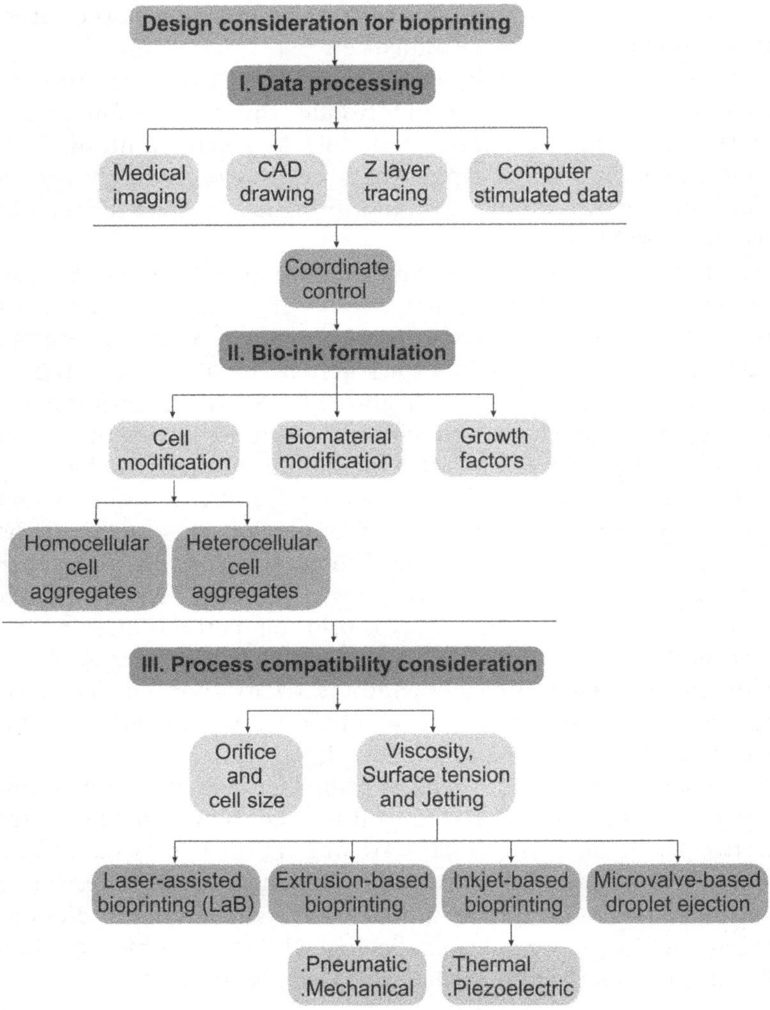

Figure 6.1 Diagram representing the various considerations when designing for bioprinting.

established to supply a variety of characterizations, such as biocompatibility and suitable physical properties, to guarantee printability and longstanding functionality after deposition.[19,24] Natural hydrogels contain polymers of ECM constituents, such as collagen, laminin, gelatin and fibronectin, as well as other natural polymers such as chitosan, alginate, fibroin and silk.[25–32] Interactions between cells and natural hydrogels have been examined in depth.[33–42]

However, less frequent than collagen, fibroin or silk proteins, naturally-derived Matrigel®,[18] decellularized adipose tissue[43] and decellularized extracellular matrix[44] have been efficaciously used for bioprinting. Matrigel is a commercialized combination of proteins isolated from mouse sarcoma cells. In its standard form it contains heparan sulfate proteoglycans, collagen, entactin, laminin and growth factors.[45] DECMs (decellularized extracellular matrices) are insulated from living tissues. The composition naturally contains a diverse array of ECM components.[40,44,46]

Unlike natural polymers, synthetic polymers like polyesters are produced by chemical synthesis so that more manageable chemical and mechanical properties are achievable.[47,48] However, their interactions with cells have not yet been considered analytically.[21] Some scientists have used a blend of natural and synthetic polymers.[49–54] Various hybrid inks have been suggested to improve the properties of the bioink, allowing more reproducible printing, better shape reliability and faster printing, and modified mechanical properties of the fabricated scaffolds.[19,55–59]

Some researchers are using an altered kind of hydrogel material 'fugitive ink' along with standard bioinks to support their bioprinted constructions. After printing is complete, the fugitive inks are converted from a gel to a liquid by changing the temperature and then the resulting liquid can be removed, leaving the bioprinted construction behind. Fugitive inks are used in two methods, either around the construction (as a supportive bath) to support its structure in open space,[60–62] or inside the construction to allow the formation of interior channels.[63,64]

6.2.2 Cells

In addition to biomaterials, encapsulated cells are another critical component of bioinks. Bioprinted cells must proliferate to form an extremely mimetic tissue or organ on a macro scale.[65] Two of the main issues to be considered when selecting cells for bioprinting are how thoroughly the bioprinted cells can mimic the biological nature of cells *in vivo*, and to what degree the bioprinted cells can sustain or improve their *in vivo* functions.[66]

Synthetic tissues are seeded by either printing functional primary cells with supporting cells or printing progenitors or stem cells for additional differentiation.[67,68] Direct printing of primary cells can rapidly increase the complexity of bioprinting. Because multiple types of cells embedded within the same or different hydrogels need to be printed in parallel, many bioinks need to be prepared for each print. Printing with stem cells will usually reduce the total number of bioinks used for a given print, but also adds its own set of complications. Additional bioink formulations with different growth factor and small molecule signals may be desirable to attempt to guide site-specific differentiation. Even without this kind of approach, there is added difficulty in post-printing culture as growth factors and other differentiation stimulators must be deposited precisely to ensure the control of differentiation, especially when vascularization is desired.[20,69,70]

6.3 Bioprinting Techniques

As yet, no single bioprinting procedure enables the fabrication of all scales and complexities of artificial tissues. The three major bioprinting techniques each have precise physical properties and limitations. Figure 6.2 presents a

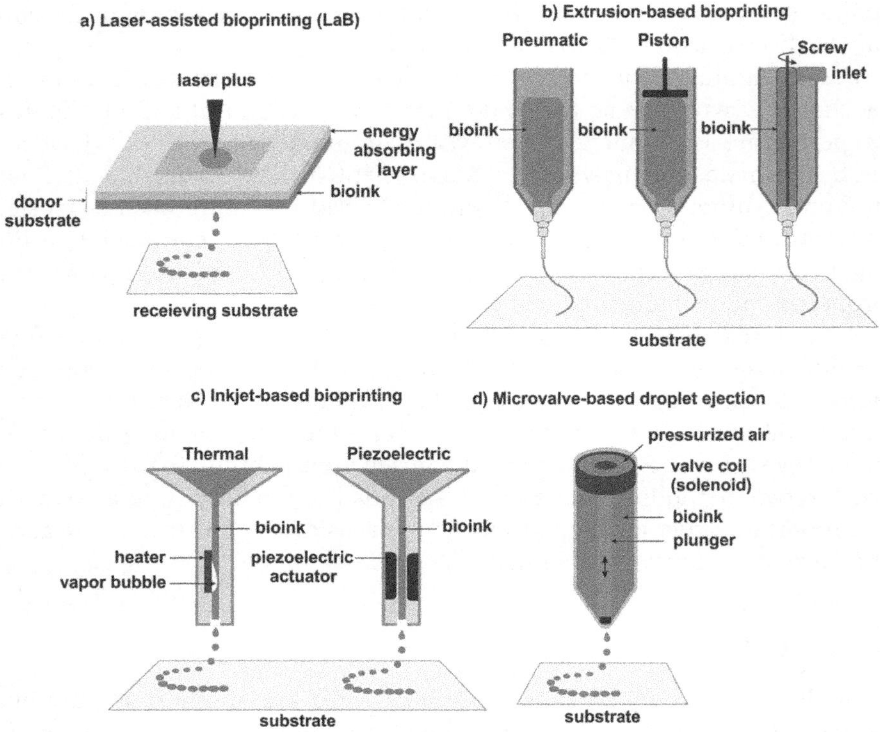

Figure 6.2 Representation of the most frequently used approaches to 3D bioprinting. a) Laser-assisted bioprinting (LaB), b) extrusion-based bioprinting, c) inkjet-based bioprinting, and d) microvalve-based droplet ejection. Reprinted from *Biomaterials*, **134**, M. K. Włodarczyk-Biegun and A. del Campo, 3D bioprinting of structural proteins, 180–201, Copyright 2017, with permission from Elsevier.

general comparison of these methods. These approaches can also be classified based on the smallest printable unit in which bioinks can either be printed layer-wise, continuously patterned, or distributed drop by drop.[71] 3D bioprinting can also be categorized into scaffold-free or scaffold-dependent approaches.[4] Most approaches to 3D bioprinting are scaffold-dependent, where there is a necessity for biomaterials, *e.g.* bioinks or scaffolds. Nevertheless, scaffold-dependent 3D bioprinting faces many problems and restrictions, such as immunogenicity of biomaterial, cost of the bioinks used, sluggish speeds and toxicity of degradation yields.[72]

6.3.1 Laser-assisted Printing

Figure 6.2a is a schematic of laser-assisted printing. This system contains a donor layer as its critical part, which responds to laser motivation. The donor layer involves a 'ribbon' construction made up of an energy-absorbing layer (*e.g.* gold or titanium) on the top and a layer of bioink solution suspended on the bottom. A concentrated laser pulse is applied to excite a small zone of the absorbing layer during printing.[73] This laser pulse generates a high-pressure bubble at the border of the bioink layer by evaporating a part of the donor layer and impels the suspended bioink. The dropping bioink droplet is collected on the delivery substrate and consequently crosslinked. The procedure is normally fast, supporting a medium-to-high density of cells.[66] Compared to inkjet printing, in laser-assisted printing there is no direct contact between the dispenser and the bioinks. This non-contact printing technique can elude mechanical tension to the cells, which results in high cell viability (typically higher than 95%).[74] Furthermore, highly viscous bioinks can be printed by laser-assisted printing. Despite these attractive features of laser bioprinting the side-effects of laser introduction on the cell are not yet completely known.[75] The other disadvantages of this method include: (i) high cost compared to other nozzle-based printing techniques due to expensive laser diodes (diodes with high resolution and intensity are more expensive); (ii) control complexity of laser printing system, and (iii) restrictive technique's adoption. Some laser printing prototype was settled by merging optical laser sources with a lens.[76] Researchers also built a high-yield laser printing system.[77]

6.3.2 Extrusion Printing

Microextrusion is the most common material-distributing technique used for bioprinting. Most up-to-date commercial bioprinters, including the NovoGen and Bioplotter (EnvisionTec, Gladbeck, Germany) 3D Bioprinting platform (Organovo, San Diego, USA), are founded on extrusion method.[78] In this method, cells and hydrogel originators are distributed from a syringe tank or micronozzle comprising cell suspension by a syringe pump plunger or pneumatic microvalve, respectively.[79] An air-force pump or a mechanical screw piston is used in extrusion printing to dispense bioinks, as shown in Figure 6.2b. Because of the single nozzle feature, the stages (and/or the nozzles) for microextrusion bioprinters need to be capable of movement along all three axes to render it possible to create 3D tissues. Extrusion printing can print continuous cylinder-shaped lines rather than a single bioink droplet by applying a constant force. It is also capable of printing almost all sorts of hydrogel solutions of different viscosity besides aggregates with great cell density.[66,80] The disadvantage of extrusion printing is that it exposes the cells to larger mechanical stresses, which can reduce the cell viability. Microextrusion bioprinters confer medium cost and the ability to encapsulate high densities of cells including spheroids, but the printing speeds are relatively low.[66,81–83]

6.3.3 Inkjet Printing

The 3D inkjet bioprinting method is comparable with conventional 2D inkjet printing and was the first bioprinting method. A hydrogel pre-polymer solution with encapsulated cells (bioink) is kept in the ink container.[84] As a bioink source during the printing procedure, the container is then joined to a printer head. The printer heads are deformed by a thermal or piezoelectric activator and squeezed to make droplets of a manageable size, as shown in Figure 6.2c. There is quite a small nozzle deformation created by either thermal or piezoelectric in current printer heads. In both cases, the stages in which the bioinks are deposited transfer in the *z*-axis to assemble 3D substances.

Inkjet printing approaches are usually cost-efficient and fast, but are limited by the fact that the bioink has to be in a relatively low-viscosity form to enable droplet formation and thus low cell concentration.[66,85] Despite good mixing of bioinks, cells begin to settle in the container, over the whole printing procedure, increasing the viscosity of the bioink and usually blocking the printer head.[65,86]

6.3.4 Hybrid Bioprinting

Although above-mentioned three printing approaches are most generally used by bioprinting scientists, there are a few novel printing procedures which remain under exploration. Nowadays, diverse 3D printing methods have been established based on fused deposition modeling (FDM), stereolithography (SLA), selective laser sintering (SLS), and PolyJet technology.[20,87–91]

Using novel categorize of self-assembling peptide-based hydrogels is another approach in bioprinting technology which gives rise to stiffness and developed structural integrity in constructed materials.[92] Organ biomanufacturing using accurate layer-by-layer employment of self-assembled tissue spheroids in sprayed tissue fusion accommodating hydrogels is called "directed tissue self-assembly" (Figure 6.3).[93]

6.4 Combining 3D Printing with Electrospun Scaffolds

3D bioprinting has also been used in combination with electrospinning procedure. Melt-plotting and electrospinning machines have been integrated with extrusion and inkjet bioprinting methods to produce multiple fragments.[44,94–96] These machines can deposit multiple materials, potentially including multiple cell types.[97]

Almost all tissues are composed of various types of cells placed inside sophisticated biochemistries and geometries of the ECMs; while electrospinning is restricted in unidirectional arrangement of nanofibers. 3D printing is able to create more- complex constructions that better imitate *in vivo* micro-environments.[98–100] Nevertheless, with the fiber diameters ranging in the nanometer or micrometer scale, electrospinning has been claimed among the most hopeful engineering methods for the construction of polymeric fibrillar networks, as tissue-engineered vascular graft (TEVGs).[101,102] This method is amenable to fabricate tubular shapes and accompanying to all the techniques of cell seeding.[103] Also, with this method, it is possible

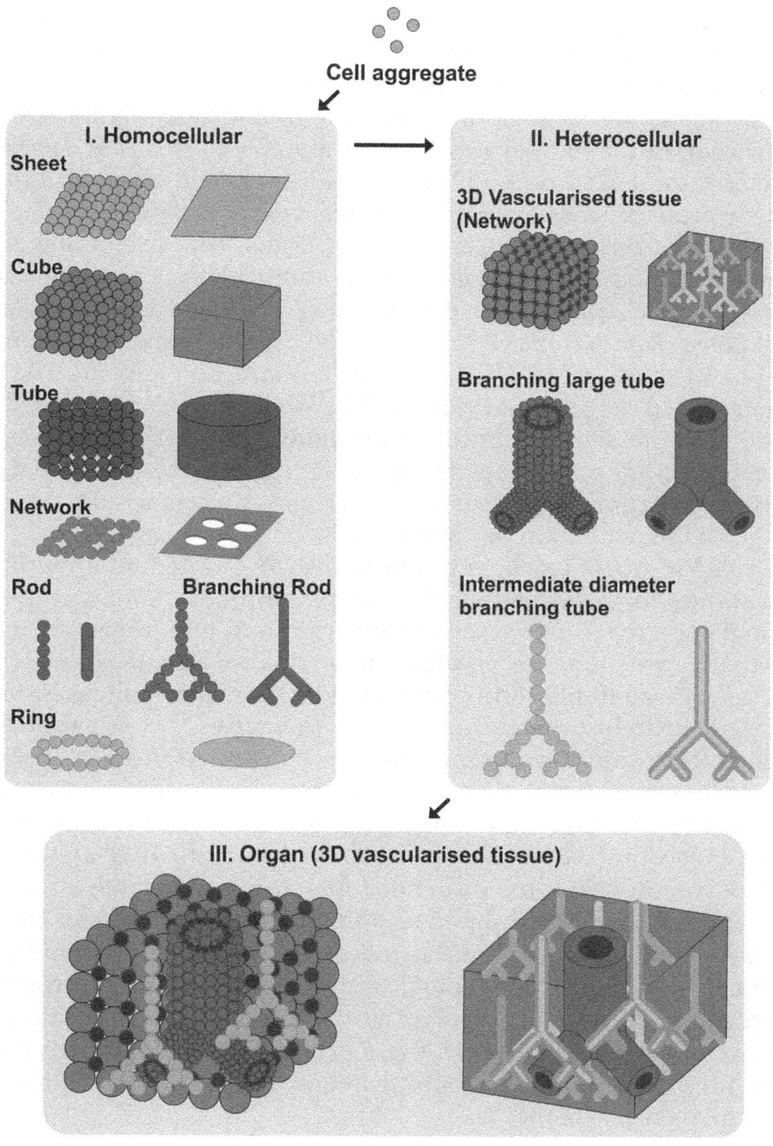

Figure 6.3 A guideline for organ printing. Reprinted from Biomaterials, **30**, V. Mironov, R. P. Visconti, V. Kasyanov, G. Forgacs, C. J. Drake, R. R. Markwald, Organ printing: Tissue spheroids as building blocks, 2164–2174, Copyright 2009, with permission from Elsevier.

to produce interconnected porous constructions presenting required physical characteristics in natural environments as a high surface area/volume ratio and a high permeability.[104] Moreover, fibers are tractable to be functionalized with different combinations and growth factors without damaging their biological activity and task.[101] Furthermore, electrospun platforms might therefore constitute a suitable candidate for TEVGs fabrication showing a permissive influence on vascular remodeling of both the cellular and extracellular constituents of a graft.[105] In spite of several efforts experimental struggles spend in this field, the capacity to mimic biomechanical features and behavior of the natural vessel wall is a problem.[106–109] The mechanical characteristics of electrospun fibrous networks are still unsuitable to withstand the pressure loads quickly after implantation in the body.[110]

To overcome this problem, recently a hybrid system combining electrospinning and bioprinting to construct bioresorbable platforms for vascular tissue engineering was developed. Spadaccio *et al.*[111] bioprinted a single-layer helical poly-ε-caprolactone (PCL) coil on the exterior surface as a reinforcement of an heparin-releasing PLLA tubular electrospun scaffold. They used computer-aided tissue engineering (CATE), which has been described to be an interesting method to fabricate three-dimensional (3D) constructs.[112,113] Biofunctionalization with heparin created a drug delivery system that could overcome the thrombogenesis of TEVG and simultaneously improved endothelial differentiation was provided.[101]

Electrospun fibers can be introduced into 3D printed microporous tissue scaffolds for investigating the neural stem cell (NSC) and primary neuronal cell growth. However, the intricacy and high cost concomitant prevent the wide-ranging application of this technique. Furthermore, the disadvantage of platforms fabricated by these mechanical systems is decreased flexibility, which more limits their uses.[84,114] A hybrid inkjet printing/electrospinning system that can be used to produce viable tissues was described for cartilage tissue engineering applications. In order to construct a five-layer tissue platform of 1 mm thickness, electrospinning of polycaprolactone fibers was alternated with inkjet printing of rabbit elastic chondrocytes suspended in a fibrin–collagen hydrogel. The printed hybrid scaffold exhibited viability of more than 80% for chondrocytes one week after printing. Furthermore, the cells proliferation and maintenance of their basic biological properties within the printed layered platform and also formation of cartilage-like tissues both *in vitro* and *in vivo* were observed. Moreover, these printed hybrid constructs demonstrated improved physical properties compared to printed alginate or fibrin–collagen gels alone.[50]

3D printing is also able to develop retinal ganglion cells (RGC) disposal in association with a radial electrospun platform designed to control axon direction. Kador *et al.*[115] studied effects of thermal inkjet printing on neurite growth and survival of primary RGCs *in vitro* when printed in different buffers, at different ejection energies and at several cell densities. Finally, initial patterns were printed integrate areas for the fovea onto electrospun scaffolds. They confirmed the ability to associate these two approaches which incorporates cell direction to the scaffold canter. Their results established that 3D printing method may be associated with intricate electrospun scaffolds in the design of future retinal models or therapies.

As mentioned before 3D printed scaffolds can be used for the engineering of complex tissues such as cartilage, vessels and nerves. Nevertheless, the 3D printing methods often provide inadequate control over the resolution and resulted mechanical properties because of limited choice of printable inks. On the other hand, naturally-derived 3D printed platforms, in spite of the ability to imitate the natural properties of tissue, still lack constructional integrity and suitable mechanical characteristics for use *in vivo*, consequently restraining their usage in load-bearing tissue engineering applications, such as bone and cartilage.[50] Utilizing a novel multi-head deposition method to fabricate sustainable products supplies the ability to combine synthetic biomaterials, which have more advanced mechanical properties than natural polymers. These platforms also benefits from the promising situation for cell growth supplied by traditional nature-based biomaterials.[50]

To address the aforementioned limitations, Se-Jun Lee *et al.*[16] used PCL and PCL/gelatin composite materials in their electrospinning system and PEG-DA for the SL 3D printing. They studied how the combination of gelatin into PCL fibers can influence neurite growth and differentiation compared to PCL fibers alone. The incorporation of electrospun fibers into the 3D scaffolds resulted in improved mechanical properties. Furthermore, it was found that 3D printed structures with PCL/gelatin fibers produced superior NSC differentiation compared to scaffolds with PCL alone fibers. Also, highly lined up PCL/gelatin fibers within the 3D printed scaffold improved the neurite outgrowth and directional control of primary cortical neurons.

FDM and electrospinning (ES) were used by Naghieh *et al.*[116] to develop hierarchical scaffolds containing poly(lactic acid) (PLA) micro struts and nanocomposite gelatin-forsterite fibrous layers respectively. Results showed that elastic modulus of PLA/gelatin-forsterite platforms, produced by a hybrid method, was considerably higher than that of pure one (about 52%).

Electrospinning and 3D printing methods were broadly used to produce bone tissue engineering scaffolds. Nevertheless, it is very difficult to infiltrate the cells into the nanoporous construction of traditional electrospun scaffold by electrospinning method. On the other hand 3D printing method has the disadvantage of low printing resolution. In order to unravel these problems, Yu[117] produced 3D composite scaffold by infused PCL/Gelatin distributed nanofibers into the networks of PCL printing platform. They approved the good potential of electrospinning/3D printing composite scaffold to use in bone tissue engineering. Due to the microporous structure of electrospun scaffold, MC3T3-E1 cells showed a better proliferation and infiltration on the composite scaffold than PCL printing one.

Another purpose of using hybrid scaffolds in tissue engineering is construction of functionally classified products which can mimic the complex distributions of the structure and functionality of the human tissues. These hybrid procedures have been reported to be used in numerous applications and the ability to print a huge range of materials.[118] Also, aligned nanogrooves and assemblies have been revealed to effect cell proliferation and morphology.[103] Yang Wu *et al.*[119] developed a hybrid procedure which combined electrospinning and electrospraying and was successfully used to produce platforms for tissue engineering application, principally for the tendon-bone crossing point. In the fabrication procedure, one substance was produced *via* electrospinning, and the other was produced *via* electrospraying. In this study, polycaprolactone (PCL) and hydroxyapatite (HA) were selected for printing.

6.5 Application Areas of 3D Bioprinting

Clinical uses for 3D printing are growing fast and are anticipated to revolutionize medical care. Medical applications for 3D printing, both concrete and budding, can be divided into numerous wide-ranging sorts, comprising: tissue and organ creation; fabrication of modified prosthetics, implants, and organs-on-a-chip models; and pharmaceutical research regarding drug screening, delivery, and discovery and cancer research.[120,121] Despite all complications, a broad variation of tissues have been successfully bioprinted, for example skin,[122] blood vessel,[123] bone[124,125] or tissues that do not requires vascularization like cartilage.[3,75,82,126–131]

6.6 Potential Markets for 3D Bioprinting

As stated by Grand View Research Inc., the universal 3D bioprinting market is anticipated to reach US%1.82 billion by 2022. Substitute for organ transplantation Increasing frequency of chronic diseases like chronic kidney disease (CKD) is estimated to increase the market progress. Developing medical usages of 3D bioprinting such as drug and toxicity testing, tissue regeneration and user product testing are estimated to promote the market positively. Additional key conclusions from the investigation propose:

- As nozzle-based bioprinters allow working in sterilized condition, they conquered the technology section.
- Magnetic printers are anticipated to witness profitable development over the prediction time. The medical applications of these printers can include human cell modeling, recognition of cardiovascular toxicity and smooth muscle regeneration.
- The medical using section is predictable to achieve significant portion of over 32.0% by 2022 which includes applications such as organ transplants, medicines and toxicity recognition.
- Dental usage section is anticipated to witness beneficial progress during the prediction span. Increasing consciousness about oral sanitation accompanied by increasing request for cosmetic dentistry are anticipated to raise the demand for dental applications.

- With above 35.0% share, North America conquered the local 3D bioprinting market and is anticipated to develop at a compound annual growth rate (CAGR) of 10.0%. Increasing salary attached with Helpful government enterprises to progress healthcare structure are essential dynamic element for this area's development.
- Asia Pacific is estimated to witness rewarding progression during the prediction span. Encouraging government funding to broaden healthcare supply with financial growths in developing countries such as China and India are anticipated to raise the request for 3D bioprinting.
- Important companies working in the 3D bioprinting manufacturing comprise Cyfuse Biomedical, BioBots, Luxexcel Group BV, Organovo Holding Inc., Voxeljet, TeVido BioDevices, Envision TEC, Aspect Biosystems, 3Dynamics Systems, Stratasys Ltd., Oceanz, Materialise NV, Bio3D Technologies, Solidscape and others.
- These companies are participated in different strategic advantages such as new product promotions, acquisitions and strategic associations and unions to achieve the control over the opponents. For example, In June 2015, Orgonovo cooperated with L'Oréal to improve skin tissue using NovoGen Bioprinting system.[132]

6.7 Conclusions and Future Perspectives

The field of 3D bioprinting and electrospinning is in improving evolution. The prospects and range of uses for bioprinting are increasing to cover all subfields of biomedical fabrications, and especially regenerative medicine. Reaching a final aim to print complete living organs appears to be more achievable than ever, though pertinent attempts have yet to be overcome to reach this dream. Now, there are around 15 diverse tissue types under research in the bioprinting field; though, there exist an amount of other tissues in the human body that have not been affected yet. Furthermore, bioprinting of new forms of organs, like bionic organs or organs that do not exist in nature, will be one of the novel guidelines in bioprinting.

The majority of bioprinting examination has progressed around homocellular tissue fabrication using a single cell type; nevertheless, natural tissues have a heterocellular and extremely complex nature. Therefore combination of bioprinting procedure with other procedures such as electrospinning and microfluidics is required for constructing physiologically simulated, complex and longstanding feasible target micro-organ arrangements. The combination of electrospun fibers into the 3D productions give rise to improved mechanical characterizations, developed cell differentiation and proliferation.

We envision that the use of hybrid electrospun/3D printed scaffolds will become more prominent for tissue models in drug testing and cancer therapy *in vitro* and organ transplantation in disease therapy.

References

1. I. T. Ozbolat, Bioprinting scale-up tissue and organ constructs for transplantation, *Trends Biotechnol.*, 2015, **33**(7), 395–400.
2. Y. Zhang, *et al.*, *In vitro* study of directly bioprinted perfusable vasculature conduits, *Biomater. Sci.*, 2015, **3**(1), 134–143.
3. I. T. Ozbolat, Scaffold-based or scaffold-free bioprinting: competing or complementing approaches? *J. Nanotechnol. Eng. Med.*, 2015, **6**(2), 024701.
4. B. C. Gross, *et al.*, *Evaluation of 3D Printing and its Potential Impact on Biotechnology and the Chemical Sciences*, ACS Publications, 2014.
5. R. Kolakovic, *et al.*, Printing technologies in fabrication of drug delivery systems, *Expert Opin. Drug Delivery*, 2013, **10**(12), 1711–1723.
6. S. E. Moulton and G. G. Wallace, 3-dimensional (3D) fabricated polymer based drug delivery systems, *J. Controlled Release*, 2014, **193**, 27–34.
7. I. D. Ursan, L. Chiu and A. Pierce, Three-dimensional drug printing: a structured review, *J. Am. Pharm. Assoc.*, 2013, **53**(2), 136–144.
8. L. K. Prasad and H. Smyth, 3D Printing technologies for drug delivery: a review, *Drug Dev. Ind. Pharm.*, 2016, **42**(7), 1019–1031.
9. P. Katakam, *et al.*, Top-down and bottom-up approaches in 3D printing technologies for drug delivery challenges, *Crit. Rev. Ther. Drug Carrier Syst.*, 2015, **32**(1), 61–87.
10. H. Yu, *et al.*, Detection of the mcr-1 colistin resistance gene in carbapenem-resistant Enterobacteriaceae from different hospitals in China, *Antimicrob. Agents Chemother.*, 2016, **60**(8), 5033–5035.
11. J. He, *et al.*, Electrohydrodynamic 3D printing of microscale poly(ε-caprolactone) scaffolds with multi-walled carbon nanotubes, *Biofabrication*, 2017, **9**(1), 015007.
12. S.-H. Park, *et al.*, Creation of a hybrid scaffold with dual configuration of aligned and random electrospun fibers, *ACS Appl. Mater. Interfaces*, 2016, **8**(4), 2826–2832.
13. J. Walser, *et al.*, Direct electrospinning of 3D auricle-shaped scaffolds for tissue engineering applications, *Biofabrication*, 2016, **8**(2), 025007.
14. J. Walser and S. J. Ferguson, Oriented nanofibrous membranes for tissue engineering applications: electrospinning with secondary field control, *J. Mech. Behav. Biomed. Mater.*, 2016, **58**, 188–198.
15. B. Derby, Printing and prototyping of tissues and scaffolds, *Science*, 2012, **338**(6109), 921–926.
16. S.-J. Lee, *et al.*, Fabrication of a highly aligned neural scaffold *via* a table top stereolithography 3D printing and electrospinning, *Tissue Eng., Part A*, 2017, **23**(11–12), 491–502.
17. F. Pati, J. Gantelius and H. A. Svahn, 3D bioprinting of tissue/organ models, *Angew. Chem., Int. Ed.*, 2016, **55**(15), 4650–4665.
18. J. Snyder, *et al.*, Bioprinting cell-laden matrigel for radioprotection study of liver by pro-drug conversion in a dual-tissue microfluidic chip, *Biofabrication*, 2011, **3**(3), 034112.

19. Y. S. Zhang, *et al.*, 3D bioprinting for tissue and organ fabrication, *Ann. Biomed. Eng.*, 2017, **45**(1), 148–163.
20. C. Mandrycky, *et al.*, 3D bioprinting for engineering complex tissues, *Biotechnol. Adv.*, 2016, **34**(4), 422–434.
21. P. Zorlutuna, N. E. Vrana and A. Khademhosseini, The expanding world of tissue engineering: the building blocks and new applications of tissue engineered constructs, *IEEE Rev. Biomed. Eng.*, 2013, **6**, 47–62.
22. A. R. Akkineni, *et al.*, A versatile method for combining different biopolymers in a core/shell fashion by 3D plotting to achieve mechanically robust constructs, *Biofabrication*, 2016, **8**(4), 045001.
23. Y. B. Kim, H. Lee and G. H. Kim, Strategy to achieve highly porous/biocompatible macroscale cell blocks, using a collagen/genipin-bioink and an optimal 3D printing process, *ACS Appl. Mater. Interfaces*, 2016, **8**(47), 32230–32240.
24. J. Leijten, *et al.*, Spatially and temporally controlled hydrogels for tissue engineering, *Mater. Sci. Eng., R*, 2017, **119**, 1–35.
25. R. Suntivich, *et al.*, Inkjet printing of silk nest arrays for cell hosting, *Biomacromolecules*, 2014, **15**(4), 1428–1435.
26. H. Tao, *et al.*, Inkjet printing of regenerated silk fibroin: From printable forms to printable functions, *Adv. Mater.*, 2015, **27**(29), 4273–4279.
27. D. A. Gregory, *et al.*, Reactive inkjet printing of biocompatible enzyme powered silk micro-rockets, *Small*, 2016, **12**(30), 4048–4055.
28. C. Henmi, M. Nakamura, Y Nishiyama, K. Yamaguchi, S. Mochizuki, K. Takiura and H. Nakagawa, Development of an effective three dimensional fabrication technique using inkjet technology for tissue model samples, *AATEX*, 2007, **14**, 689–692.
29. A. M. Compaan, K. Christensen and Y. Huang, Inkjet bioprinting of 3D silk fibroin cellular constructs using sacrificial alginate, *ACS Biomater. Sci. Eng.*, 2017, **3**(8), 1519–1526.
30. L. Sun, *et al.*, Direct-write assembly of 3d silk/hydroxyapatite scaffolds for bone co-cultures, *Adv. Healthcare Mater.*, 2012, **1**(6), 729–735.
31. S. Ghosh, *et al.*, Direct-write assembly of microperiodic silk fibroin scaffolds for tissue engineering applications, *Adv. Funct. Mater.*, 2008, **18**(13), 1883–1889.
32. S. T. Parker, *et al.*, Biocompatible silk printed optical waveguides, *Adv. Mater.*, 2009, **21**(23), 2411–2415.
33. V. Lee, *et al.*, Design and fabrication of human skin by three-dimensional bioprinting, *Tissue Eng., Part C*, 2013, **20**(6), 473–484.
34. J. Y. Park, *et al.*, A comparative study on collagen type I and hyaluronic acid dependent cell behavior for osteochondral tissue bioprinting, *Biofabrication*, 2014, **6**(3), 035004.
35. M. Lei and X. Wang, Biodegradable polymers and stem cells for bioprinting, *Molecules*, 2016, **21**(5), 539.
36. I. T. Ozbolat, W. Peng and V. Ozbolat, Application areas of 3D bioprinting, *Drug discovery today*, 2016, **21**(8), 1257–1271.
37. A. B. Dababneh and I. T. Ozbolat, Bioprinting technology: a current state-of-the-art review, *J. Manuf. Sci. Eng.*, 2014, **136**(6), 061016.

38. C. Di Bella, *et al.*, 3D bioprinting of cartilage for orthopedic surgeons: reading between the lines, *Front. Surg.*, 2015, **2**, 39.
39. C. M. Smith, *et al.*, Three-dimensional bioassembly tool for generating viable tissue-engineered constructs, *Tissue Eng.*, 2004, **10**(9–10), 1566–1576.
40. M. K. Włodarczyk-Biegun and A. del Campo, 3D bioprinting of structural proteins, *Biomaterials*, 2017, **134**, 180–201.
41. B. Duan, State-of-the-art review of 3D bioprinting for cardiovascular tissue engineering, *Ann. Biomed. Eng.*, 2017, **45**(1), 195–209.
42. J. Malda, *et al.*, 25th anniversary article: engineering hydrogels for biofabrication, *Adv. Mater.*, 2013, **25**(36), 5011–5028.
43. F. Pati, *et al.*, Biomimetic 3D tissue printing for soft tissue regeneration, *Biomaterials*, 2015, **62**, 164–175.
44. F. Pati, *et al.*, Printing three-dimensional tissue analogues with decellularized extracellular matrix bioink, *Nat. Commun.*, 2014, **5**, 3935.
45. C. S. Hughes, L. M. Postovit and G. A. Lajoie, Matrigel: a complex protein mixture required for optimal growth of cell culture, *Proteomics*, 2010, **10**(9), 1886–1890.
46. P. M. Crapo, T. W. Gilbert and S. F. Badylak, An overview of tissue and whole organ decellularization processes, *Biomaterials*, 2011, **32**(12), 3233–3243.
47. J. Zhu and R. E. Marchant, Design properties of hydrogel tissue-engineering scaffolds, *Expert Rev. Med. Devices*, 2011, **8**(5), 607–626.
48. R. Balint, N. J. Cassidy and S. H. Cartmell, Conductive polymers: towards a smart biomaterial for tissue engineering, *Acta Biomater.*, 2014, **10**(6), 2341–2353.
49. W. Schuurman, *et al.*, Bioprinting of hybrid tissue constructs with tailorable mechanical properties, *Biofabrication*, 2011, **3**(2), 021001.
50. T. Xu, *et al.*, Hybrid printing of mechanically and biologically improved constructs for cartilage tissue engineering applications, *Biofabrication*, 2012, **5**(1), 015001.
51. I. Donderwinkel, J. C. van Hest and N. R. Cameron, Bio-inks for 3D bioprinting: recent advances and future prospects, *Polym. Chem.*, 2017, **8**(31), 4451–4471.
52. A. E. Jakus, *et al.*, Three-dimensional printing of high-content graphene scaffolds for electronic and biomedical applications, *ACS Nano*, 2015, **9**(4), 4636–4648.
53. P. Bajaj, *et al.*, 3D biofabrication strategies for tissue engineering and regenerative medicine, *Annu. Rev. Biomed. Eng.*, 2014, **16**, 247–276.
54. G. Villar, A. D. Graham and H. Bayley, A tissue-like printed material, *Science*, 2013, **340**(6128), 48–52.
55. N. E. Sanjana and S. B. Fuller, A fast flexible ink-jet printing method for patterning dissociated neurons in culture, *J. Neurosci. Methods*, 2004, **136**(2), 151–163.
56. S. Das, *et al.*, Bioprintable, cell-laden silk fibroin–gelatin hydrogel supporting multilineage differentiation of stem cells for fabrication of three-dimensional tissue constructs, *Acta Biomater.*, 2015, **11**, 233–246.

57. S. Das, *et al.*, Enhanced redifferentiation of chondrocytes on microperiodic silk/gelatin scaffolds: toward tailor-made tissue engineering, *Biomacromolecules*, 2013, **14**(2), 311–321.
58. S. Li, *et al.*, Direct fabrication of a hybrid cell/hydrogel construct by a double-nozzle assembling technology, *J. Bioact. Compat. Polym.*, 2009, **24**(3), 249–265.
59. W. Xu, *et al.*, Rapid prototyping three-dimensional cell/gelatin/fibrinogen constructs for medical regeneration, *J. Bioact. Compat. Polym.*, 2007, **22**(4), 363–377.
60. S.-H. Kim, J. Turnbull and S. Guimond, Extracellular matrix and cell signalling: the dynamic cooperation of integrin, proteoglycan and growth factor receptor, *J. Endocrinol.*, 2011, **209**(2), 139–151.
61. R. O. Hynes, The extracellular matrix: not just pretty fibrils, *Science*, 2009, **326**(5957), 1216–1219.
62. S. F. Badylak, The extracellular matrix as a scaffold for tissue reconstruction, in *Seminars in Cell & Developmental Biology*, Elsevier, 2002.
63. L. Cai and S. C. Heilshorn, Designing ECM-mimetic materials using protein engineering, *Acta Biomater.*, 2014, **10**(4), 1751–1760.
64. D. J. Cornelissen, A. Faulkner-Jones and W. Shu, Current developments in 3D bioprinting for tissue engineering, *Curr. Opin. Biomed. Eng.*, 2017, **2**, 76–82.
65. Y. S. Zhang, *et al.*, Three-dimensional bioprinting strategies for tissue engineering, *Cold Spring Harbor Perspect. Med.*, 2017, a025718.
66. S. V. Murphy and A. Atala, 3D bioprinting of tissues and organs, *Nat. Biotechnol.*, 2014, **32**(8), 773–785.
67. D. F. D. Campos, *et al.*, Three-dimensional printing of stem cell-laden hydrogels submerged in a hydrophobic high-density fluid, *Biofabrication*, 2012, **5**(1), 015003.
68. M. Gruene, *et al.*, Laser printing of three-dimensional multicellular arrays for studies of cell–cell and cell–environment interactions, *Tissue Eng., Part C*, 2011, **17**(10), 973–982.
69. C. M. Smith, *et al.*, Characterizing environmental factors that impact the viability of tissue-engineered constructs fabricated by a direct-write bioassembly tool, *Tissue Eng.*, 2007, **13**(2), 373–383.
70. I. T. Ozbolat and Y. Yu, Bioprinting toward organ fabrication: challenges and future trends, *IEEE Trans. Biomed. Eng.*, 2013, **60**(3), 691–699.
71. A. Blaeser, D. F. D. Campos and H. Fischer, 3D bioprinting of cell-laden hydrogels for advanced tissue engineering, *Curr. Opin. Biomed. Eng.*, 2017, **2**, 58–66.
72. C. S. Ong, T. Fukunishi, A. Nashed, A. Blazeski, H. Zhang, S. Hardy, D. DiSilvestre, L. Vricella, J. Conte, L. Tung and G. Tomaselli, Creation of cardiac tissue exhibiting mechanical integration of spheroids using 3D bioprinting, *J. Visualized Exp.*, 2017, **125**, DOI: 10.3791/55438.
73. L. Koch, *et al.*, Skin tissue generation by laser cell printing, *Biotechnol. Bioeng.*, 2012, **109**(7), 1855–1863.

74. S. Michael, *et al.*, Tissue engineered skin substitutes created by laser-assisted bioprinting form skin-like structures in the dorsal skin fold chamber in mice, *PLoS One*, 2013, **8**(3), e57741.
75. B. Duan, *et al.*, 3D bioprinting of heterogeneous aortic valve conduits with alginate/gelatin hydrogels, *J. Biomed. Mater. Res., Part A*, 2013, **101**(5), 1255–1264.
76. Y. Nahmias, *et al.*, Laser-guided direct writing for three-dimensional tissue engineering, *Biotechnol. Bioeng.*, 2005, **92**(2), 129–136.
77. B. Guillotin, *et al.*, Laser assisted bioprinting of engineered tissue with high cell density and microscale organization, *Biomaterials*, 2010, **31**(28), 7250–7256.
78. D. F. Duarte Campos, *et al.*, The stiffness and structure of three-dimensional printed hydrogels direct the differentiation of mesenchymal stromal cells toward adipogenic and osteogenic lineages, *Tissue Eng., Part A*, 2014, **21**(3–4), 740–756.
79. A. Arslan-Yildiz, *et al.*, Towards artificial tissue models: past, present, and future of 3D bioprinting, *Biofabrication*, 2016, **8**(1), 014103.
80. T. J. Hinton, *et al.*, Three-dimensional printing of complex biological structures by freeform reversible embedding of suspended hydrogels, *Sci. Adv.*, 2015, **1**(9), e1500758.
81. S. Khalil and W. Sun, Biopolymer deposition for freeform fabrication of hydrogel tissue constructs, *Mater. Sci. Eng., C*, 2007, **27**(3), 469–478.
82. S. Rhee, *et al.*, 3D bioprinting of spatially heterogeneous collagen constructs for cartilage tissue engineering, *ACS Biomater. Sci. Eng.*, 2016, **2**(10), 1800–1805.
83. M. Yeo, *et al.*, An innovative collagen-based cell-printing method for obtaining human adipose stem cell-laden structures consisting of core–sheath structures for tissue engineering, *Biomacromolecules*, 2016, **17**(4), 1365–1375.
84. E. Roth, *et al.*, Inkjet printing for high-throughput cell patterning, *Biomaterials*, 2004, **25**(17), 3707–3715.
85. M. Singh, *et al.*, Inkjet printing—process and its applications, *Adv. Mater.*, 2010, **22**(6), 673–685.
86. K. Arai, *et al.*, Three-dimensional inkjet biofabrication based on designed images, *Biofabrication*, 2011, **3**(3), 034113.
87. J. S. Miller, *et al.*, Rapid casting of patterned vascular networks for perfusable engineered 3D tissues, *Nat. Mater.*, 2012, **11**(9), 768.
88. R. Gauvin, *et al.*, Microfabrication of complex porous tissue engineering scaffolds using 3D projection stereolithography, *Biomaterials*, 2012, **33**(15), 3824–3834.
89. B. Horstkotte, O. Elsholz and V. Cerdá, Review on automation using multisyringe flow injection analysis, *J. Flow Injection Anal.*, 2005, **22**(2), 99.
90. C. Calderilla, *et al.*, 3D printed device including disk-based solid-phase extraction for the automated speciation of iron using the multisyringe flow injection analysis technique, *Talanta*, 2017, **175**, 463–469.

91. J. R. Tumbleston, *et al.*, Continuous liquid interface production of 3D objects, *Science*, 2015, **347**(6228), 1349–1352.
92. B. Raphael, *et al.*, 3D cell bioprinting of self-assembling peptide-based hydrogels, *Mater. Lett.*, 2017, **190**, 103–106.
93. V. Mironov, *et al.*, Organ printing: tissue spheroids as building blocks, *Biomaterials*, 2009, **30**(12), 2164–2174.
94. J.-H. Shim, *et al.*, Development of a hybrid scaffold with synthetic biomaterials and hydrogel using solid freeform fabrication technology, *Biofabrication*, 2011, **3**(3), 034102.
95. H.-W. Kang, *et al.*, A 3D bioprinting system to produce human-scale tissue constructs with structural integrity, *Nat. Biotechnol.*, 2016, **34**(3), 312–319.
96. T. Xu, *et al.*, Complex heterogeneous tissue constructs containing multiple cell types prepared by inkjet printing technology, *Biomaterials*, 2013, **34**(1), 130–139.
97. I. T. Ozbolat, H. Chen and Y. Yu, Development of 'Multi-arm Bioprinter'for hybrid biofabrication of tissue engineering constructs, *Robot. Comput. Integr. Manuf.*, 2014, **30**(3), 295–304.
98. J. Wu, *et al.*, Biomimetic nanofibrous scaffolds for neural tissue engineering and drug development, *Drug Discovery Today*, 2017, **22**(9), 1375–1384.
99. F.-Y. Hsieh and S.-H. Hsu, 3D bioprinting: a new insight into the therapeutic strategy of neural tissue regeneration, *Organogenesis*, 2015, **11**(4), 153–158.
100. W. Zhu, *et al.*, Highly aligned nanocomposite scaffolds by electrospinning and electrospraying for neural tissue regeneration, *Nanomedicine*, 2015, **11**(3), 693–704.
101. C. Spadaccio, *et al.*, Preliminary *in vivo* evaluation of a hybrid armored vascular graft combining electrospinning and additive manufacturing techniques, *Drug Target Insights*, 2016, **10**(suppl. 1), 1.
102. X. Mo, *et al.*, Electrospun P (LLA-CL) nanofiber: a biomimetic extracellular matrix for smooth muscle cell and endothelial cell proliferation, *Biomaterials*, 2004, **25**(10), 1883–1890.
103. C. Xu, *et al.*, Electrospun nanofiber fabrication as synthetic extracellular matrix and its potential for vascular tissue engineering, *Tissue Eng.*, 2004, **10**(7–8), 1160–1168.
104. A. S. Badami, *et al.*, Effect of fiber diameter on spreading, proliferation, and differentiation of osteoblastic cells on electrospun poly(lactic acid) substrates, *Biomaterials*, 2006, **27**(4), 596–606.
105. C. K. Hashi, *et al.*, Antithrombogenic property of bone marrow mesenchymal stem cells in nanofibrous vascular grafts, *Proc. Natl. Acad. Sci. U. S. A.*, 2007, **104**(29), 11915–11920.
106. C. P. Barnes, *et al.*, Nanofiber technology: designing the next generation of tissue engineering scaffolds, *Adv. Drug Delivery Rev.*, 2007, **59**(14), 1413–1433.

107. R. Murugan and S. Ramakrishna, Nano-featured scaffolds for tissue engineering: a review of spinning methodologies, *Tissue Eng.*, 2006, **12**(3), 435–447.
108. S. J. Lee, *et al.*, *In vitro* evaluation of electrospun nanofiber scaffolds for vascular graft application, *J. Biomed. Mater. Res., Part A*, 2007, **83**(4), 999–1008.
109. T. K. Merceron, *et al.*, A 3D bioprinted complex structure for engineering the muscle-tendon unit, *Biofabrication*, 2015, **7**(3), 035003.
110. S. H. Kim, *et al.*, Hierarchical assembly of nanostructured organosilicate networks *via* stereocomplexation of block copolymers, *Nano Lett.*, 2008, **8**(1), 294–301.
111. M. Centola, *et al.*, Combining electrospinning and fused deposition modeling for the fabrication of a hybrid vascular graft, *Biofabrication*, 2010, **2**(1), 014102.
112. S. Giannitelli, *et al.*, Current trends in the design of scaffolds for computer-aided tissue engineering, *Acta Biomater.*, 2014, **10**(2), 580–594.
113. S. Giannitelli, *et al.*, Combined additive manufacturing approaches in tissue engineering, *Acta Biomater.*, 2015, **24**, 1–11.
114. W. C. Wilson and T. Boland, Cell and organ printing 1: protein and cell printers, *Anat. Rec.*, 2003, **272**(2), 491–496.
115. K. E. Kador, *et al.*, Control of retinal ganglion cell positioning and neurite growth: combining 3D printing with radial electrospun scaffolds, *Tissue Eng., Part A*, 2016, **22**(3–4), 286–294.
116. S. Naghieh, *et al.*, Combinational processing of 3D printing and electrospinning of hierarchical poly(lactic acid)/gelatin-forsterite scaffolds as a biocomposite: mechanical and biological assessment, *Mater. Des.*, 2017, **133**, 128–135.
117. Y. Yu, *et al.*, Fabrication and characterization of electrospinning/3D printing bone tissue engineering scaffold, *RSC Adv.*, 2016, **6**(112), 110557–110565.
118. W. E. Teo and S. Ramakrishna, A review on electrospinning design and nanofibre assemblies, *Nanotechnology*, 2006, **17**(14), R89.
119. Y. Wu, J. Fuh, Y. S. Wong and J. Sun, A hybrid electrospinning and electrospraying 3D printing for tissue engineered scaffolds, *Rapid Prototyp. J.*, 2017, **23**(6), 1011–1019.
120. C. L. Ventola, Medical applications for 3D printing: current and projected uses, *Pharm. Ther.*, 2014, **39**(10), 704.
121. K. Jakab, *et al.*, Tissue engineering by self-assembly and bio-printing of living cells, *Biofabrication*, 2010, **2**(2), 022001.
122. J. Du and X. Liu, 3D bio-printing: a promising technique to fabricate dermal equivalent for skin wound healing, *Int. J. Med. Med. Sci.*, 2017, **4**(3), 11–17.
123. M. Itoh, *et al.*, Scaffold-free tubular tissues created by a bio-3D printer undergo remodeling and endothelialization when implanted in rat aortae, *PLoS One*, 2015, **10**(9), e0136681.

124. A. C. Daly, *et al.*, 3D bioprinting of developmentally inspired templates for whole bone organ engineering, *Adv. Healthcare Mater.*, 2016, **5**(18), 2353–2362.
125. B. Byambaa, *et al.*, Bioprinted osteogenic and vasculogenic patterns for engineering 3D bone tissue, *Adv. Healthcare Mater.*, 2017, **6**(16), DOI: 10.1002/adhm.201700015.
126. L. Hockaday, *et al.*, Rapid 3D printing of anatomically accurate and mechanically heterogeneous aortic valve hydrogel scaffolds, *Biofabrication*, 2012, **4**(3), 035005.
127. J. Visser, *et al.*, Biofabrication of multi-material anatomically shaped tissue constructs, *Biofabrication*, 2013, **5**(3), 035007.
128. B. Duan, *et al.*, Three-dimensional printed trileaflet valve conduits using biological hydrogels and human valve interstitial cells, *Acta Biomater.*, 2014, **10**(5), 1836–1846.
129. T. Billiet, *et al.*, The 3D printing of gelatin methacrylamide cell-laden tissue-engineered constructs with high cell viability, *Biomaterials*, 2014, **35**(1), 49–62.
130. N. Brandenberg and M. P. Lutolf, *In situ* patterning of microfluidic networks in 3D cell-laden hydrogels, *Adv. Mater.*, 2016, **28**(34), 7450–7456.
131. X. Ma, *et al.*, Deterministically patterned biomimetic human iPSC-derived hepatic model *via* rapid 3D bioprinting, *Proc. Natl. Acad. Sci. U. S. A.*, 2016, **113**(8), 2206–2211.
132. 3D Bioprinting Market Analysis, by Technology, by Application, Market Size, Application Analysis, Competitive Strategies and Forecasts To 2022, San Francisco, Grand View Research, Inc., 2017, Report Code: BT 4971, www.marketsandmarkets.com.

CHAPTER 7

Electrospinning: Large-scale Industrial Applications of Superhydrophobic Surfaces in Filtration/Sorbents Applications

S. T. ARUNA

Surface Engineering Division, Council of Scientific and Industrial Research-National Aerospace Laboratories, Bangalore, 560 017, Karnataka, India
*E-mail: aruna_reddy@nal.res.in

7.1 Introduction

Electrospinning, a derivative of electrospraying, is a century-old technique that utilizes electrostatic forces to generate polymer droplets. In recent years, electrospinning has been recognized as an efficient technique for the fabrication of polymer nanofibers and, as the name implies, it shares characteristics of electrospraying and conventional solution dry spinning of fibers. It is a fiber production method which uses high voltages to draw charged threads of polymer solutions or polymer melts to form fibers with diameters in the order of a few hundred nanometers and lengths running to several kilometers.

Soft Matter Series No. 7
Electrospinning: From Basic Research to Commercialization
Edited by Erich Kny, Kajal Ghosal and Sabu Thomas

Published by the Royal Society of Chemistry, www.rsc.org

The origin of this technique dates back to 1600 when W. Gilbert, an English physician, noticed that a water drop sitting on a dry surface deformed into a conical shape when a piece of rubbed amber was held in proximity to it. This was the first experimental observation of deformed liquid–air interfaces under the influence of an external electric field.[1–7] In 1897, the term 'electrospinning' was coined by Raleigh. In late 1930s, Petryanov-Sokolov and collaborators carried out experiments that rapidly led to the first industrial facility for the production of electrospun fibrous materials, called Petryanov filters for gas masks.[8] In the 1990s, Reneker produced fibers with a diameter of less than 5 μm using high voltages to charge the polymer dispersion.[9–13] Tucker *et al.*[14] have narrated the history, science and technology of the electrospinning process from 1600 to 1995, and a large number of review articles have been published on various aspects of electrospinning.[15–20]

7.1.1 Principles of Electrospinning

The electrospinning technique is based on overcoming a liquid's surface tension by means of an externally applied voltage. When the electric field force overcomes the surface tension of the precursor solution, a Taylor cone is formed and thin fibers originate from the tip of this cone (Figure 7.1a). Spinneret and collector are maintained at opposite charges. A high-voltage power supply remains in contact with the metallic spinneret. Viscous liquid from the syringe with an optimum flow rate is slowly collected on the spinneret and forms spherical droplets controlled by surface tension. The equipment used for electrospinning basically consists of a tank of polymer dispersion with a pump, a high-voltage power source, a nozzle and a conductive collector. Figure 7.1b and c show, respectively, photographs of a lab-scale electrospinning facility along with its important components such as (i) plate collector, (ii) drum collector, (iii) rpm meter for drum collector, (iv) syringe pump, (v) syringe pump controller and (vi) discharge rod. The steps involved in the electrospinning process are as follows: (i) synthesis of polymer precursor solution with suitable viscosity and (ii) electrospinning the solution onto a plate or drum collector and optimizing the parameters such as voltage and distance between the collector and the syringe.

7.1.2 Factors Controlling Electrospinning

Various parameters control the fiber formation, the most influential factors being viscosity or polymer concentration, solvent, flow rate, distance between tip and collector, applied voltage and temperature. The complete list of parameters affecting the diameter of fibers is shown in Table 7.1.[21] However, only a few parameters can be controlled and altered.

Studies have shown that by decreasing the distance between nozzle and collector, wet electrospun fibers with a beaded structure are obtained.[22,23] It is obvious that for aqueous polymer dispersions a higher distance has to be maintained to dry the fiber.[23] Polymer concentration or viscosity plays an important role that can influence the size, alignment and defects of the

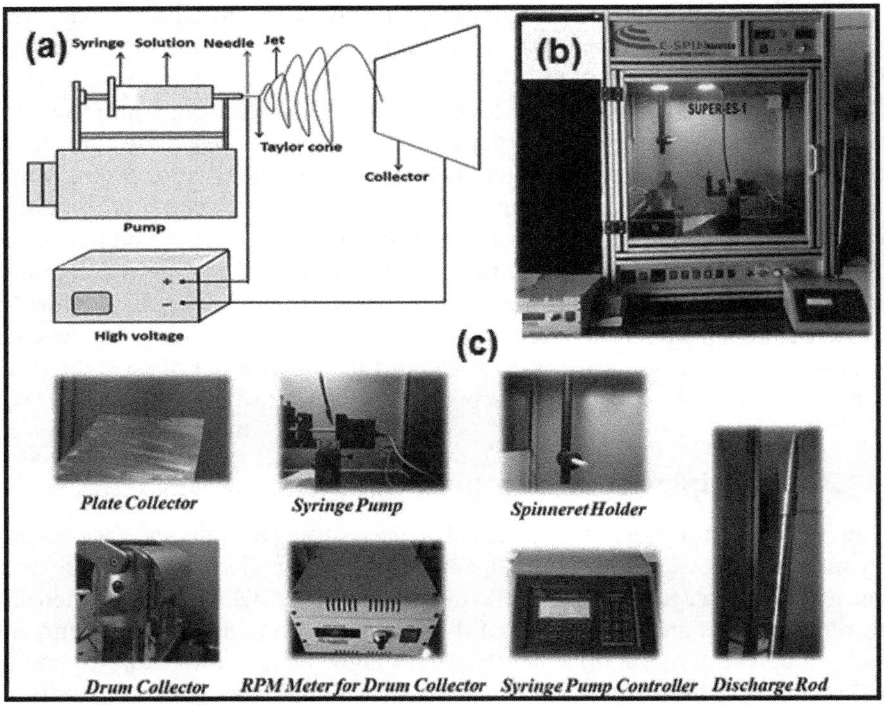

Figure 7.1 (a) Schematic showing the principle of electrospinning and photographs showing (b) the electrospinning facility along with (c) various components of the set-up.

Table 7.1 Different parameters used in electrospinning. Reprinted from *Polymer*, **46**, S.-H. Tan, R. Inai, M. Kotaki, S. Ramakrishna, Systematic parameter study for ultra-fine fiber fabrication *via* electrospinning process, 6128–6134. Copyright 2005, with permission from Elsevier.

Solution properties	Processing conditions	Operating conditions
Viscosity	Applied voltage	Temperature
Polymer concentration	Needle to collector distance	Humidity
Molecular weight of polymer	Volume feed rate	Atmospheric pressure
Electrical conductivity	Needle diameter	
Elasticity		
Surface tension		

fibers. Higher polymer concentration can lead to large fiber diameter and a much lower number of beads. Lower polymer concentration affects the overall fiber production by forming a greater number of beads and fibers with lower diameters. The most widely used polymers in electrospinning include polyvinyl alcohol (PVA), polyvinylpyrrolidone (PVP), polyacrylonitrile (PAN), polyamide-6, polyacrylic acid, polyurethane, polyethylene oxide, poly(trimethylene terephthalate), gelatin, collagen, chitosan and silk. Figure 7.2 is a photograph of electrospun PVP fibers in the author's laboratory.

Figure 7.2 Photograph showing electrospun polyvinylpyrrolidone (PVP) fibers.

The solvent used in the electrospinning technique is considered to be very important as it is directly related to surface tension and evaporation rate. The most frequently used solvents include formic acid, chloroform, tetrahydrofuran, trifluoroacetic acid, acetic acid, isopropanol, dimethyl formamide, water, dichloromethane, methanol and ethanol. When a solvent with average surface tension and volatility is used, the spinning of fibers is less challenging. Also, solvents with very low surface tension result in no fiber formation at all. Water is a proven good solvent for various polymers but due to its very high surface tension it is very challenging to overcome the surface tension to draw fibers. Flow rate is related directly to fiber diameter. Fibers with more uniform diameters can be obtained with a lower flow rate. Beading takes place with higher flow rates as the fibers do not have sufficient time to dry and this can also result in the formation of electrosprayed particles.[24] Modifications to fiber diameter are also achievable by increasing the distance between the tip and collector. Increase in distance gives enough time for stretching of the jet and delayed flight time, so as to obtain thinner fibers. Voltage is one of the most important factors that influences morphology and defects in fibers. Fibers formed in the presence of a low voltage are generally reported as smooth and beadless fibers, whereas beaded and rougher fibers are observed in the presence of a high voltage. Temperature is inversely related to fiber diameter, *i.e.* an increase in temperature decreases the fiber diameter and *vice versa*.

7.1.3 Advantages of Electrospinning

The advantages of electrospinning include high surface area-to-volume and length-to-diameter ratios, uniform pore size, ease of fabrication, higher efficiency, low cost, high yield and high degree of reproducibility of the product. When the diameters of fiber materials are reduced from micro- to nanometers it will result in a large increase in the surface area-to-volume ratio, which can be as much as 103 times that of a microfiber. As a result of this several properties are improved, including flexibility in surface functionalities, and superior mechanical properties (*e.g.* stiffness and tensile strength). In addition, properties such as degree of hydrophobicity, conductivity and percolation limit can also be tailored by changing the electrospinning parameters and chemical modification techniques.

7.1.4 Applications of Electrospinning

Electrospun nanofibers can find use in wide range of fields such as the biomedical industry, filtration applications, supporting or electrode materials in energy harvesting devices, protective clothing, composite materials, sensors and many more. Since 1990, numerous modifications have been performed on electrospun fibers to obtain morphologies such as core–shell, nanofibrous mats, ribbons and many more, drawn out of polymer solution for different applications.[25,26] A few applications are shown in Figure 7.3 and are explained in brief. There are many review articles published on different applications of electrospun nanofibers and the details are presented in Table 7.2.[27–32]

Physical entanglement, porous morphology and high surface-to-volume ratio of fibers have gained more attention in the area of filtration. A fibrous material's ability to capture micron and sub-micron level impurities has proved beneficial in industries producing clean compressed air.[33] Gibson *et al.*[34] prepared nanofiber mats for trapping airborne particles with sizes ranging from 0.5 to 200 μm. Fibers exhibit higher surface reception and thus facilitate the fabrication of highly sensitive sensors.[35–37] Due to their high surface area-to-volume ratio, electrospun nanofiber fabrics show greater efficiency in capturing sub-micron level particulates and also prove to be useful as lightweight protective clothing.[38]

Electrospun nanofibers are considered as one of the most promising materials in the field of energy systems, to meet the increasing demand for low-cost

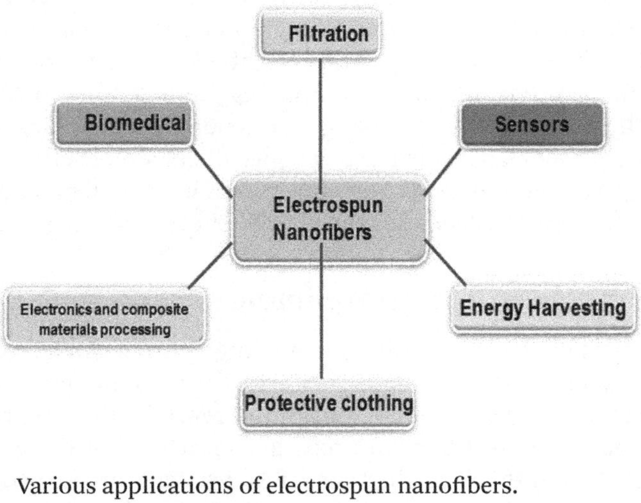

Figure 7.3 Various applications of electrospun nanofibers.

Table 7.2 Review articles on applications of electrospun fibers.

S. no	Application	Salient features	Reference
1.	Glucose sensor	Increased lifetime	27
2.	Biosensor	Faster, simpler, and less expensive detection methods for monitoring food/agricultural products	28
3.	Gas sensor	Ultrahigh sensitivity, very short response and recovery time, good reversibility, reproducibility, stability of sensors based on acoustic wave, resistivity, photoelectricity and optical activity	29
4.	Energy devices	Solar cells, fuel cells, nanogenerators, hydrogen energy harvesting and storage-enhanced efficiency; lithium-ion batteries, and supercapacitors – high specific capacitance and better cycling stability	30
5.	Electronic components and devices	Insulators, separators, electrolytes, optoelectronics	32

materials and higher efficiency. Due to their unique one-dimensional structure, surface reception, porosity, homogeneity and high electronic conductivity, fibers are used as cathode materials, catalysts and supports.[39–41]

Nanofibers have gained more attention for having solution to overcome limitations in biomedical field than any other field. Targeted drug delivery can be achieved with nanofibers as carrier of drugs or composite fibers consisting of drug components along with a carrier by which the drugs reach the affected area without affecting other mechanisms and parts.[42,43]

Composite materials with fiber as one of their part can be used in photocatalysis application by alteration of band gap, degrading dye in the presence of visible light more efficiently and in lesser time.[44,45] Nanofibers with conducting nature holds promising role in fabrication of various miniature electronic devices.[46] Besides, this technique is widely utilized for generating membranes for membrane distillation and nanofiltration processes. Electrospun nanofibrous membranes can be used for environmental remediation like removal of oil spills and waste water treatment. Superhydrophobic electrospun nanofibrous membranes (ENMs) with a fully interconnected porous web structure and high void volume fraction have been proposed for membrane distillation (MD) for desalination and heavy metal removal applications.

7.2 Electrospun Superhydrophobic Nanofibers

Over the past decade, superhydrophobic (SH) surfaces have attracted considerable attention owing to their potential applications in diverse areas, such as self-cleaning, anticorrosion, drag-reduction coatings, microfluidic devices and intelligent membranes.[47–51] Superhydrophobic self-cleaning, also called the lotus effect, utilizes the right combinations of surface chemistry and topology to form a very high contact angle on a surface and drive water droplets away from it.[52] Superhydrophobic surfaces exhibit extraordinarily high water contact angles (WCAs, >150°) and low contact angle hysteresis between advancing and receding contact angles (<5°). For the first time, Neinhuis and Barthlott discovered superhydrophobicity and patented it in 1998 as the lotus effect.[53] The lotus effect has been attributed to the combination of low surface energy and hierarchical micro–nano roughness. Using the principle of lotus effect a plethora of superhydrophobic surfaces have been fabricated. Superhydrophobic surfaces can be produced in two ways (i) by creating a rough surface and modifying the existing surface with low surface energy materials and (ii) creating nano–micron roughness on a low surface energy material.

In the past one decade many reviews describing the different synthesis strategies to fabricate artificial superhydrophobic surfaces and applications have been published.[54–60] Inspired by the self-cleaning behavior of lotus leaves in nature, surfaces with superhydrophobic features have been produced in three ways:[61,62] (i) the bottom up approach – involving methods like colloidal assembly, self-assembly, sol–gel, chemical deposition, chemical vapor deposition, electrodeposition, phase separation, electrospinning *etc.* (ii) top down approach – involving methods like plasma method, sputtering, lithography, template method, electrospinning, solution immersion, crystallization control, layer-by-layer deposition *etc.* and (iii) combination approach involving both bottom up and top down approaches. A high quality superhydrophobic surface will have a uniform layer of pinned or trapped air that produces a very high contact angle (>150°), a very low roll-off angle (<10°), small contact angle hysteresis and will produce an optical mirror effect when the surface is submerged in water due to total internal reflections of light reflecting off the pinned air layer (Figure 7.4a and b). Among

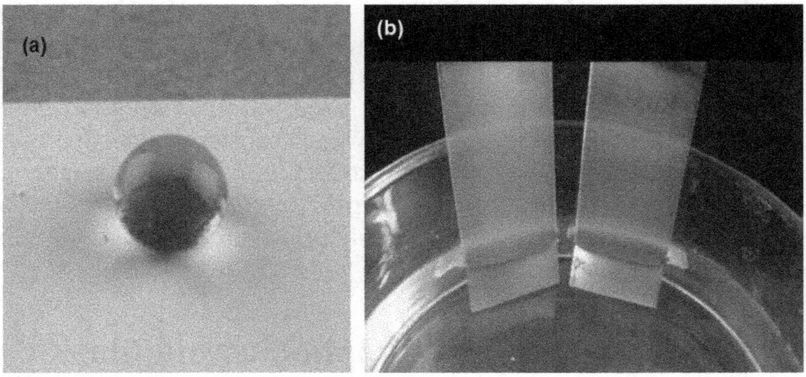

Figure 7.4 Photographs showing (a) water contact angle on a superhydrophobic surface, (b) total internal reflection layer caused due to superhydrophobic film when dipped in water.

all these methods, electrospinning technique has been widely explored to fabricate superhydrophobic fibers. In electrospun fibers, smaller fiber diameters contribute to superhydrophobicity due to the unique surface roughness/texture of the fibers. Some nanomaterials have also been applied to the post-treatment of electrospun fibers that contribute to additional roughness and facilitate the creation of superhydrophobic nano fibers.[52] A literature survey on electrospinning and superhydrophobicity using Scopus analysis clearly shows an increasing trend in the number of papers published over the years and China is the leading country that publishes papers in this area (Figure 7.5).

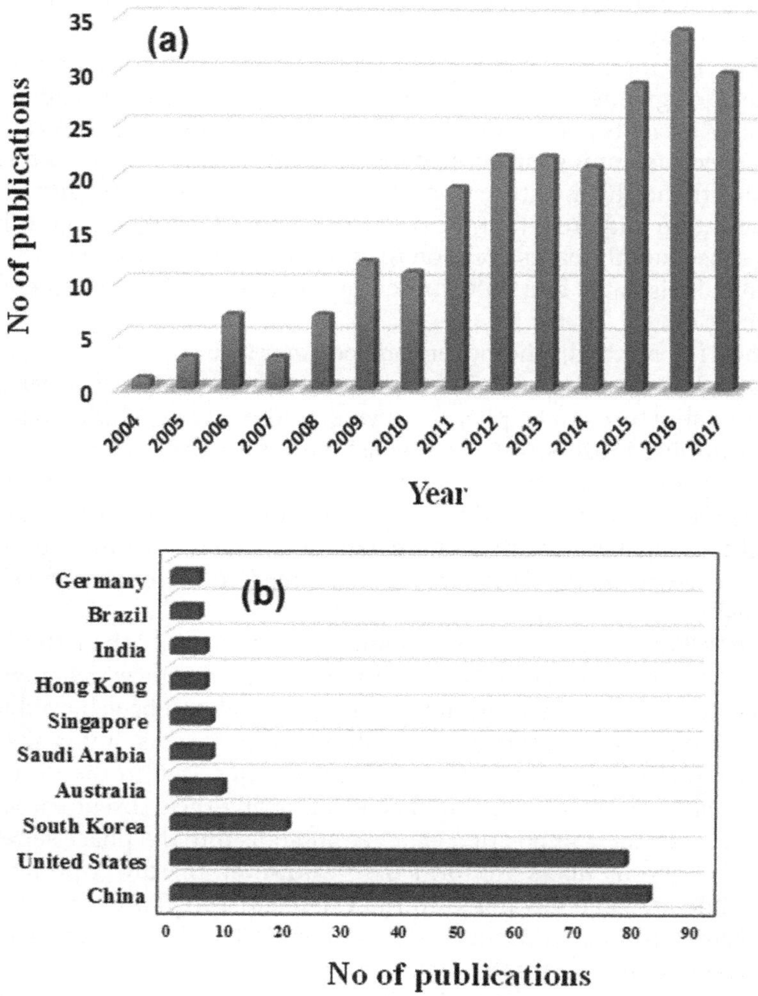

Figure 7.5 Scopus analysis using the search words superhydrophobic and electrospinning as on August 2017: (a) publications by year and (b) top countries publishing papers on electrospun superhydrophobic fibers.

7.3 Fabrication of Electrospun Superhydrophobic Porous Nanofibers

In general, the fabrication of electrospun nanofibers involves the synthesis of the polymer solution followed by electrospinning by varying the parameters like voltage and the distance between the needle and the collector. For sorbent and filtration applications it is very important to fabricate porous nanofibers. Phase separation is a common strategy adopted to produce porous structures in the field of polymer membranes. The same strategy can be employed for electrospinning fibrous nanofibers. Phase separation can be achieved in three different ways: (i) vapor-induced phase separation (VIPS), (ii) non-solvent-induced phase separation (NIPS) and (iii) thermally induced phase separation (TIPS).

In the VIPS method, usually water vapors are condensed as non-solvents that cause originally homogeneous solutions to phase separate and after the removal of solvents porous structures are obtained. The same VIPS can occur during electrospinning process. When solvent evaporates, the jet surface cools down and the moisture from the atmosphere condenses and initiates the nucleation and growth of the water-rich phase floating on the solution, which subsequently leaves pores on the fiber surface.[63–68] However, the pores are generally isolated and only formed on the fiber surface due to moisture condensation.

In the TIPS method, a homogeneous polymer solution is first prepared at an elevated temperature and then rapidly cooled to a temperature where the solvent quality turns to be poor to polymer and phase separation occurs. In electrospinning process, TIPS can be achieved by maintaining the collector at relatively low temperature than the syringe to induce phase separation and there by create porous nanofibers.[69] Although this method produces pores throughout the fibers, the fibers have to undergo a post-treatment to remove residual solvents as the fibers are not completely dried during the electrospinning process.

In the NIPS method, also known as immersion precipitation method, a cast polymer solution is immersed in a non-solvent bath during which phase separation occurs.[70] In electrospinning technique, NIPS can be achieved by premixing appropriate amounts of non-solvents in polymer solution that does not cause phase separation before electrospinning. Due to the differences in the volatility of solvent and non-solvent, during the subsequent solidification process, the composition changes and falls into the phase separation region and porous fibers are formed.[71–73] However, the fibers produced by this method are generally not uniform. Chen and Tung have developed one step electrospinning to produce non-solvent induced macroporous fibers with ultrahigh oil adsorption capability.[74]

Generally, to induce superhydrophobic properties to electrospun fibers, nanoparticles/CNTs are added or special pretreatments applied. The preparation of superhydrophobic electrospun nanofibers are discussed under sorption and filtration subheadings.

7.4 Characterization of Electrospun Superhydrophobic Nanofibers

In order to qualify the electrospun nanofibers for sorbent/filtration application, particularly oil spill clean-up a series of tests have to be carried out and details are provided below:

7.4.1 Contact Angle and Sliding Angle Measurements

The superhydrophobic nature of the electrospun fibers is evaluated by measuring the contact angle and sliding angle using the standard goniometer. To accurately measure the contact angles the fiber mats should be flat and firm. To achieve this, the glass slides are attached to the aluminum foil on the plate collector of the electrospinning set-up and fibers are deposited onto the foil and also on the glass slides as continuous mats. The fibers adhere well to the glass slides and excess mat material is trimmed off from the edges of the glass slides. To compare to the intrinsic flat surface contact angle of the polymer system, a glass slide is generally coated with the same polymer system by evaporating the solvent from the glass slide, to form a smooth non-porous coating of the polymer on the slide.

7.4.2 Mechanical Robustness and Stability of Membrane

Good mechanical strength and stability in harsh conditions are essential for a good membrane for practical application. It should exhibit high tensile modulus. The standard tensile strength studies are carried out for the fibers using universal tensile testing machine.

7.4.3 Measurement of Oil Sorption Capacity

The oil sorption capacity of an electrospun membrane is determined as per ASTM F726-60 standard using different oils as discussed in the literature.[74] In a typical experiment, 10 g of oil and 150 ml of water are added into a 200 ml glass bottle. Due to the lower density of oil, it will immediately float on water. The fiber which acts as adsorbent is gently placed on the oil/water mixture. The sorbent being superhydrophobic will attract oil and oil gets adsorbed. The wet sorbent is taken with the help of tweezers and it is hung for 3 min in air to remove the excess oil before weighing. The experiments are repeated at least three times and with different oils. The oil sorption capacity is quantified by the weight ratio as oil absorbency (g g^{-1}) from the following equation:

$$Q_{\mathrm{oil}} = \frac{m_{\mathrm{s}} - m_{\mathrm{i}}}{m_{\mathrm{i}}}$$

where Q_{oil} is the oil sorption capacity (g g^{-1}) of the sample, m_{s} is the weight of the wet sorbent after draining (g), and m_{i} is the initial weight of the dried sorbent before sorption (g).

7.4.4 Desorption and Reusability Test

In order to measure the reusability of the electrospun nanofiber sorbents, their desorption abilities have to be investigated. The sorbent fiber mats with oil are compressed between two plates and the sorbent fiber mats are rinsed with alcohol and pure water to remove the residual oil and dried in an oven at 60 °C for 10 min. The dried sorbents are then reused for oil sorption, and the sorption/desorption cycles are continued until the sorbent is damaged.

7.4.5 Assessment of Oil–Water Separation for Filtration Membranes

The oil–water separation of a filtration membrane is determined as follows: the as-prepared electrospun fibrous membrane is fixed between two glass funnels with effective separation area of 56.25×10^{-6} m^2 (Figure 7.6).[75] A mixture of 10 ml of oil and 10 ml of water is poured slowly into the upper test glass vessel through the as-prepared membrane and the separation is achieved, driven solely by gravity. Then the separated oil and water are collected respectively with the beaker and the glass vessel.

The flux (F) of liquid through the membrane is calculated from the following equation:

$$F = \frac{V}{A\Delta t}$$

Here, V is defined as the volume of filtrated liquid, A (m^2) being the effective (surface) area of the separation membrane in the set-up and t is the filtration time. The 'separation efficiency' (η) of liquid through the membrane is obtained using the following equation:

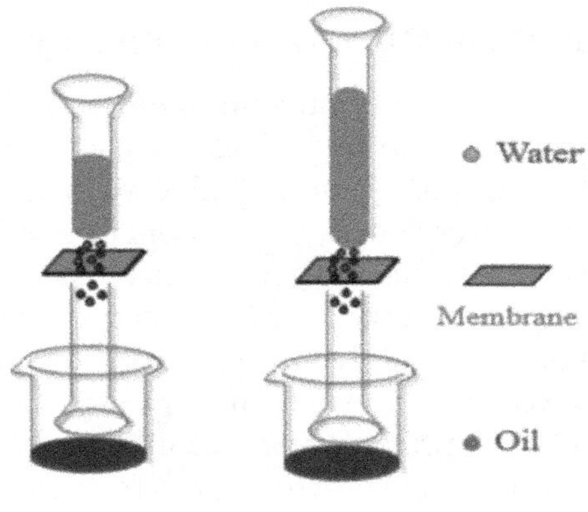

Figure 7.6 Schematic of oil–water separation set-up. Reprinted from *Separation and Purification Technology*, **177**, W. Ma, Z. Guo, J. Zhao, Q. Yu, F. Wang, J. Han, H. Pan, J. Yao, Q. Zhang, S. K. Samal, S. C. De Smedt, C. Huang, Polyimide/cellulose acetate core/shell electrospun fibrous membranes for oil-water separation, 71–85, Copyright 2017, with permission from Elsevier.

$$\eta = \frac{V_1}{V_0} \times 100$$

where V_0 and V_1 are the volume of the water before and after the separation processes, respectively.

7.4.6 The Evaluation of Porosity (P) of the Fibrous Membrane

A manually deposited polymer film is obtained by completely evaporating a 1.5 ml of 1 wt% polymer solution on a microscopic glass slide. The film thickness is measured using a micrometer screw gauge. The porosity (P) of the fibrous membrane is calculated as follows: [75]

$$\rho_{\text{membrane}} = \frac{\text{mass}_{\text{membrane}}}{\text{area}_{\text{membrane}} \times \text{thickness}_{\text{membrane}}}$$

$$\rho_{\text{film}} = \frac{\text{mass}_{\text{film}}}{\text{area}_{\text{film}} \times \text{thickness}_{\text{film}}}$$

$$P = \left(1 - \frac{\rho_{\text{membrane}}}{\rho_{\text{film}}}\right) \times 100\%$$

where ρ_{membrane} and ρ_{film} are the density of the electrospun fibrous membrane and the corresponding manually deposited 'polymer film', respectively.

7.5 Literature on Electrospun Superhydrophobic Nanofibers for Sorption and Filtration Applications

Oil spills that occur due to manmade disasters are of great concern worldwide as it results in both immediate and long-term environmental damage and can last for decades after the spill occurs. Besides, oil spilled by pipelines or offshore oil rigs and damaged tankers coats everything it touches including birds and fishes and becomes an unpleasant but long-term part of every ecosystem it enters. Due to the increased amount of industrial oily wastewater, as well as frequent oil spill accidents like the Deepwater Horizon oil spill in the Gulf of Mexico in 2010, oil–water separation has become a worldwide challenge.[76–81] The most widely used methods for cleaning up of oil spills along with their advantages and disadvantages are illustrated in Table 7.3.[82–85]

It is possible to collect oil and completely remove the oil from oil spill site using sorbent materials and cross-flow filtration using membranes.[86,87] Membrane technology is considered to be the most promising approach to treat oily wastewater due to the high separation efficiency and relatively simple operational procedures. Polymeric nanofibers have been used in air filtration applications for more than seven decades. However, due to their poor bulk mechanical properties they are organized over a filtration medium substrate.

Table 7.3 Methods used for cleaning up oil spills.

S. no.	Method	Advantages	Disadvantages	Reference
1.	Bioremediation using biological agents/ microorganisms	Economical	Slow process	82
2.	Controlled burning	Minimizes the adverse effect of the oil on the environment	It can be done only in low wind and can cause air pollution and sinking of semi-solid or semi-liquid residues	83
3.	Dispersants to break and form water soluble micelles	Smaller oil particles are more easily biodegraded	Increased toxicity and may kill fish eggs	84
4.	Skimming	Greater overall oil separation efficiency	Requires calm waters at all times and removes a mixture of oil and water	85

Smaller fiber diameters (<0.5 μm) enable slip flows at fiber surfaces, causing an increase in the interception and inertial impaction efficiencies of these composite filter media and thereby enhancing the filtration efficiency at the same pressure drop. By tailoring the electrospinning parameters both membranes and absorbents can be prepared with nanostructured geometries, selective wettability and reduced manufacturing costs and can be used for oil spill clean-up.[73] Due to its inherent adaptability, electrospinning technique can be employed for the preparation of non-polar polymeric absorbents and non-polar and polar polymeric selective filtration membranes.[73] Different strategies can be used for oil spill clean-up using electrospun nanofibers such as using absorbents of hydrophobic-oleophilic nanofibrous membranes and cross-flow filtration membranes which are either hydrophobic or hydrophilic nanofibrous membranes.

7.5.1 Electrospun Nanofibers as Sorbent Materials

The use of sorbents is generally considered to be one of the most efficient techniques for the removal of oil and polycyclic aromatic hydrocarbons from waters.[88,89] An ideal sorbent material for oil spill clean-up should exhibit high uptake capacity and high rate of uptake, hydrophobicity, retention over time, buoyancy, durability in aqueous media, reusability or biodegradability, and recoverability of oil. Albeit several natural materials such as rice straw, cotton fibers, recycled wool, activated carbon have been proposed as sorbents for oil removal, polymeric materials based on non-woven polypropylene (PP) are the most generally used commercial sorbents in oil spill clean-up. The properties of PP sorbents closely match those of an ideal oil sorbent.[90–92] Most sorbents, however, end up in landfills or in incineration after a single use. These possibilities increase the oil recovery cost and also introduce another source of pollution.

Recently, superhydrophobic and superoleophilic sorbents have attracted much attention for oil clean-up because of their wide application. There are three main classes of oil sorbent materials: inorganic mineral products – silica aerogels, organic natural products – cotton and rice husk, and synthetic organic products – polytetrafluoroethylene (PTFE), polypropylene (PP) and polyurethane (PU). Among them, organic synthetic fibers play an important role in oil spill clean-up due to their low cost, tunable oleophilic and hydrophobic properties and availability on a large scale.

To date, a large number of fabrication methods have been used for synthesizing sorbents for oil spill clean-up such as template-based techniques, self-assembly of block copolymers, layer-by-layer assembly, *etc.* These methods are of not practical use as they involve multistep and have limitation in substrate sides. Recently, electrospinning technique has been widely exploited to fabricate superhydrophobic electrospun fibers for a variety of applications. A large number of papers have been devoted to oil spill clean-up, desalination, waste water treatment *etc.* Electrospun superhydrophobic sorbent fibers are attractive for oil spill clean-up as they enable complete removal of the oil from the oil spill state. These fibers exhibit hydrophobicity and oleophilicity, high rate of uptake, higher retention time, and easy recovery.

7.5.1.1 Polystyrene-based Electrospun Fibers

Wu *et al.*[93] fabricated electrospun superhydrophobic-oleophilic polystyrene (PS) fibers by controlling the fiber diameter and creating porous morphology which facilitated high oil sorption capacity and thereby enhancing the oil/water selective adsorption. The detailed procedure for the preparation was as follows: 2 ml of 20 wt% precursor solution of PS (M_w = 350 000) in DMF was placed in a 5 ml syringe. To form PS fibers with different diameters (0.565, 0.667, 0.918, 1.473, 2.454, and 3.716 μm), metal needles with different inner diameters (0.5, 0.6, 0.7, 0.9, 1.2, and 1.6 mm) were used (Figure 7.6). The fibers were collected on a stainless steel plate covered with a sheet of aluminum foil. The fibers were electrospun by maintaining a distance of 25 cm between the needle tip and collector and maintaining a voltage of 12 kV. The PS fibrous films with porous structure on fibers were electrospun by changing the solvent from DMF to THF and maintaining the same parameters (Figure 7.7). The porous fiber film exhibited higher and lower water contact and oil contact angles respectively which may be attributed to higher surface roughness.

The PS fibrous film exhibited higher sorption capacity due to the combined superhydrophobic and oleophilic nature of the film which was caused by lower surface tension, voids between the interconnected fibers and due

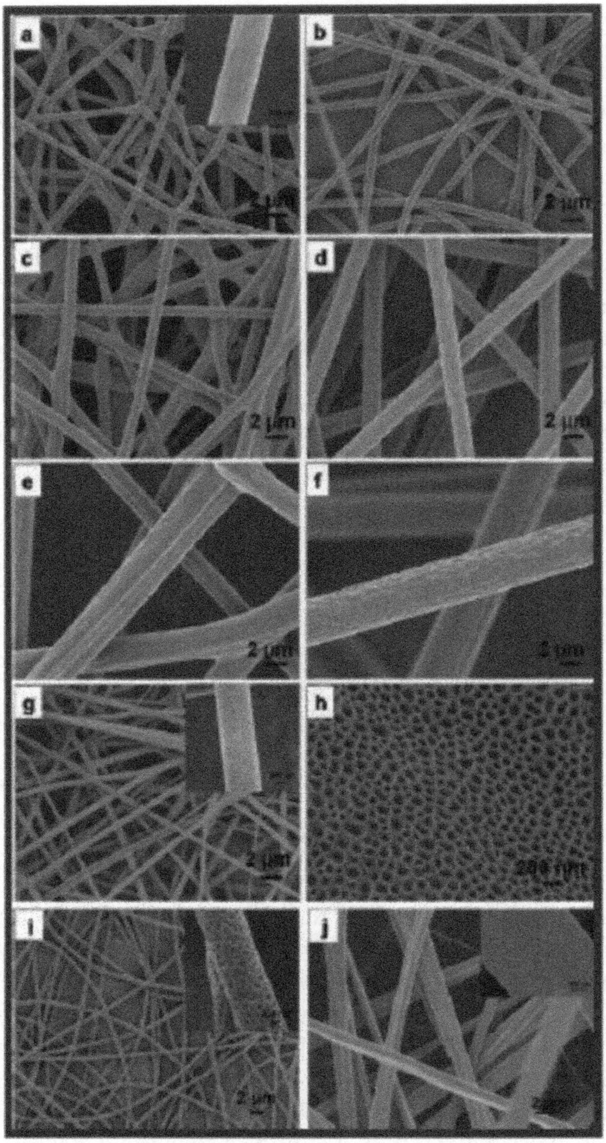

Figure 7.7 SEM images of PS fibrous films. (a)–(f) SEM images of as-prepared samples without a porous structure. To form PS fibers with different diameters, different metal needles of 0.5, 0.6, 0.7, 0.9, 1.2 and 1.6 mm inner diameter were used (the samples were named F5, F6, F7, F9, F12 and F16). The average diameters of these fibers were 0.565, 0.667, 0.918, 1.473, 2.454 and 3.716 μm, respectively. Inset to (a) shows the magnified morphology of F5. (g)–(j) Images of porous PS fibrous films (called PF5, PF7 and PF12). The fiber diameters are approximately 0.540, 1.020 and 2.135 μm. Insets to (g), (i) and (j) show the magnified morphology of the samples. (h) Amplification of PF5 sample fiber. There are nanopores on the surface of PF5, and pore sizes range from 100 to 200 nm. Reprinted with permission from J. Wu, N. Wang, L. Wang, H. Dong, Y. Zhao and L. Jiang, *ACS Appl. Mater. Interfaces*, 2012, **4**, 3207, Copyright 2012 American Chemical Society.

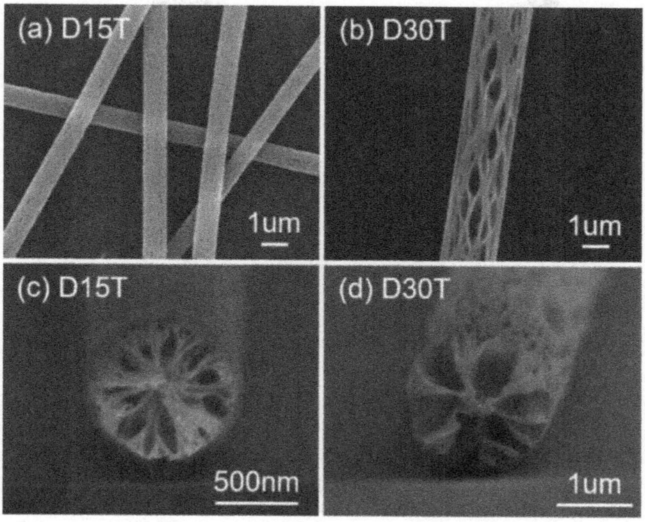

Figure 7.8 SEM micrographs of the surfaces (a, b) and cross-sections (c, d) of the as-spun D15T and D30T fibers. Reprinted with permission from P.-Y. Chen and S.-H. Tung, *Macromolecules*, 2017, **50** (6), 2528, Copyright 2017 American Chemical Society.

to high surface area. The mechanism of oil adsorption has been understood as follows: (i) initially during oil sorption onto the film surfaces, the oleophilic interaction and van der Waals forces play an important role between the oil and film's surface. (ii) Chemical compatibility between oils and PS films leads to minimum energy barrier for oil to penetrate to the fiber voids or fibrous structures.

Chen and Tung[74] mixed polystyrene in cosolvents of chlorobenzene and dimethylsulfoxide followed by magnetic stirring at 50 °C for 12 h to obtain homogeneous solutions followed by electrospinning using a single-capillary spinneret using 4–10 kV and 15 cm distance. For electrospinning solution, tetrabutylammonium perchlorate (TBAP) and DMSO was added to enhance the conductivity of the solutions. The SEM images of D15T and D30T fibers electrospun from the cosolvents of 15% and 30% DMSO and 1% TBAP are shown in Figure 7.8. It was found that fibers with 1% TBAP were of smaller diameter due to increased solvent conductivity. By combining the polar non-solvent and the organic salt, PS fibers were electrospun to obtain fibers with smaller diameters and higher porosity. The maximum oil adsorption capacities of the fibrous sorbents for four different oils with different viscosities are displayed in Figure 7.9. The amount of oil the sorbents can adsorb is highly dependent on the type of oils. Higher the viscosity, higher is the adsorption which may be attributed to better adhesion of high viscous oils to the fibers. In fibrous sorbents, the adsorbed oil is mainly distributed on the fiber surface and is trapped in the void space between the fibers and is driven by the capillary force to reduce the surface free energy of the fibers.

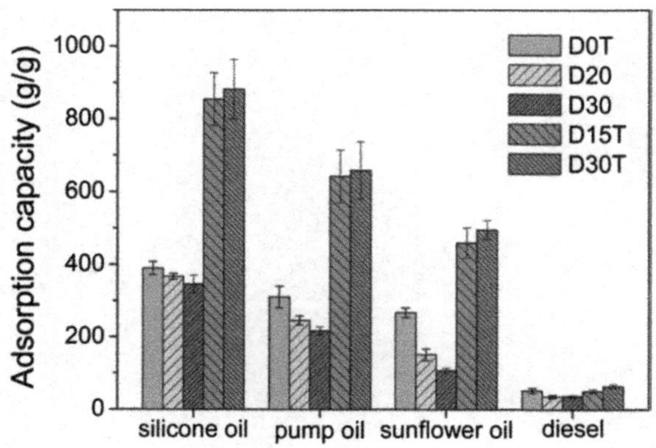

Figure 7.9 Maximum oil adsorption capacities of the fibrous sorbents for different oils. Reprinted with permission from P.-Y. Chen and S.-H. Tung, *Macromolecules*, 2017, **50** (6), 2528, Copyright 2017 American Chemical Society.

Thus, by simply tuning the fraction of the non-solvent and organic salt in solutions, fibrous sorbents with extraordinary oil adsorption capacity and controllable adsorption rate can be successfully fabricated through the NIPS assisted electrospinning technique. The adsorption capacity of porous PS sorbent for certain oil spill reached to about 100 g of oil per gram of sorbent and thus can be employed for efficient adsorption of oil/water system and can be widely used in technological areas.[93] However, due to the poor mechanical properties of PS sorbents they are prone to damage during transportation or post-treatment processes. Therefore, various materials like polyacrylonitrile (PAN), iron oxide (Fe_3O_4) and polyvinylidene fluoride (PVDF) have been introduced to modify the mechanical strength of PS sorbents, but the incorporation of these materials decreased the oil sorption capacity of the resulting sorbents.[94]

Carbon nanotubes (CNTs) have shown excellent potential for oil–water separation due to their high porosity, large specific surface area, low density, hydrophobic and oleophilic nature, and ease of functionalization.[95] By taking advantage of PS polymer and CNTs' properties, Wu *et al.*[96] have fabricated a superhydrophobic PS-CNTs composite sorbent using electrospinning technique to investigate their oil sorption capacity. A well-aligned PS-CNTs sorbent, which is superhydrophobic-oleophilic, was fabricated by simple electrospinning using PS, DMF and 1*H*,1*H*,2*H*,2*H*-perfluorooctyltriethoxysilane (FTES) functionalized multiwalled CNTs. Functionalized CNTs were employed in this study to obtain homogeneous dispersion of CNT in PS and improve the interfacial compatibility of CNTs with the polymer matrix. The PS-CNTs composite sorbent exhibited a significant increase in the porosity, surface area, and water contact angle, as well as oleophilic nature, which plays an important role in oil sorption. PS-CNT exhibited excellent oil removal efficiency even after several cycles of the oil sorption test. Thus PS-CNT can be a potential sorbent for oil spill clean-up and is cost effective, convenient and can be produced in large scale.

Another challenge has been the poor nanofiber strength and resiliency after oil sorption due to which the reusability is very poor. This has been overcome by the development of core–shell configuration of fibers using coaxial electrospinning method.[75] Ma *et al.*[75] used two syringe-like apparatus and a spinneret with an inner diameter of 0.51 mm needle coaxially placed inside an outer diameter of 1.2 mm one to prepare core–shell configured electrospun nanofibers. The solution concentrations of polyamide acid (PAA) in *N*,*N*-dimethylacetamide (DMAc (core)) and cellulose acetate (CA) in a mixed solvent of dichloromethane (DCM)/acetone (75 : 25 v/v) (shell) were optimized at 3 wt% and 7.5% w/v, respectively. The feed rate of the inner and outer dope was set at 0.5 and 1 ml h^{-1}, respectively. Both PAA and CA solutions were delivered *via* the coaxial needle with a programmable syringe pump. The distance between the needle tip and the collector was 13 cm and a 25 kV voltage was applied to the needle. A rotating roller (1500 rpm) controlled by a stepping motor was used as collector. Non-woven sheets of the electrospun nanofibers were obtained at ambient temperature of 25 °C with a relative humidity of 40% and were subsequently imidizated by annealing at different temperatures (150°C/2 h, 200°C/3 h, 250°C/2 h and 260°C/1 h) followed by modification of the surface of the fibers with fluorinated polybenzoxazine (F-PB), in the presence of silica nanoparticles (SNPs) (Figure 7.10a). The TEM image of PI/CA (Figure 7.10b) fibers showed a clear core/shell structure and the outer diameter of the fibers was approximately 200 nm with a wall thickness of approximately 5 nm, indicating that the PI nanofibers were coated by a thin CA layer: it looks very much like the inclusion of a highly flexible PI wire (core) into the CA fiber (shell). Therefore, the good mechanical performance of PI/CA nanofibers could be explained (Figure 7.10b) by the designed core/sheath structure. Therefore, the core/shell structure PI/CA nanofibers inherited both the high mechanical property of a PI fiber core and the surface roughness of a CA shell. After the surface modification with F-PB/SNPs, the membranes become superhydrophobic and superoleophilic.

However, the oil sorption capacity of the majority of these samples is only 10 g of oil per gram of sorbent, which needs to be improved for practical application.

7.5.2 Electrospun Nanofibers for Cross-flow Filtration

Membrane technology based cross-flow filtration is a highly efficient method for environmental remediation. In recent years, much attention is given to the fabrication of extremely efficient nanofibrous membranes with advantages of high filtration efficiency due to very high surface area-to-volume ratios. The synthesis of superhydrophobic and superhydrophilic membranes is of paramount importance for environmental remediation. To ensure the efficiency of the filtration process, the selection of a suitable

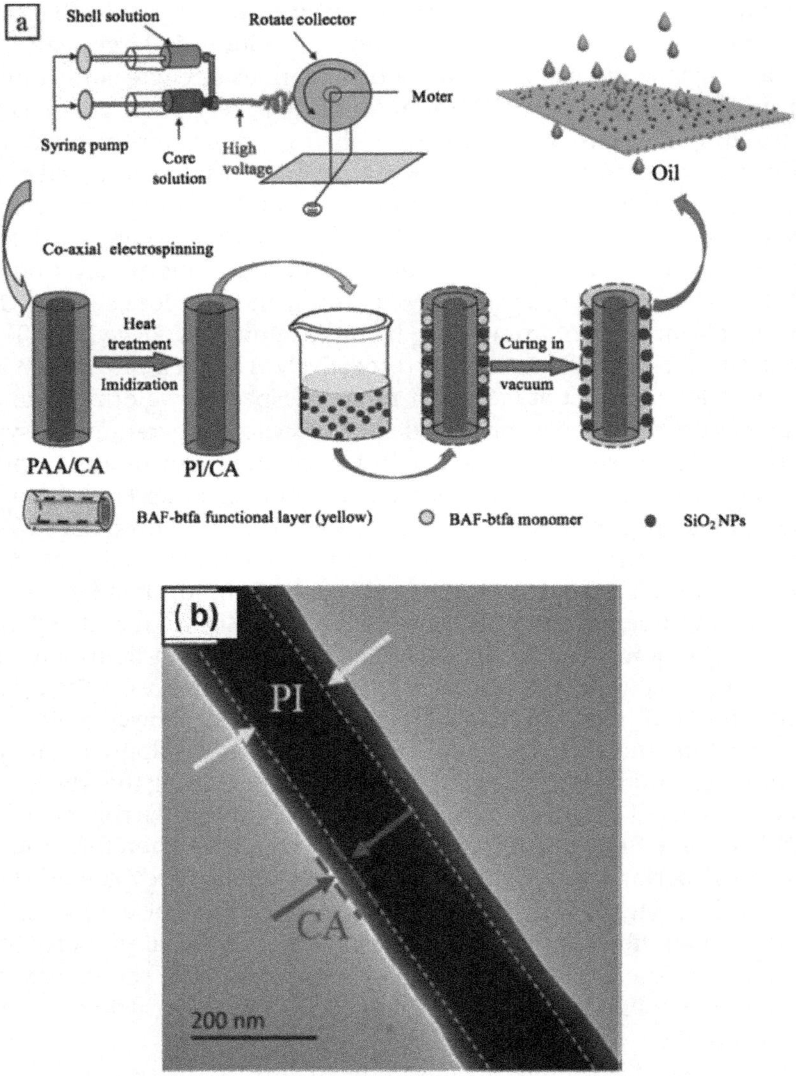

Figure 7.10 (a) Schematic showing the steps involved in the fabrication of core–shell electrospun nanofiber F-PB/SNP modified PI/CA nanofibrous membranes prepared through coaxial electrospinning: cellulose acetate (CA-shell), polyamide acid (PAA-core), polyimide (PI-core) and (b) TEM image of a (flexible) PI/CA nanofiber (ODA/PDA molar ratio of 10 : 0). Reprinted from *Separation and Purification Technology*, **177**, W. Ma, Z. Guo, J. Zhao, Q. Yu, F. Wang, J. Han, H. Pan, J. Yao, Q. Zhang, S. K. Samal, S. C. De Smedt, C. Huang, Polyimide/cellulose acetate core/shell electrospun fibrous membranes for oil–water separation, 71–85, Copyright 2017, with permission from Elsevier.

membrane is essential. There are some key parameters related to the membrane, which are dependent on the characteristics of the feed stream used in the system. Primarily, this depends upon the pore size, porosity, surface energy, and also on the nature of the membrane, whether it is hydrophobic or hydrophilic. The use of nanofiber webs as a filtering medium is well established. Due to the small size of the fibers London-Van Der Waals forces are an important method of adhesion between the fibers and the captured materials. The review by Wang *et al.* provides the state-of-the art developments on electrospun nanofiber materials as versatile medium for effective oil/water separation.[97] Table 7.4 lists some of the recent electrospun superhydrophobic nanofibers along with their features and applications published during 2016/17.[74,96,98–110]

A novel membrane with hierarchical structure consisting of a top layer of nano/micro-beaded porous polyvinylidene fluoride (PVDF)-silica composite and a PVDF nanofibrous intermediate layer on a non-woven support with switchable super-wettability for oil and water has been reported.[98] The surface of the composite membrane was modified to exhibit high surface free energy. The membrane was superamphiphilic in air, superoleophobic under water and superhydrophobic under oil (Figure 7.11). This membrane is versatile and can treat various types of oil/water mixtures, from simple layered solutions to emulsions (including surfactant-free or surfactant-stabilized oil-in-water and water-in-oil emulsions) without external driving force or under an ultralow pressure (0.1 bar) in a cross-flow filtration process. The membrane exhibited a flux up to 2000 l m^{-2} h and high separation efficiency (>99.99%). The membrane also exhibited excellent robustness under harsh conditions, including strong acidic or alkaline solutions, hot water and petroleum and presents remarkable antifouling and easy-cleaning properties and can be a promising membrane for industrial application if it is mass produced.

Superhydrophobic, reversibly elastic, moldable, and electrospun (SupREME) fibers exhibiting multimodal functions for arranging spatial responses of substances with distinct affinities to oil phases have been fabricated by coaxially electrospinning polysulfone and poly(glycerol sebacate) (PGS), followed by a thermal process.[103] The multimodal functions of SupREME fibers include superhydrophobicity/oleophilicity, reversible elasticity, flexibility, oil absorbing property and sealing property.[103] Haddad *et al.*[111] used design of experiment for the fabrication of PVDF-nano Al_2O_3 electrospun mats. Solution parameters were found to have a much greater effect compared to the process parameters considered. PVDF concentration was also an important factor for all the responses, and the ratio of acetone/DMF determined the fiber diameter and the WCA. The review article by Ma *et al.*[112] discusses in detail the preparation of electrospun superhydrophobic/superoleophilic fibers/surfaces/fibrous membranes for oil/water separation, including the preparation of electrospun fibers for oil absorption and oil filtration. The review article also focuses on the recent developments on superoleophobic/superhydrophilic electrospun fibers and their application for oil–water separation.[112]

Table 7.4 Different electrospun nanofiber filter membranes for various applications.

S. no.	System	Parameters used	Properties achieved	Reference
1.	Polystyrene + solvents tetrabutylammonium perchlorate	4–10 kV; 15 cm distance	Higher adsorption capacity	74
2.	Polystyrene/CNT	13–15 kV; 15 cm distance	Good mechanical properties and reusability for oil removal	96
3.	Polyvinylidene fluoride (PVDF)–silica composite nano/micro-beaded top layer and a PVDF nanofibrous intermediate layer on a non-woven support	28–30 kV; 7 cm distance; spinneret speed 3 cm min^{-1}	In-air super-amphiphilic, underwater super-oleophobic, under-oil super-hydrophobic Excellent mechanical properties, good flexibility, robustness in oil	98
4.	PVDF-*co*-hexafluoropropylene + TiO_2	18–24 kV; 10–15 cm distance	High flux membrane distillation	99
5.	PVDF-nano alumina particles + cationic surfactant (HTAB)	10–18 kV; 15 cm distance	Heavy metal lead (Pb) removal	100
6.	PTFE@PVA hybrid nanofibrous calcination at 380 °C	16 kV; 15 cm distance	Higher tensile strength (19.7 MPa)	101
7.	Aminated polyacrylonitrile + Ag nanoclusters	25 kV; 15 cm distance	Liquid entry pressure: 175 ± 3 kPa, similar to the conventional PVDF commercial membrane	102
8.	Poly(glycerol sebacate) PGS (core) and PSF (sheath), followed by a thermal baking process	14 kV; 17 cm distance	Multimodal functions as absorbents for selectively separating oil-based substances and as sealants for temporarily blocking the passage/leakage of aqueous phases	103
9.	Blend electrospinning of poly(vinylidene fluoride) (PVDF) mixed with silane coupling agent modified SiO_2 nanoparticles	16 kV; 15 cm distance	Contact angle 160°	104
10.	Expanded polystyrene (EPS)/zeolite	10–20 kV	Contact angle >150°	105
11.	Cellulose acetate nanofibers (CA-NF) modified with perfluorooctyltriethoxysilane	15 kV; 15 cm distance	Durable, robust and water resistant	106

12.	3D polystyrene superhydrophobic miniature box	25 kV; 15 cm distance	Effectively separates heavy oil–water and light oil–water mixtures by a gravity-driven process, without requiring a tilting device	107
13.	PVDF-HFP/SiNPs hybrid membrane	16 kV; 15 cm distance	Highest permeate flux: 48.6 kg m^{-2} Rejection of NaCl: 99.99%	108
14.	PDMS/PMMA/THF/DMF	7 kV; 16 cm distance	Contact angle 163°, tensile strength of 1833.89 kPa, average membrane flux of 39.61 l/m^2h, excellent salt rejection efficiency (99.96%) Membrane distillation	109
15.	Poly(vinylidene fluoride-*co*-hexafluoropropene) (PVDF-HFP) membrane + PDMS/PVDF + polymeric organosilicon compound	18 kV; 15 cm distance	Contact angle 155.4°, dyeing wastewater treatment + antifouling property	110

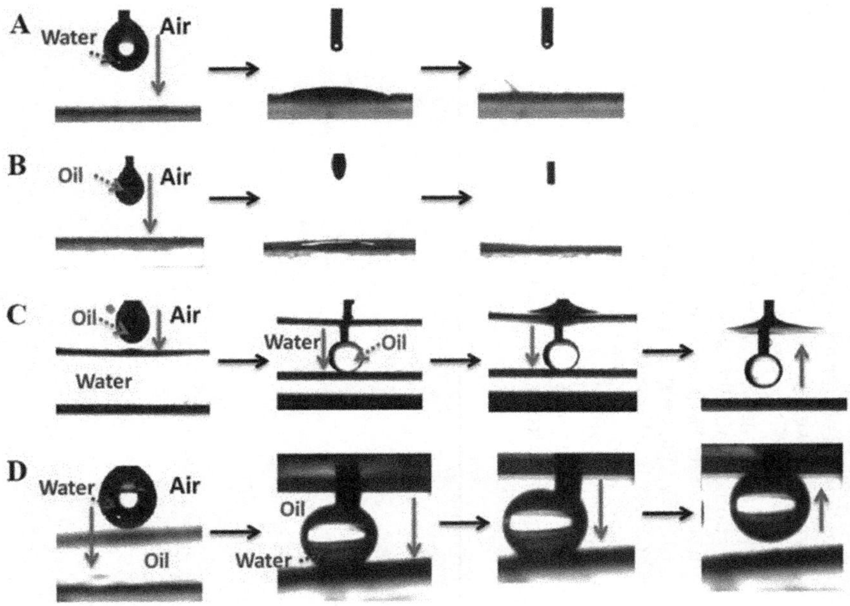

Figure 7.11 Behaviors of (A) a water droplet and (B) an oil droplet in air, (C) an underwater oil droplet, and (D) an under-oil water droplet on the M-PVDF membrane surface. Reprinted from *Journal of Membrane Science*, **543**, Y. Liao, M. Tian, R. Wang, A high-performance and robust membrane with switchable super-wettability for oil/water separation under ultralow pressure, 123–132, Copyright 2017, with permission from Elsevier.

7.6 Industrial Applications of Electrospun Superhydrophobic Nanofibers

Out of the many papers reported in recent years, two of the reports seem to be interesting and promising in view of the simplicity and industrial application of electrospun nanofibers for oil spill clean-up.[102,107] The traditional method of removal of oil spill involves the collection of the bulk of oil/water first and then pour it over to a 2D superhydrophobic filtration membrane. On the other hand, 3D superhydrophobic absorption sponges require a squeezing process after each absorption cycle, which is inconvenient and time-consuming. To solve these problems, Li *et al.*[107] have designed a methodology for continuous oil spill removal from seawater. Polystyrene nanofibers were electrospun using a solution containing 12 wt% polystyrene in ethyl acetate on a stainless steel mesh (300-mesh size, 8 cm × 8 cm) using an applied voltage of 25 kV and distance of 15 cm and then folded in the form of a box (4 cm × 4 cm) under an infrared lamp. The as-prepared coated mesh was superoleophilic and superhydrophobic with contact angle of 154 ± 1° and sliding angle of 4 ± 1° in air. 3D superhydrophobic box can separate oils under harsh conditions. In addition, Li *et al.*[107] have performed a continuous

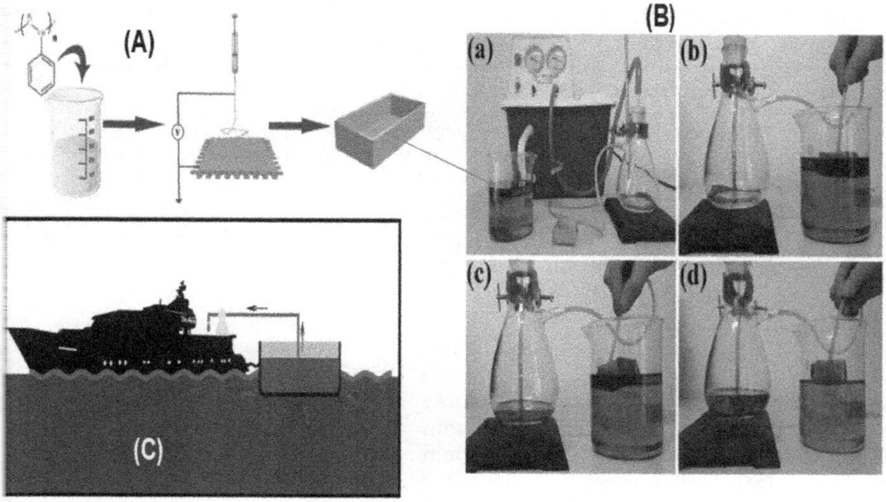

Figure 7.12 (A) Schematic for the fabrication of the PS-coated mesh by electrospinning. (B) Continuous oil removal process from water using kerosene as an example. (C) Scheme illustrating a ship with large oil collection equipment (superhydrophobic mesh collector) to clear the large area where the oil spill floats (red crosshatched area) on the sea surface. Reprinted from ref. 107 with permission from the Royal Society of Chemistry.

oil/water separation using the superhydrophobic and superoleophilic 3D miniature box (Figure 7.12), thus demonstrating its potential applications in oil spill clean-up. This design may be the most suitable methodology to solve the issue of pollution due to frequent oil spills on sea surfaces.

Li *et al.*[102] have reported the synthesis of nanofibrous membrane containing immobilized Ag nanocluster for oil/water separation. To achieve this, a combination of the amination of electrospun polyacrylonitrile (APAN) nanofibers and immobilization of a Ag nanocluster with an electroless plating technique, followed by surface modification using *n*-hexadecyl mercaptan has been used. By introducing the hierarchically rough structure and low surface energy, APAN nanofibrous membranes exhibited superhydrophobicity with a high water contact angle of 171.1 ± 2.3° and superhydrophilicity with a WCA of 0° and low water contact angle hysteresis. The selective wettability of the surface was manipulated by varying the electroless Ni plating time and most importantly it exhibited extremely high liquid entry pressure of water. The developed membrane exhibited excellent oil–water separation efficiency in both hyper saline environment and broad pH range conditions along with excellent recyclability. Thus this system is a promising candidate for remediation of industrial oil-contaminated water treatment and marine spilt oil clean-up.

In recent years the electrospinning versatility has started to gain industrial market. There has been a transition of electrospinning technology from

Table 7.5 List of companies selling pilot plant/industrial-scale electrospinning machines.

S. no.	Company name	Country	Website
1.	Elmarco	Czech Republic	http://www.elmarco.com/
2.	SPUR	Czech Republic	http://www.spur-nanotechnologies.cz/
3.	MECC Co. Ltd	Japan	http://www.mecc.co.jp.en/
4.	Fuence	Japan	http://www.fuence.co.jp/en
5.	ANSTCO (Asian Nanostructures Technology Company)	Iran	http://anstco.com/
6.	Fnm Co. (Fanavaran Nano-Meghyas)	Iran	http://en.fnm.ir/
7.	Inovenso	Turkey	http://www.inovenso.com
8.	Yflow	Spain	http://www.yflow.com/
9.	Bioinicia	Spain	http://bioinicia.com/
10.	IME Technologies	Netherlands	http://www.imetechnologies.com
11.	Tong Li Tech Co. Ltd	China	http://www.electrospinning.com/
12.	Linari Engineering S.R.L.	Italy	http://www.linaribiomedical.com/
13.	TOPTEC	Korea	http://www.toptec.co.kr
14.	NanoStatics	USA	http://www.nanostatics.com

academic to industrial sector as evident from the review article by Persano *et al.*[113] The multi-needle and needle-free electrospinning technologies have been used to realize large-scale electrospinning. The production rate has increased from a mere mg per hour to pilot scale and current industrial-scale production is in the range of kg per hour or m min^{-1} level. Some of the leading companies manufacturing electrospun products are DuPont, Ahlstrom, Donaldson *etc.* It is estimated that more than 100 companies are selling electrospinning set-ups or products based on electrospun nanofibers. Countries like the USA, Japan, Czech Republic and Korea pioneered industrialization of the electrospinning technology. Elmarco has built industrial-scale production equipment without nozzles, needles or spinnerets using Nanospider™ technology. This technology is based upon the possibility to create Taylor cones and the subsequent flow of material not only from the tip of a capillary, but also from a thin film of a polymer solution. Table 7.5 lists some of the companies involved in manufacturing pilot/industrial-scale electrospinning equipment.

Of particular interest from the industrial point of view is the peculiar capability of electrospinning to realize complex bi- and tri-dimensional architectures in a single run, and engineering the collector geometry is strategic in this respect. For scale up process, conveyor belts made of insulating or metallic materials sliding on grounded metallic guides can be used as collectors.

This strategy in addition to fiber alignment, offers significant possibilities of scaling up the material production. The integration of motorized conveyor guides can indeed allow continuous operation processing over different substrates.[114] Within the environment protection area, filtration, and purification materials based on nanofiber mats are undoubtedly the currently most industrially advanced products.

Among commercially available industrial-scale electrospinning set-up, Inovenso Ltd. supplies an apparatus producing fibers with diameter down to 50 nm with a throughput area up to 5000 m^2 daily or to production rates up to about 200 g h^{-1} for nanofiber coatings.[115] Various electrospinning industrial plants, some with production capability up to several kg h^{-1} have been reported in the literature.[116,117] It should be noted that most of these applications have not reached their industry level, but just at a laboratory research and development stage. However, their promising potential is believed to be attracting attentions and investments from academia, governments, and industry all over the world.

Despite the many positive prospects more intense efforts are still required. There is a need for (i) increasing the relative low output production rate of commercial electrospinning set-ups; (ii) fabrication of reproducible and standardized hybrid functional nanofibers (iii) and to improve the dispersion of nanomaterials into the bulk or onto surface of the functional nanofibers.

Although multi-spinneret techniques can yield high mass throughput, the cost of this new equipment may prevent its increased use in manufacturing. The fabrication of 3D polymer nanofiber/nets using a recently developed advanced electrospinning/netting technique holds future prospects as it offers one step fabrication of 3D nanofiber/nets.[118] Finally, with respect to filtration applications, potential development in this area may include employing various polymers and/or post-treatment methods to control the formation of pores on fibers, along with a better understanding of transport mechanism through fibers with microporous rough surfaces.[119]

7.7 Concluding Remarks and Future Prospects

Electrospinning is a versatile and facile technique for the fabrication of superhydrophobic nanofibers. A large number of superhydrophobic nanofibers and composites based on polystyrene, PAN, PVDF, CA and fillers like alumina, silica, CNTs have been prepared by electrospinning. These nanofibers have shown excellent sorption properties and is having potential application for oil spill clean-up. Electrospun superhydrophobic nanofibers have shown excellent properties for oil spill clean-up, desalination and removal of pollutants. Some of the nanofibers have shown excellent properties and have great potential for industrial applications. More intense efforts should be focused to commercialize electrospun superhydrophobic nanofibers and industry-academia should join hands in this endeavor.

Acknowledgements

The author acknowledges the encouragement and support received from Director, CSIR-NAL and Head, Surface Engineering Division, CSIR-NAL. The author would also like to thank Mr Jerald and Ms Lashmi for help in formatting the references and tables. Special thanks to Mr B. Shivaram for providing the required references.

References

1. W. Gilbert and D. Magnete, *Translated 1893 from Latin to English by Paul Fleury Mottelay*, Dover Publications, New York, 1991.
2. A. Nollet, *Recherches Sur Les Causes Particulieres des Phenomenes Electriques*, les Freres Guerin, Paris, 1749.
3. J. F. Cooley, US Pat., 692 63, 1902.
4. J. M. Morton, US Pat., 0 705 691, 1902.
5. J. Zeleny, *Phys. Rev.*, 1914, **3**, 69.
6. J. Zeleny, *Phys. Rev.*, 1917, **10**, 1.
7. A. Formhals, US Pat., 1 975 504, 1934.
8. J. S. Kim and D. H. Reneker, *Polym. Eng. Sci.*, 1999, **39**, 849.
9. D. H. Reneker and A. L. Yarin, *Polymer*, 2008, **49**, 2387.
10. X. Fang and D. H. Reneker, *J. Macromol. Sci., Part B: Phys.*, 1997, **36**, 169.
11. H. R. Darrell and C. Iksoo, *Nanotechnology*, 1996, **7**, 216.
12. G. Srinivasan and D. H. Reneker, *Polym. Int.*, 1995, **36**, 195.
13. J. Doshi and D. H. Reneker, *J. Electrost.*, 1995, **35**, 151.
14. N. Tucker, J. J. Stanger, M. P. Staiger, H. Razzaq and K. Hofman, *J. Eng. Fibers Fabr.*, 2012, **7**, 63.
15. M. Mirjalili and S. Zohoori, *J. Nanostruct. Chem.*, 2016, **6**, 207.
16. X. Wang, B. Ding, G. Sun, M. Wang and J. Yu, *Prog. Mater. Sci.*, 2013, **58**, 1173.
17. W. E. Teo and S. Ramakrishna, *Nanotechnology*, 2006, **17**, 14.
18. F. E. Ahmed, B. S. Lalia and R. Hashaikeh, *Desalination*, 2015, **356**, 15.
19. S. S. Ray, S.-S. Chen, C.-W. Li, N. C. Nguyen and H. T. Nguyen, *RSC Adv.*, 2016, **6**, 85495.
20. A. Haider, S. Haider and I.-K. Kang, *Arabian J. Chem.*, 2015, DOI: 10.1016/j.arabjc.2015.11.015.
21. S.-H. Tan, R. Inai, M. Kotaki and S. Ramakrishna, *Polymer*, 2005, **46**, 6128.
22. S. Megelski, J. S. Stephens, D. B. Chase and J. F. Rabolt, *Macromolecules*, 2002, **35**, 8456.
23. C. J. Buchko, L. C. Chen, Y. Shen and D. C. Martin, *Polymer*, 1999, **40**, 7397.
24. J. Miao, M. Miyauchi, T. J. Simmons, J. S. Dordick and R. J. Linhardt, *J. Nanosci. Nanotechnol.*, 2010, **10**, 5507.
25. P. Gibson, S. Gibson and H. D. Rivin, *Colloids Surf., A*, 2001, **187**(188), 469.

26. J. M. Deitzel, J. Kleinmeyer, J. K. Hirvonen and T. N. C. Beck, *Polymer*, 2001, **42**, 8163.
27. A. Senthamizhan, B. Balusamy and T. Uya, *Anal. Bioanal. Chem.*, 2016, **408**, 1285.
28. L. A. Mercante, V. P. Scagion, F. L. Migliorini, L. H. C. Mattoso and D. S. Correa, *Trends Anal. Chem.*, 2017, **91**, 91.
29. B. Ding, M. Wang, J. Yu and G. Sun, *Sensors*, 2009, **9**, 1609–1624.
30. X. Shi, W. Zhou, D. Ma, Q. Ma, D. Bridges, Y. Ma and A. Hu, *J. Nanomater.*, 2015, **140716**, 20.
31. J. Miao, M. Miyauchi, T. J. Simmons, J. S. Dordick and R. J. Linhardt, *J. Nanosci. Nanotechnol.*, 2010, **10**, 5507–5519.
32. M. Afshari, *Electrospun Nanofibers*, Woodhead Publishing, Amsterdam, 2016, vol. 1.
33. K. Graham, M. Ouyang, T. Raether, T. Grafe, B. McDonald and P. Knauf, *Fifteenth Annual Technical Conference & Expo of the American Filtration & Separations Society*, Galveston, April 2002.
34. P. Gibson, H. S. Gibson and D. Rivin, *Colloids Surf., A*, 2001, **187**(188), 469.
35. S. J. Kwoun, R. M. Lec, B. Han and F. K. Ko, *Proceedings of the 2000 IEEE/EIA International Frequency Control Symposium and Exhibition*, 2000, pp. 52–57.
36. S. J. Kwoun, R. M. Lec, B. Han and F. K. Ko, *Proceedings of the 2001 IEEE/EIA International Frequency Control Symposium and Exhibition*, 2001, pp. 9–10.
37. S. H. Lee, B. C. Ku, X. Wang, L. A. Samuelson and J. Kumar, *Mater. Res. Soc. Symp. Proc.*, 2001, **708**, 403.
38. D. Smith and D. H. Reneker, PCT/US, 00/27737, 2001.
39. L. Dimesso, C. Spanheimer, T. Nguyen, R. Hausbrand and W. Jaegarmann, *EPJ Web of Conferences*, EDP sciences, 2012.
40. T. A. Arun, A. A. Madhavan, D. K. Chacko, G. S. Anjusree, T. G. Deepak, S. Thomas, S. V. Nair and A. S. Nair, *Dalton Trans.*, 2014, **43**, 4830.
41. M. Yanilmaz and X. Zhang, *Polymers*, 2015, 7, 629.
42. R. Augustine, P. Dan, A. Sosnik, N. Kalarikkal, N. Tran, B. Vincent, S. Thomas, P. Menu and D. Rouxel, *Nano Res.*, 2017, **10**, 3358.
43. B. Joseph, A. George, S. Gopi, N. Kalarikkal and S. Thomas, *Int. J. Pharm.*, 2017, **524**, 454.
44. D. A. Gopakumar, D. Pasquini, M. A. Henrique, L. C. De Morais, Y. Grohens and S. Thomas, *ACS Sustainable Chem. Eng.*, 2017, **5**, 2026.
45. T. Krishnamoorthy, V. Thavasi, G. M. Subodh and S. Ramakrishna, *Energy Environ. Sci.*, 2011, **4**(8), 2807.
46. Z.-M. Huang, Y.-Z. Zhang, M. Kotaki and S. Ramakrishna, *Compos. Sci. Technol.*, 2003, **63**, 2223.
47. B. Ding, M. Wang, X. Wang, J. Yu and G. Sun, *Mater. Today*, 2010, **13**, 16.
48. H. Y. Erbil, A. L. Demirel, Y. Avci and O. Mert, *Science*, 2003, **299**, 1377.
49. V. A. Lifton, J. A. Taylor, B. Vyas, P. Kolodner, R. Cirelli, N. Basavanhally, A. Papazian, R. Frahm, S. Simon and T. Krupenkin, *Appl. Phys. Lett.*, 2008, **93**, 043112.

50. M. Callies and D. Quere, *Soft Matter*, 2005, **1**, 55.
51. H. Q. Cao, H. Zheng, K. Y. Liu and J. H. Warner, *ChemPhysChem*, 2010, **11**, 489.
52. M. Ma, M. Gupta, Z. Li, L. Zhai, K. K. Gleason, R. E. Cohen, M. F. Rubner and G. C. Rutledge, *Adv. Mater.*, 2007, **19**, 255.
53. W. Barthlott and C. Neinhuis, *Planta*, 1997, **202**, 1.
54. L. Feng, S. Li, H. Li, J. Zhai, Y. Song, L. Jiang and D. Zhu, *Angew. Chem., Int. Ed.*, 2002, **41**, 1221.
55. S. S. Latthe, A. B. Gurav, C. S. Maruti and R. S. Vhatkar, *J. Surf. Eng. Mater. Adv. Technol.*, 2012, **2**, 76.
56. X. M. Li, D. Reinhoudt and M. C. Calama, *Chem. Soc. Rev.*, 2007, **36**(8), 1350.
57. J. Genzer and K. Efimenko, *Biofouling*, 2006, **22**(5), 339.
58. L. Gao, T. J. McCarthy and X. Zhang, *Langmuir*, 2009, **25**(24), 14100.
59. P. Roach, N. J. Shirtcliffe and M. I. Newton, *Soft Matter*, 2008, **4**(2), 224.
60. K. Liu, X. Yao and L. Jiang, *Chem. Soc. Rev.*, 2010, **39**(8), 3240.
61. Z. Guo, W. Liu and B. L. Su, *J. Colloid Interface Sci.*, 2011, **353**(2), 335.
62. J. T. Simpson, S. R. Hunter and T. Aytug, *Rep. Prog. Phys.*, 2015, **78**(8), 086501.
63. J. Wu, N. Wang, L. Wang, H. Dong, Y. Zhao and L. Jiang, *ACS Appl. Mater. Interfaces*, 2012, **4**, 3207.
64. S. Megelski, J. S. Stephens, D. B. Chase and J. F. Rabolt, *Macromolecules*, 2002, **35**, 8456.
65. P. Lu and Y. Xia, *Langmuir*, 2013, **29**, 7070.
66. R. M. Nezarati, M. B. Eifert and E. Cosgriff-Hernandez, *Tissue Eng.*, 2013, **19**, 810.
67. C. L. Casper, J. S. Stephens, N. G. Tassi, D. B. Chase and J. F. Rabolt, *Macromolecules*, 2004, **37**, 573.
68. J. Y. Park and I. H. Lee, *J. Nanosci. Nanotechnol.*, 2010, **10**, 3473.
69. J. T. McCann, M. Marquez and Y. Xia, *J. Am. Chem. Soc.*, 2006, **128**, 1436.
70. D. J. Lin, H. H. Chang, T. C. Chen, Y. C. Lee and L. P. Cheng, *Eur. Polym. J.*, 2006, **42**, 1581.
71. Z. Qi, H. Yu, Y. Chen and M. Zhu, *Mater. Lett.*, 2009, **63**, 415.
72. Z. Wei, Q. Zhang, L. Wang, X. Wang, S. Long and J. Yang, *Colloid Polym. Sci.*, 2013, **291**, 1293.
73. K. A. G. Katsogiannis, G. T. Vladisavljević and S. Georgiadou, *Eur. Polym. J.*, 2015, **69**, 284.
74. P.-Y. Chen and S.-H. Tung, *Macromolecules*, 2017, **50**(6), 2528.
75. W. Ma, Z. Guo, J. Zhao, Q. Yu, F. Wang, J. Han, H. Pan, J. Yao, Q. Zhang, S. K. Samal, S. C. De Smedt and C. Huang, *Sep. Purif. Technol.*, 2017, **177**, 71.
76. X. Feng and L. Jiang, *Adv. Mater.*, 2006, **18**, 3063.
77. K. Liu, X. Yao and L. Jiang, *Chem. Soc. Rev.*, 2010, **39**, 3240.
78. Z. Xue, S. Wang, L. Lin, L. Chen, M. Liu, L. Feng and L. Jiang, *Adv. Mater.*, 2011, **23**, 4270.
79. B. Wang and Z. Guo, *Chem. Commun.*, 2013, **49**, 9416.

80. J. Zhang and S. Seeger, *Adv. Funct. Mater.*, 2011, **21**, 4699.
81. S.-J. Choi, T.-H. Kwon, H. Im, D.-I. Moon, D. J. Baek, M.-L. Seol, J. P. Duarte and Y.-K. Choi, *ACS Appl. Mater. Interfaces*, 2011, **3**, 4552.
82. *Before and after an Oil Spill: The Arthur Kill*, ed. J. Burge, The State University, Rutgers, 1994.
83. A. A. Allen and R. J. Ferek, *Advantages and Disadvantages of Burning Spilled Oil, International Oil Spill Conference Proceedings*, Washington, March 1993, pp. 765–772.
84. S. Kleindienst, J. H. Paul and S. B. Joye, *Nat. Rev. Microbiol.*, 2015, **13**, 388.
85. M. Fingas, *Oil Spill Science and Technology*, Elsevier, Amsterdam, 2017.
86. A. Raza, J. Ge, Y. Si, J. Yu, G. Sun and B. Ding, in *Nanostructure Science and Technology*, ed. B. Ding and J. Yu, Springer-Verlag, Berlin Heidelberg, 2014, vol. 1, ch. 17, pp. 433–447.
87. A. Bayat, S. F. Aghamiri, A. Moheb and G. R. Vakili-Nezhaad, *Chem. Eng. Technol.*, 2005, **28**(12), 1525.
88. X. F. Sun, R. Sun and J. X. Sun, *J. Agric. Food Chem.*, 2002, **50**, 6428.
89. G. Deschamps, H. Caruel, M. E. Borredon, C. Bonnin and C. Vignoles, *Environ. Sci. Technol.*, 2003, **37**, 1013.
90. M. M. Radetic, D. M. Jocic, P. M. Jovancic, Z. L. Petrovic and H. F. Thomas, *Environ. Sci. Technol.*, 2003, **37**, 1008.
91. Q. F. Wei, R. R. Mather, A. F. Fotheringham and R. D. Yang, *Mar. Pollut. Bull.*, 2003, **46**, 780.
92. A. Bayat, S. F. Aghamiri, A. Moheb and G. R. Vakili, *Chem. Eng. Technol.*, 2005, **28**, 1525.
93. J. Wu, N. Wang, L. Wang, H. Dong, Y. Zhao and L. Jiang, *ACS Appl. Mater. Interfaces*, 2012, **4**, 3207.
94. Z. Jiang, L. D. Tijing, A. Amarjargal, C. H. Park, K.-J. An, H. K. Shon and C. S. Kim, *Composites, Part B*, 2015, **77**, 311.
95. S. Gupta and N.-H. Tai, *J. Mater. Chem. A*, 2014, **4**, 1550.
96. J. Wu, A. K. An, J. Guo, E.-J. Lee, M. U. Farid and S. Jeong, *Chem. Eng. J.*, 2017, **314**, 526.
97. X. Wang, J. Yu, G. Sun and B. Ding, *Mater. Today*, 2016, **19**, 403.
98. Y. Liao, M. Tian and R. Wang, *J. Membr. Sci.*, 2017, **543**, 123.
99. S. Mahdi, S. Shahabadia, H. Rabieec, S. M. Seyedib, A. Mokhtared and J. A. Branta, *J. Membr. Sci.*, 2017, **537**, 140.
100. H. Attia, S. Alexander, C. J. Wright and N. Hilal, *Desalination*, 2017, **420**, 318.
101. W. Qing, X. Shi, Y. Deng, W. Zhang, J. Wang and C. Y. Tang, *J. Membr. Sci.*, 2017, **540**, 354.
102. X. Li, M. Wang, C. Wang, C. Cheng and X. Wang, *ACS Appl. Mater. Interfaces*, 2014, **6**(17), 15272.
103. S. Lee, B. Kim, S.-H. Kim, E. Kim and J.-H. Jang, *Adv. Funct. Mater.*, 2017, **25**, 1702310.
104. H. Sun, Y. Xu, Y. Zhou, W. Gao, H. Zhao and W. Wang, *J. Appl. Polym. Sci.*, 2017, **134**(13), 44501.

105. A. S. Oluwagbemiga, D. E. O. Dare, A. M. M. Titus, A. A. Kehinde and P. O. Aiyedun, *Phys. Chem. Earth*, 2016, **92**, 7.
106. O. Arslan, Z. Aytac and T. Uyar, *ACS Appl. Mater. Interfaces*, 2016, **8**(30), 19747.
107. J. Li, P. Guan, M. Li, Y. Zhang, P. Cheng and R. Jia, *New J. Chem.*, 2017, **41**(12), 4862.
108. D. Hou, D. Lin, C. Ding, C. Wang and D. Wang, *Sep. Purif. Technol.*, 2017, **189**, 82.
109. L. F. Ren, F. Xia, J. Shao, X. Zhang and J. Li, *Desalination*, 2017, **404**, 155.
110. A. K. An, E. J. Guo, E.-J. Lee, S. Jeong, Y. Zhao, Z. Wang and T. Leiknes, *J. Membr. Sci.*, 2017, **525**, 57.
111. Z. A. A. Haddad, E. Svinterikosa and I. Zuburtikudis, *Colloids Surf., A*, 2017, **523**, 81.
112. W. Ma, Q. Zhang, D. Hua, R. Xiong, J. Zhao, W. Rao, S. Huang, X. Zhan, F. Chen and C. Huang, *RSC Adv.*, 2016, **6**(16), 12868.
113. L. Persano, A. Camposeo, C. Tekmen and D. Pisignano, *Macromol. Mater. Eng.*, 2013, **298**, 504.
114. W. Dayong, H. Wang and J. Cao, WO Pat., 48644, 2010.
115. Inovenso Ltd., *Industrial Machines*, http://www.inovenso.com/products/industrial-nanofiber-machines, accessed September, 2017.
116. A. Varesano, F. Rombaldoni, G. Mazzuchetti, C. Tonina and R. Comotto, *Polym. Int.*, 2010, **59**, 1606.
117. C. J. Luo, S. D. Stoyanov, E. Stride, E. Pelanb and M. Edirisinghe, *Chem. Soc. Rev.*, 2012, **41**, 4708.
118. X. Wang, B. Ding, G. Sun, M. Wang and J. Yu, *Prog. Mater. Sci.*, 2013, **58**, 1173.
119. B. Kannan, H. Cha and I. C. Hosie, in *Nano-size Polymers*, ed. S. Fakirov, Springer, Cham, 2016, vol. 1, pp. 309–342.

CHAPTER 8

Large-scale Production of Electrospun-based Mat to Explore in Electronics and Sensors

MAILIN MISSON[a] AND HU ZHANG*[b]

[a]Bioprocess Engineering Research Group, Biotechnology Research Institute, Universiti Malaysia Sabah, Jalan UMS, 88400, Kota Kinabalu, Sabah, Malaysia; [b]School of Chemical Engineering, The University of Adelaide, Adelaide, SA 5000, Australia
*E-mail: hu.zhang@adelaide.edu.au

8.1 Introduction

Nanofiber is a nanostructured material ranging from micro- to nano-scale size fabricated from synthetic or natural polymer or a combination of both. Nanofiber has been widely used in voluminous applications driven by its simplicity and cost-effective preparation.[1] A surface modification introduces functional groups on nanofiber which beneficial for biomolecules adsorption like DNAs, proteins and enzymes. Besides a large surface area that allows sufficient molecule binding, nanofiber is also associated with high porosity and interconnectivity to endow low hindrance for mass transfer.[2] Due to these distinguished properties, nanofiber is promising to be incorporated in electronic devices as catalyst and also can be used for biodetection

Soft Matter Series No. 7
Electrospinning: From Basic Research to Commercialization
Edited by Erich Kny, Kajal Ghosal and Sabu Thomas

Published by the Royal Society of Chemistry, www.rsc.org

and biosensing. Nanofiber has been recognized able to improve drug encapsulation and control release efficiency that is appropriate for biodetection and biosensing applications.

8.1.1 Fabrication of Nanofibers Using Electrospinning

Electrospinning is a traditional procedure used to generate polymer nanofibers using electrostatic force to generate high repulsive forces. The electrospinning device consists of three major components: (i) a high-voltage power supply, (ii) a syringe with a control pump connected to a needle to which a high voltage is supplied, and (iii) a grounded collector (Figure 8.1). The principle behind the electrospinning technique is 'electrostatic interactions'.[1] When a high voltage is applied (~10–30 kV), the polymer solution at the tip of the needle becomes electrostatically charged and forms a cone-like structure called a 'Taylor cone'. The polymer solution has its own surface tension inside the syringe. The presence of an electric field creates a repulsive force and overcomes the surface tension of the solution leading to a polymer jet eruption from the spinneret tip.[3] The stream of the polymer solution is converted into continuous uniform fibers under optimum conditions and collected at the grounded surface. The viscosity of the polymer, conductivity, applied electric voltage and needle tip to collector distance are among the important operating conditions that could affect the structural and

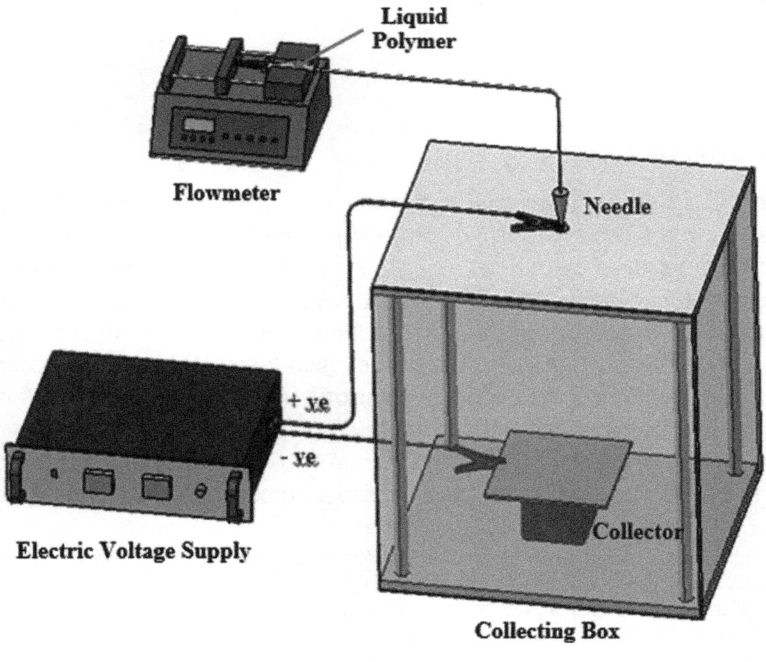

Figure 8.1 Schematic of the major components of the electrospinning device.

physicochemical properties of electrospun nanofibers. Formation of beads in nanofiber morphology can be reduced by increasing the distances from needle tip to collector.

8.1.2 Characteristics and Properties of Nanofibers

Development of electrospun nanofiber with excellent quality is a growing biotechnological interest. The performance and functionality of nanofibers are generally influenced by their surface structure. The presence of defects on a nanofiber surface can reduce the surface area of the materials for biomolecule attachment. A homogeneous and droplet-free fiber mat spread on the collector surface inspected through visual observations (Figure 8.2a). Condensed and non-homogenous fibrous mats on the collector indicate poor quality of fibers (Figure 8.2b). The homogeneous fiber mat is considered as a good quality as illustrated in Figure 8.3a. Meanwhile, the morphological examination of the uniformity of fibrous mats and the defects, including the formation of bead (Figure 8.3b) and a coil-like geometry (Figure 8.3c), can be further performed using microscopies such as a basic microscope, a scanning electron microscope (SEM) and a transmission electron microscopy (TEM).

During the fabrication of nanofibers using an electrospinning device, a range of operating variables can affect the quality and physicochemical properties of electrospun fiber mats. The polymer concentration (10–30%, w/v), electric voltage (20–30 kV) and the distance between needle tips and collector (10–25 cm) have been reported among the important variables need to be considered and optimized.[4] Among these three parameters, the concentration of polymer was found to have a strong influence towards the production of nanofibers of excellent quality. Regardless of the applied voltage or

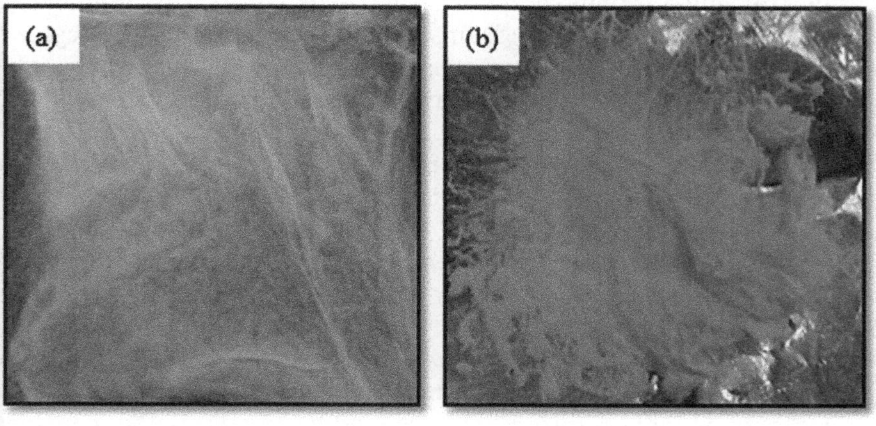

Figure 8.2 Visual observations of fibrous mats on a collector: (a) homogeneous and droplet-free fibrous mats, (b) condensed and non-homogeneous fibrous mats.

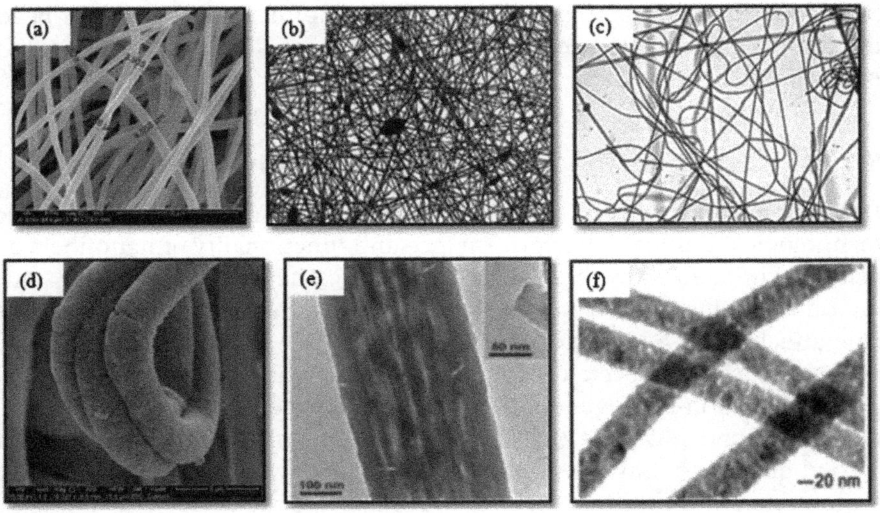

Figure 8.3 (a) SEM images of uniform size and ordered alignment of fibrous mats, (b) microscopic images at 40× objective magnification that indicate NF with bead formation and (c) disordered alignment forming a coil-like geometry, (d) structural morphology of NF after surface modification, (e) TEM image of polyvinyl alcohol (PVA) NF with G-PBASE embedded in the sidewall[41] and (f) TEM image of TiO_2 nanofibers prepared by calcining sample in air at 500 °C for 3 h.[42] Adapted from ref. 41 with permission from John Wiley and Sons, © 2010 WILEY-VCH Verlag GmbH & Co. KGaA, Weinheim[41] and D. Li, Y. Xia, *Nano Letters*, 2003, **3** (4), 555–560, Copyright 2003 American Chemical Society.[42]

distance, desirable nanofibers made from polystyrene as a precursor were obtained at 20% (w/v) concentration. If the concentration was too low, the polymer remained as liquid droplets at the collector instead of forming fiber mats. The reason behind this circumstance is associated with the degree of surface tension of the polymer. During the electrospinning process, fiber mats are produced at the collector after an electric field charges the polymer solution and creates a repulsive force. It will overcome the surface tension that leads to a polymer jet eruption from the spinneret tip.[3] Failure of the synthesis at a low polymer concentration is probably due to high surface tension associated with low viscosity of the polymer.[5] Meanwhile, condensed fibers with major defects were obtained at high concentrations of the precursor, resulting in fibers fusing together into a network rather than individual fibers and generating rough and thick mats.

An innovative core/shell nanofiber with control over morphology and function is promising with electrospinning. The possibility for fine-tuning the properties of nanofibers through surface modification is another beneficial feature in nanofiber development. The modification introduces chemical changes on the nanofiber surface, which in some cases creates destructive effects on fragile structures (Figure 8.3d). Advancements in technologies drive the development of nanofibers by providing useful insight into understanding the structure and physicochemical properties of materials. TEM imaging can elucidate the fundamental mechanisms and interfacial reactions associated with the materials, for instance, the successful synthesis of polyvinyl alcohol (PVA) nanofibers with G-PBASE embedded in the sidewall (Figure 8.3e) and a titanium oxide-filled nanofiber (Figure 8.3f).

8.2 Lab-scale Production

The simplicity and scalability of the nanofiber production process makes nanofiber an ideal choice for various applications including filtration, protective materials, biosensors, and biocomposites. The nanofiber market will reach nearly $2.0 billion in 2020 from its 2015 value of US$383.7 million (figures from BCC Research, http://www.bccresearch.com, in May 2016). Many different polymers have been successfully subjected to electrospinning, including polystyrene,[6] polyethylene terephthalate (PET)[7] and polyvinvylcarbazole.[8] The most common method for lab-scale production of nanofiber is using electrospinning. Compared to other approaches, the electrospinning technique is widely used due to its simplicity and reproducibility, less consumption of chemical precursors and economical effectiveness.

8.2.1 Nanofiber-based Electronics

Nanofibers have unique properties like high tensile strength, large surface area and porosity which suitable for electronic and sensor applications. There has been an increase in the demand for a power source of portable devices in the research of microfluidic fuel cells (MFC). Carbon nitride (CN_x) nanofibers have been used as a cathode catalyst for the application of catalysts or catalyst supports. Fuel cells show remarkable efficiency, clean emissions and a high-power density, making them an appropriate alternative power source for applications such as electric vehicles, small portable electronics and distributed power generators.[9] CN_x nanofibers as cathode catalysts permit a wide range of anolyte (formic acid) and catholyte ($KMnO_4$) flow rates during MFC operation. Optimum flow rates of fuel and oxidant are vital to maximum MFC power density. CN_x nanofibers is also tolerant towards fuel crossover effect.[10] By electrospinning the polyacryonitrile (PAN) solution containing CN_x nanoparticles, a higher power density of 3.43 mW cm^{-2} and a current density of 9.79 mA cm^{-2} were generated, comparing with 2.72 mW cm^{-2}, 6.04 mA cm^{-2} from Au electrodes and 3.09 mW cm^{-2}, 6.18 mA cm^{-2} from Pt electrodes, respectively. CN_x nanofibers allow MFC to operate at wider ranges of flow rates for fuel and oxidant.[11] Image from SEM showed that the average diameter of the nanofibers was 75 nm. The successful dispersion of CN_x on PAN nanofibers was confirmed by the TEM micrograph.

Co-electrospinning of Nafion and polyaniline (PANI) has been carried out as a catalyst support for fuel cell electrodes. PANI is another promising conducting polymer due to its high electrical conductivity, large surface area and favorable interaction with Pt. PANI, however, has poor solubility in most organic solvents, ascribing to its short and rigid polymer chains.[9] Nafion were electrospun with a small fraction of poly(ethylene oxide) (PEO), a carrier polymer that is able to enhance chain entanglements, making both Nafion and PANI solutions electrospinnable. PANI/PEO nanofibers were observed free of beads with an average diameter of 506 ± 15 nm under SEM micrographs. Nafion–PANI assembly also showed a smooth morphology with an electrical conductivity of 0.078 S cm^{-1}. Nafion has also been developed to improve proton conductivity at a low relative humidity.[12]

Makinouchi *et al.*[12] employed a novel proton conductive nanofiber composite membrane, composed of sulfonated polyimide (SPI) nanofibers and a typical polymer electrolyte, Nafion, to improve electrolyte characteristics focusing on proton conductivity at low relative humidity. The electrospun SPI nanofibers showed much higher proton conductivity along the nanofiber axis (5.1×10^{-2} S cm^{-1} at 90 °C, 95% RH) than the corresponding SPI membrane (8.3×10^{-2} S cm^{-1}). It is assumed that the SPI nanofibers formed an effective. Furthermore, in comparison with the recast-Nafion (without nanofiber), the nanofiber composite has indicated a higher proton conductivity at all range of tested temperatures between 30 and 90 °C. Water uptake is an important characteristic related to proton conductivity. Water uptake of the nanofiber composite membrane under water immersion was lower than that of the recast-Nafion. The three-dimensional nanofiber network structure helps to inhibit excessive swelling of Nafion matrix to reduce water uptake. At high relative humidity, lower water uptake was also observed making the nanofiber composite membrane more stable even during water retention.

8.2.2 Nanofiber-based Sensors

Nanofiber also has been employed in biosensing and tissue engineering. The potential of electrospun nanofibers for sensor applications is attributed by their large specific surface area, porosity and ease of functionalization. Electrospun nanofiber has been employed for disease treatment. Effective delivery of insulin or anti-diabetic drugs has been a constant challenge in pharmaceutical industry. The preparation of gelatin/insulin fibers has been reported for transbuccal insulin delivery.[13] Gelatin was electrospun into fibers and converted into a semi-interpenetrating network using polyethylene glycol diacrylate (PEG-DA) as a crosslinker. Extended release of insulin from the scaffold was observed for 4 h. The released insulin was found had successfully triggered AKT phosphorylation and resulted in enhanced transbuccal transport of insulin. The transmucosal delivery of linagliptin for the treatment of type 2 diabetes mellitus has also been reported using ultrathin nanofibers.[14] Oral formulation of an anti-diabetic drug using polymeric nanofiber was conducted by encapsulating linagliptin in poly(vinyl alcohol) (PVA) nanofiber using electrospinning technique. The encapsulation efficiency of the linagliptin was determined at 92%. In comparison with empty nanofibers, the drug-filled nanofiber was found to have higher muco-adhesion. The *in vitro* release study suggested that the nanofiber patch played an important role in control release of the drug.

Nanofibers have been recognized to play a role in drug delivery. The fabrication of biomedical core–shell fine curcumin nanofiber mats with a polymer-free core was realized using a simple coaxial spinneret electrospinning technique. This application is useful to prevent wound infections. Curcumin, a natural drug with a superior microbial activity, has been loaded inside nanofibers made up of polyvinyl alcohol (PVA) and chitosan (CS).[15] Core/shell nanofibers with a polymer-free core could improve drug bioactivity and bioavailability and prevent its initial burst of release. Bacterial inhibition was conducted against MRSA and *S. epidermidis* after six days of incubation with the observed results determined to be at 92.18 ± 0.71% and 82.08 ± 3.89%, respectively.

A curcumin based biocompatible nanofibers also had been fabricated for lead ion detection.[16] Curcumin was loaded inside a cellulose acetate nanofiber membrane by electrospinning method. Curcumin–cellulose acetate mixtures were stirred for 6 h at room temperature prior to the electrospinning process, performed in horizontal alignment using a grounded aluminum foil as a collector. The electrospun nanofiber membrane strip can be used for colorimetric sensing of lead (Pb^{2+}) with good selectivity. The visual detection limit of Pb^{2+} is 20 μM involving a color change from yellow to orange.

Electrospun poly(ε-caprolactone) (PCL) nanofibers have been used as a fibrous matrix for delivery of a reference drug, Naproxen (NAP).[17] The inclusion complex (NAP-βCD-IC) was derived by mixing NAP and beta-cyclodextrin (βCD) and inserting them into PCL *via* the electrospinning method. The formation of NAP-βCD-IC assembly promotes release of NAP from the PCL nanofibrous mat. A hydrophobic drug, hydroxypropyl-beta-cyclodextrin (HPβCD) inclusion complex (IC) of 9% sulfisoxazole (SFS), also has been inserted in hydroxypropyl cellulose (HPC) nanofibers *via* a similar technique.[10]

Enzyme immobilization through electrospun nanofibers has great potential in catalytic electrochemical biosensors. The use of carbon nanofiber-based glucose biosensors has demonstrated high sensitivity and reproducibility as well as longer lifetimes. The effective immobilization of glucose oxidase on the carbon nanofibers is important in analytical detection and food fermentation industries.[18] Immobilization of proteases also demonstrates similar properties such as extended lifetime, improved catalytic efficiency and stable half-life in both organic and aqueous solvents. Furthermore, nanofibrous enzymes have been shown to demonstrate a longer half-life than its native counterparts.[19] Nanofiber-immobilized proteases have been recognized promising for applications in wastewater treatment, detergent and leather industries.[20]

8.2.3 Electrospinning Strategies

Electrospinning generates fine and interconnected electrospun nanofibers on the ground collector. Electrospun nanofiber is normally in single fiber configuration (Figure 8.4a) with fiber diameters ranging from nanometers to a few micrometres.[25] Through advancement of technology, Nanofibers with hybrid structures such as side-by-side, core–shell and hollow structures have been innovatively invented.[28] Well-aligned and uniform side-by-side hybrid nanofiber (Figure 8.4b) allows to simultaneously carry two biomolecules on each side. This new technique can be employed using a dual-opposite-spinneret electrospinning that merges two polymer solutions into a single jet. Producing core–shell nanofibers is another innovative technique in electrospinning technology. Non-electrospinnable materials can be incorporated as a core while the shell template is made from electrospinnable polymers (Figure 8.4c). The synthesis can be done using the coaxial electrospinning process, consisting of two syringes for inner (non-electrospinnable) and outer (electrospinnable) solutions, respectively.[3] Removal of the core leaves the inner part empty generating a hollow structure inside the nanofibers. Core–shell nanofibers are suitable for biosensor application for wound healing and drug delivery. Drugs are encapsulated inside the core of electrospun nanofibers. Incorporating inorganic and organic material is a new technique to improve the robustness and mechanical strength of materials. Inserting

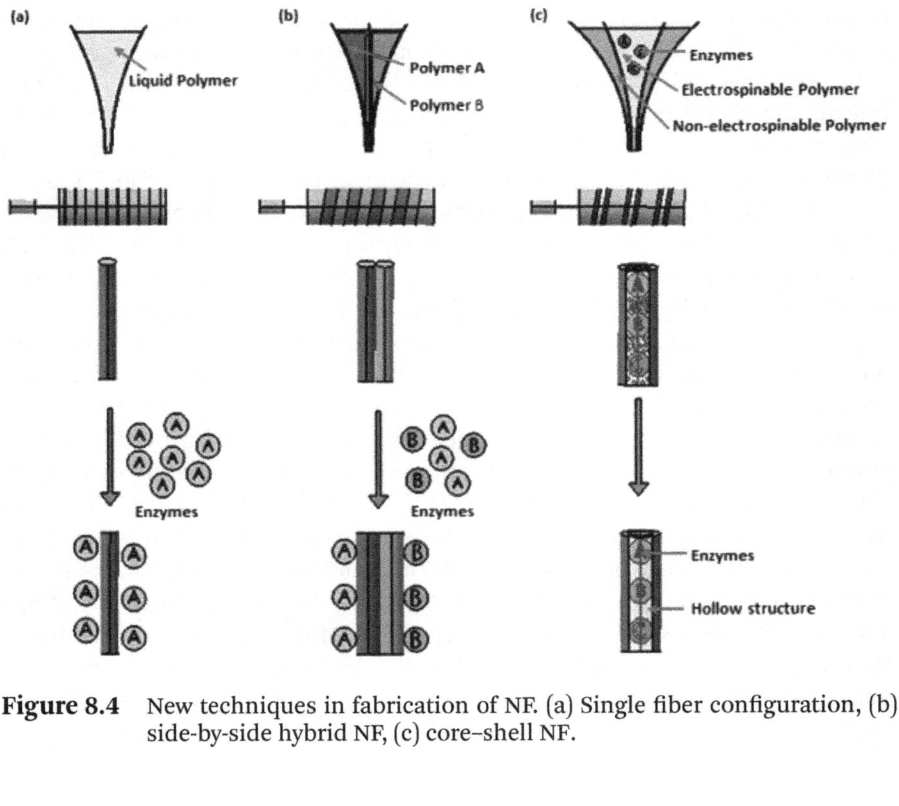

Figure 8.4 New techniques in fabrication of NF. (a) Single fiber configuration, (b) side-by-side hybrid NF, (c) core–shell NF.

silica nanoparticles inside hollow ultrafine polymer fibers is an example of the fabrication of a hybrid nanocomposite. This organic–inorganic hybrid is able to withstand shear stress during bioreaction without significant deformation.[30,31] This feature is beneficial to shelter and/or stabilize biomolecules on the hybrid nanocomposite against chemical and environmental attacks. This technique is feasible for application of nanofiber in electronic devices such as catalysts or catalyst supports.

8.2.4 Drawbacks in Lab-scale Production

Lab-scale production of nanofibers has been successful through electrospinning. However, critical challenges still remain for commercialization. Firstly, low throughput is the major hurdle for up-scaling practices. Conventional electrospinning devices have a single nozzle with a high-voltage power supply, and both efficiency and yield are very low (the needle flow rate on the order of a few μl min^{-1} and the mass production rate of 0.01–0.1 g per hour).[38] Additionally, the reproducibility of electrospun nanofibers is low, due to many interdependent parameters involved and challenges in optimization. Therefore, the productivity is uneconomical for commercial production of nanofibers. Electrospinning also depends on the dielectric properties of the polymer solution and the electrical yield. It is not applicable to use electrospinning to manufacture nanofibers from low-conductivity polymers.

Another hurdle in commercialization is optimization of rigorous parameters involved in the electrospinning process. These parameters briefly include the polymer type and solution concentration, applied voltage, temperature and effect of collector and ambient conditions. Slight alterations in these parameters will affect the morphology of the resultant nanofibers. The removal of the nanofibers from the metal-based collector without product loss is one of the key challenges in large-scale production. The collection of electrospun nanofiber requires the original thin-layer structure to be preserved. Deposit of nanofibers on a large plate or a rotating drum is usually employed for electrospinning or centrifugal spinning, while removal of the nanofibers from the surface often results in product loss.

The usage of electrospinning also poses various safety hazards. Firstly, a high electric voltage ranging from several tens of kVs is needed to power the polymer jet emitting the polymer solutions or melts. Usage of such a high voltage is a safety issue as contacting the charged jet carelessly can lead to serious injury. Secondly, the use of certain polymers may pose health hazards during the electrospinning process as it emits harmful gases either from organic solvents or polymer melts. For example, polytetrafluoroethylene (PTFE) emits gaseous products at high temperatures which may cause lung injury.[21] Concurrently, the use of certain highly flammable solvents such as dimethylformamide (DMF), hexafluoroisopropanol (HFIP) and tetrahydrofuran (THF) is a serious safety hazard.[22]

8.3 Methods of Large-scale Nanofiber Production

Large-scale manufacture of nanofibers has to overcome the drawbacks of lab-scale instruments, more importantly, practical operations and quality control have to be considered in manufacture process. The operations have to allow the operation window of process parameters to be large so that the operations can be flexible, meanwhile, the device has to be easily cleaned and maintained. Product quality control should be implemented in the device for guaranteeing uniform size and morphology of fibers. All these issues demand a thorough understanding of the principles for the process, and often multi-disciplinary collaborative inputs are essential for a large-scale manufacturing process.

8.3.1 Modified Electrospinning Process for Scalable Production of Nanofibers

The simple up-scaling method of current electrospinning process is to expand a single nozzle into multi-nozzles or to increase the number of spinning points. The maximum throughput for multi-needles is reported to be about 6.5 kg h^{-1} by Finetex Inc.[23] An industrial production system requires thousands of needles in parallel to generate a large load. All needles should be quality-reliable, operation-consistent, and maintenance-simple. Meanwhile, multi-needles have to avoid jet interference and perturbations between jets or from neighboring needles. Microfabrication technology allows fabrication of arrays of spinning sites with a high density at a low cost. Velasquez-Garcia's group[24] developed a monolithic linear array of 225 emitters (spinning sites) with a density up to 100 sites cm^{-2}. Sharp corners at long and narrow emitters had the enhanced electric field, which triggered the electro-hydrodynamic jetting. Mass flux rates reached as high as 417 g h^{-1} cm^{-2}, four times the reported rate from leading commercial electrospinning sources. However, this requires very complicated fabrication technology for the emitters, and the emitter clogging may also be unavoidable due to the minute size of these emitters.

Rather than confining the liquid stream into multi-needles or nozzles, one-dimensional spinning sites are employed such as a confined channel, the edge of a bowl, and a coil of wire. Thoppey *et al.*[25] developed an approach of using the bowl edge to spin the polymer solution (Figure 8.5a). The polymer solution formed multiple jets spontaneously at the lip of a bowl, and then the jets migrated to the edge to form equally spaced spinning sites around the bowl. A high-voltage interval was applied to induce jet formation and the voltage was then reduced for stable fiber formation. The edge electrospinning produced similar fiber morphologies but the mass production increased to 0.68 g h^{-1} for around 20 jets from the bowl edge. A replenishing source was required in this case to maintain stable jet formation and the electrical field characteristics (magnitude and gradient) had to balance well with the effective feed rate.

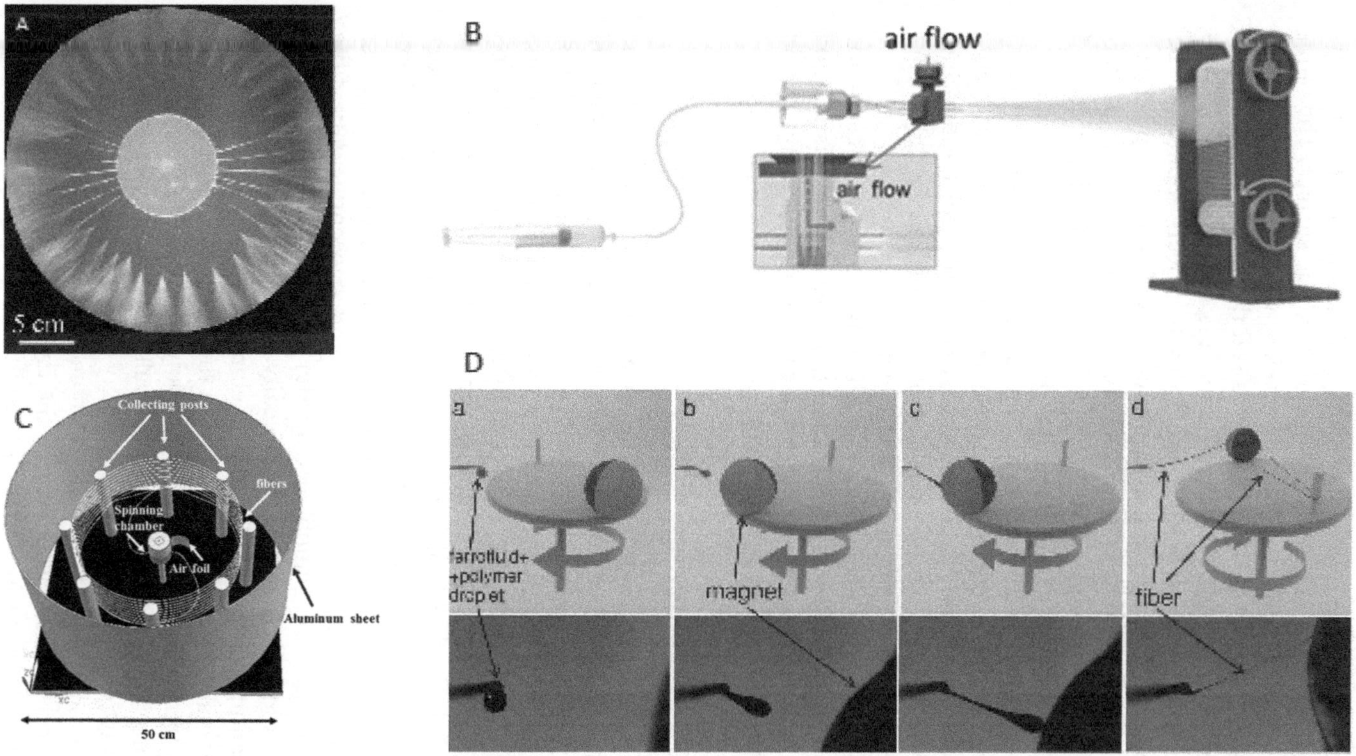

Figure 8.5 Scalable NF production technologies: (a) one-dimensional free surface electrospinning, or bowl edge spinning,[25] (b) gas-assisted spinning or blow-spinning,[36] (c) centrifugal force assisted spinning,[31] and (d) magnetic force-assisted spinning.[40] (a) Adapted from ref. 25 with permission from IOP Publishing. (b) Adapted with permission from B. Khalid, X. Bai, H. Wei, Y. W. H. Huang, Y. Cui, *Nano Letters*, 2017, **17**, 1140–1148, Copyright 2017 American Chemical Society.[36] (c) Adapted with permission from L. Ren, R. Ozisik, S. P. Kotha, P. T. Underhill, *Macromolecules*, 2015, **48**, 2593–2602, Copyright 2015 American Chemical Society.[31] (d) Adapted from ref. 40 with permission from John Wiley and Sons, © 2015 WILEY-VCH Verlag GmbH & Co. KGaA, Weinheim.[40]

Alternatively, a spinning drum is partially submerged in a bath of polymer solution. A thin layer of polymer solution covers the surface of the rotating drum that is connected to a high-voltage supply. A number of spinning jets are generated and they are distributed over the electrode surface. The number and location of the jets are distributed in their optimal conditions basing on the analysis by Lukas *et al.*[26] Wang *et al.*[27] used the needle-free in-house devices for producing ultrafine chitosan nanofibers for air separation. The needle-free technique using rotating electrodes has been applied in the commercial scale, and its representative product is Nanospider™ produced by Elmarco in Czech Republic. The product has five rotating charged electrodes and 11 counter electrodes, and a very high electricity field is generated between these electrodes for spinning nanofibers.

The above-mentioned techniques may increase the throughput for nanofiber production, but limited starting materials are available for producing nanofibers for electronic devices and sensors. For example, poly(3-hexylthiophene) widely used in electronics is not electrospinnable due to short chain entanglement and Rayleigh instability. More electrospinnable materials are in the early stage of development. Although the mixture of organic–organic or organic–inorganic has been demonstrated to be feasible by this method for nanoelectronics or nanosensors, large-scale production of such a hybrid structure still requires further development.

8.3.2 Melt Electrospinning

Melt electrospinning has been scaled up in an industrial scale and the throughput can be achieved at thousands of meters per minute. The method is to spin melt polymer chains through the spinneret at an elevated temperature, and melt polymer chains are drawn into a very thin sheet by overcoming the surface tension force by the electric force. The nanofiber formation is realized by solidification after rapid cooling of the melt polymers. The process has the advantage of solvent-free and it is environmentally friendly. The nanofiber quality is dependent on the electric field, polymer molecular weight and molecular chain conformation, and post-annealing process during the cooling period. Due to high viscosity and low conductivity of the polymer melt, the polymer nanofiber diameter is often much bigger than other methods. Additives to reduce the viscosity may help reduce the diameter down to a nanometer scale. Because fiber elongation experiences instable forces, the fiber diameter is not uniform from the process.

Excepting the centrifugal force spinning of polymer melts, multilayer coextrusion is also an emerging method for large-scale manufacture of nanofibers due to its high efficiency, versatility, simplicity and low cost. The polymer melt is forced into a series of multipliers. In each multiplier, the viscoelastic melt is cut, spread and stacked so that the layer is doubled and the fiber diameter is halved. Coextrusion through 11 multipliers allows production of 4096 layers and the individual fiber diameter can be reduced to less than 10 nm.[28] Du *et al.*[29] further refined the method by reducing the number of layers down to 1024 but ending with a special cut die. The die had seven cutting blades inside the channel and the polymer melt was split into eight separate ribbons. The polymer melt was further extensively stretched by a rotating roll for receiving the nanofibers and the fiber diameter dropped from 500 nm down to 50 nm during the last cutting step. The lab-scale unit processed 2 kg polymers per hour and produced several hundred meters of nanofibers.

The melt electrospinning has an advantage of no solvent in the process, but the starting materials must be electrospinnable and preferably has a low melt point, which may become a limiting factor for manufacturing nanofibers for sensors or electronic devices.

8.3.3 Centrifugal Force-based Methods

Centrifugal spinning is a fabrication method of cotton candy and has been adapted for nanofiber production because it has a high production rate and a low product cost, typically 1 g min^{-1} for a lab-scale system and hundreds of meters per minute for an industrial system.[30] More importantly, the technique allows more choices for the starting materials. Non-conductive solutions are also able to be spun into nanofibers by centrifugal spinning. The liquid stream is centrifuged to provide a continuous thin jet when the centrifugal force overcomes the capillary force in the orifices. The thin jet becomes elongated and thinner filaments under the action of capillary and centrifugal forces and resistance forces from inertia, elastic and viscous stresses. The filaments reach the glassy state of the polymer solution and solidify into fibers after the solvent evaporation (Figure 8.5b). Polymer components, solution viscosity, elasticity, solvent mass transfer coefficient and evaporation rate, spinning speed, and fiber collectors are the key parameters for manipulating the fiber production in the centrifugal spinning method. Ren *et al.*[31] have investigated a range of dimensionless number and found that the capillary number ($Ca = \mu V/\sigma$, μ is liquid solution viscosity, V is the characteristic velocity and σ is the surface tension of the liquid solution) was useful for controlling uniform fiber formation and the Weissenberg number ($We = \lambda V/l$, λ is the longest or characteristic relaxation time of the liquid, l is the characteristic scale of flow) for quantifying the fiber morphologies.

The Forcespinning™ apparatus uses the centrifugal force to spin the polymer solution or melt. It has the unique capacity of melt spinning with a high speed up to 25 000 rpm for the precursor solutions. Multiple spinnerets are designed to centrifugally force the melt or solution in a cavity through designed orifices. Examples of using the apparatus to generate Sn/C composite nanofibers anode for lithium ion batteries,[32] and chitosan/polyamide nanofibers to remove anionic dyes from aqueous solutions.[33]

By combining the centrifugal and electrospinning, the polymer solution is charged at the tip of a syringe-needle spinneret; meanwhile the spinneret is also driven by a motor to eject the solution due to centrifugal force. In one example, the elongated jets landed perpendicularly to 102 collecting wire electrodes spaced 1.27 cm apart. Jets were stretched axially due to solvent evaporation as well as tangentially due to electric forces when jets were attached to the wire electrodes. The final nanofibers were well aligned across the wire gap and harvested from the wire gap. This system allows improvement in the product collection efficiency and controlled orientation of the nanofibers.[34]

Because the technique allows conductive and non-conductive materials for nanofiber production, fibers can be applied widely, such as for electronic devices and sensors. Indium tin oxides, barium titanate, hematite, and titanium oxides can be mixed with a wide range of polymers to centrifugally spin into nanofibers for sensors or due to their optical, ferroelectric, or thermoelectric properties.

8.3.4 Shear Force-assisted Methods

The polymer solution may be attenuated into thinner jets by shear force. The shear force is generated from the bulk liquid or gas stream surrounding the polymer solution. The shear force from the liquid stream was applied to produce staple polymeric nanofibers by researchers at North Carolina State University. The principle is to stretch a polymeric droplet generated from a spinneret in a coagulation bath under shear stress in the bulk of anti-solvents, and the polymer molecules precipitate in the anti-solvents to form nanofibers with an average diameter of 300–500 nm and a length of 10 mm to 10 cm.[35] The barrier of the technique is mass transfer of polymer molecules from the polymer solution into the bulk solvents, and the solidification process is slow compared with melt spinning. However, this process is simple and cheap to produce nanofibers. XanoShear™ was the commercial applications of this technique by Xanofi. The production rate for a continuous process is above 3.6 kg h^{-1}.

Another source for generating shear stress is gas streams. This method has a high productivity and can be applied to polymers with low dielectric constant or electric conductivity. The gas blowing technique has been used for polymer solutions and melts. The polymer solutions or melts spin out of spinneret(s) that may be or may not be connected with the electrodes. A gas stream with a high velocity is introduced to run in parallel with the polymer stream. The stream is attenuated due to shear stress exerted from the gas stream. Hot air is often employed to melt the polymers or evaporate the solvents from the solution. Without electrodes, the polymer stream becomes thinner purely due to shear stress and the fibers may be inconsistent and thick. The method is referred as blow-spinning. Air flow velocity, temperature, spinneret and the distance between the spinneret and collector are key parameters to affect the fiber quality. Among them, spinneret design is essential for success in the process. For example, Khalid *et al.*[36] used the method to spin the polymer solutions through 22-gauge needle bend tips to produce polyacrylonitrile nanofibers for removal of particulate pollutants in the air (Figure 8.5c). A uniformly arranged transparent fiber coating with a fiber diameter of 200 nm was achieved on a 2.5 m mesh in 3 min.

Electroblowing is the combination of electrospinning and air-blowing. Polymer solutions or melts are charged before releasing from the spinneret, the polymer stream is attenuated by both shear stress and electrostatic forces to produce very fine fibers down to 100 nm in diameter. Electrospinning and electroblowing were employed to spin 2-hydroxypropyl-β-cyclodextrin as a carrier for a drug, diclofenac sodium, which has a poor water solubility, to prepare an injection. The electrospinning process lasted only 1–2 seconds despite very uniform fibers generated. After introducing a high velocity air stream at 20 l min^{-1} comparing to the liquid stream flow of 20 ml min^{-1}, the fibers were formed steadily. The fiber-assisted drug formulation resulted in a clear solution and a fast dissolution rate of the drug.[37] A higher velocity and a higher temperature were found to further decrease the fiber diameter for the polymer melts.[38] Polyacrylonitrile nanofibers were also produced by electroblowing for air filtration, and the fiber diameter was controlled at around 200 nm.[39]

The shear force-based method offers no constraints for starting materials and solvents in the spinning process. Nanofibers from this method have the great potential for electrodes in fuel cells and supercapacitors. By combining the polymers with other metal oxides, the nanofibers may also be used for chemical and biochemical sensors.

8.3.5 Magnetic Force-based Methods

Magnetospinning is a new spinning technique employing a ferrofluid under a magnetic field. The ferrofluid is a colloidal dispersion of magnetic nanoparticles. The ferrofluid surface forms a spike as a magnet approaches the ferrofluid, and the spike is drawn towards the magnet to form a liquid bridge. The spike is induced due to energetically favorable balance between gravity, surface tension and magnetic force. The liquid bridge can be manipulated to become thinner to allow solvent evaporation and the nanofibers are produced. Tokarev *et al.*[40] employed a time-varying magnetic field by rotating the magnet at an angular velocity of 50–100 rpm to stretch the ferrofluid droplets into a very thin liquid bridge (Fd). After the solvent evaporated, the liquid bridge turned into a fine fiber. In this process, polymer concentration, angular velocity, magnetic nanoparticle concentration, magnetic field gradient, and solvent evaporation rate were key parameters for forming fibers, controlling fiber diameter, and the fiber quality. Dilute polymer concentration resulted in a too low viscosity to maintain a stable liquid bridge, while a too fast angular velocity led to no attachment of the droplet into the magnet. The magnetic nanoparticle concentration was reduced to 1 wt% by increasing the magnetic field gradient. This method is scalable, simple, and applicable to conductive and non-conductive polymers. However, magnetic nanoparticles are present in the final nanofiber products, and the operational window for the rotational speed and polymer concentration is also quite narrow. The magnetospinning method does not rely on the conductivity of staring materials, and it can be applicable to a wide range of polymers. However, the resultant nanofibers contain magnetic nanoparticles. These nanofibers may be useful for developing magnetic sensors.

8.4 Concluding Remarks

Nanofibers have attracted great interest in diverse applications in biomedical, pharmaceutical, chemical, and energy industries. Scalable techniques are developed to meet the growing demands of nanofiber market. Among them, electrospinning, shear force-assisted spinning, centrifugal force-assisted spinning, and magnetic force-assisted spinning are very promising. Some of these techniques have been transferred into manufacturing instruments from companies such as Elmarco and Inovenso, and some of these products include Nanospider and Nanospinner416. No device information is available from a few companies which only provide electrospun products, such as Dupont. Most of commercially available nanofiber-based products are applied in the air/water filtration, such as nTEX series by NXTGEN NANOFIBER, Exceed@™ and Zeus Filtriq™ PTFE Non-Woven Membrane, while very few products are in the medical device area, such as Nanosan series developed by SNS Nano Fiber Technology LLC.

The perspectives of the nanofibers for electronics and sensors show bright future as well as exciting challenges. Assembly of individual nanofibers into a device requires more investigation for easy operation, low maintenance, and cheap price. Although the development of nanofiber has been extensively studied, including characterizing the properties and investigating the potential applications, there are still very few of commercial products for electronic devices and sensors are available at the moment. A material science company, Axium Nanofibers LLC, commercializes nanomaterials for broad applications including the electronic device for energy storage. Through advanced nanofiber platform, Axium research focus has been the commercialization of a breakthrough lithium ion battery. Meanwhile, AZ Electronic Materials company produces materials for integrated circuits, flat panel displays and light-emitting diodes. Axium and Az recently collaborate for the synthesis of hybrid nanofibers that can enhance charge transport and ionic conductivity. With such efforts from material scientists in collaboration with chemical engineers, product commercialization can be accelerated. More products are expected to enter into the market in the next few years.

Acknowledgements

MM gratefully acknowledges the financial support from the Universiti Malaysia Sabah and the Malaysian Government. H. Z. thanks for the support from ARC Project (DP160104632) and The Medical Advancement Without Animal (MAWA) Trust.

References

1. S. Thenmozhi, N. Dharmaraj, K. Kadirvelu and H. Y. Kim, *Mater. Sci. Eng., B*, 2017, **217**, 36–48.
2. M. Misson, H. Zhang and B. Jin, *J. R. Soc., Interface*, 2015, **12**(102), 1–8.
3. C. Feng, K. C. Khulbe, T. Matsuura, S. Tabe and A. F. Ismail, *Sep. Purif. Technol.*, 2013, **102**, 118–135.
4. M. Misson, S. Dai, B. Jin, B. H. Chen and H. Zhang, *J. Biotechnol.*, 2016, **222**, 56–64.
5. J. M. Deitzel, J. Kleinmeyer, D. Harris and N. C. Beck Tan, *Polymer*, 2001, **42**(1), 261–272.
6. S. Megelski, J. S. Stephens, D. B. Chase and J. F. Rabolt, *Macromolecules*, 2002, **35**(22), 8456–8466.
7. H. R. Darrell and C. Iksoo, *Nanotechnology*, 1996, 7(3), 216.
8. M. Bognitzki, W. Czado, T. Frese, A. Schaper, M. Hellwig, M. Steinhart, A. Greiner and J. H. Wendorff, *Adv. Mater.*, 2001, **13**(1), 70–72.
9. S. K. Simotwo and V. Kalra, *Electrochim. Acta*, 2016, **198**, 156–164.
10. Z. Aytac, H. S. Sen, E. Durgun and T. Uyar, *Colloids Surf., B*, 2015, **128**, 331–338.
11. A. Jindal, S. Basu, N. Chauhan, T. Ukai, D. S. Kumar and K. T. Samudhyatha, *J. Power Sources*, 2017, **342**, 165–174.
12. T. Makinouchi, M. Tanaka and H. Kawakami, *J. Membr. Sci.*, 2017, **530**, 65–72.
13. L. Xu, N. Sheybani, S. Ren, G. L. Bowlin, W. A. Yeudall and H. Yang, *Pharm. Res.*, 2015, **32**(1), 275–285.
14. V. Modgill, T. Garg, A. K. Goyal and G. Rath, *Curr. Drug Delivery*, 2015, **323–332**(12), 323–332.
15. R. Sedghi and A. Shaabani, *Polymer*, 2016, **101**, 151–157.
16. S. Raj and D. R. Shankaran, *Sens. Actuators, B*, 2016, **226**, 318–325.
17. M. F. Canbolat, A. Celebioglu and T. Uyar, *Colloids Surf., B*, 2014, **115**, 15–21.
18. V. Vamvakaki, K. Tsagaraki and N. Chaniotakis, *Anal. Chem.*, 2006, **78**(15), 5538–5542.
19. H. Jia, *Methods Mol. Biol.*, 2011, **743**, 205–212.
20. M. F. Najafi, D. Deobagkar and D. Deobagkar, *Electron. J. Biotechnol.*, 2005, **8**(2), 197–203.
21. D. J. Shusterman, *Occup. Med.*, 1993, **8**(3), 519–531.
22. J. H. Wendorff, S. Agarwal and A. Greiner, *Electrospinning: Materials, Processing, and Applications*, Wiley, New Jersey, 2012.
23. K. Kim, *Electronic Spinning Apparatus and a Process of Preparing Nonwoven Fabric Using the Same*, 2008.
24. P. De Leon, F. A. Hill, E. V. Heubel and L. F. Velasquez-Garcia, *Nanotechnology*, 2016, **26**, 225301.
25. N. Thoppey, J. R. Bochinski, L. I. Clark and R. I. Gorga, *Nanotechnology*, 2011, **22**, 345301.

26. D. Lukas, A. Sarkar and P. Pokorny, *J. Appl. Phys.*, 2008, **103**, 084309.
27. L. Wang, C. Zhang, F. Gao and G. Pan, *RSC Adv.*, 2016, **6**, 105988.
28. M. Ponting, A. Hiltner and E. C. Baer, *Macromol. Symp.*, 2010, **294**(1), 19–23.
29. J. Du, D. Liu, S. Chen, D. Wan and H. Pu, *Polymer*, 2016, **120**, 209–213.
30. R. Nava, L. Cremar, V. Agubra, J. Sanchez, M. Alcoulabi and K. Lozano, *ACS Appl. Mater. Interfaces*, 2016, **8**, 29365–29372.
31. L. Ren, R. Ozisik, S. P. Kotha and P. T. Underhill, *Macromolecules*, 2015, **48**, 2593–2602.
32. V. Agubra, L. Zuniga, D. D. Garza, L. Gallegos, M. Pokhrel and M. Alcoutlabi, *Solid State Ionics*, 2016, **218**, 72–82.
33. G. L. Dotto, J. M. N. Santos, E. H. Tanabe, D. A. Bertuol, E. L. Foletto, E. C. Lima and F. A. Pavan, *J. Cleaner Prod.*, 2017, **144**, 120–129.
34. A. Erickson, D. Edmondson, F.-C. Chang, D. Wood, A. Gong, S. L. Levengood and M. Zhang, *Carbohydr. Polym.*, 2015, **124**, 467–474.
35. O. Delev, S. Smoukou and M. Marquez, *Nanospinning of Polymer Fibers from Sheared Solutions*, 2010.
36. B. Khalid, X. Bai, H. Wei, Y. Huang, W. Hu and Y. Cui, *Nano Lett.*, 2017, **17**, 1140–1148.
37. A. Balogh, T. Horvathova, Z. Fulop, T. Loftsson, A. H. Harasztos, G. Marosi and Z. K. Nagy, *J. Drug Delivery Sci. Technol.*, 2015, **26**, 28–34.
38. E. Zhayev, D. Cho and Y. L. Joo, *Polymer*, 2010, **51**, 4140–4144.
39. C. Liu, P.-C. Hsu, H.-W. Lee, M. Ye, G. Zheng, N. Liu, W. Li and Y. Cui, *Nat. Commun.*, 2015, **6**, 6205.
40. A. Tokarev, O. Trotsenko, I. M. Griffiths, H. A. Stone and S. Minko, *Adv. Mater.*, 2015, **27**, 3560–3565.
41. Q. Bao, H. Zhang, J.-X. Yang, S. Wang, D. Y. Tang, R. Jose, S. Ramakrishna, C. T. Lim and K. P. Loh, *Adv. Funct. Mater.*, 2010, **20**(5), 782–791.
42. D. Li and Y. Xia, *Nano Lett.*, 2003, **3**(4), 555–560.

CHAPTER 9

Electrospun Materials for Proton Exchange Membrane Fuel Cells and Water Electrolysis

S. CAVALIERE*, M. ZATOŃ, F. FARINA, D. JONES AND J. ROZIÈRE

Institut Charles Gerhardt Montpellier, UMR UM CNRS 5253, Laboratoire des Agrégats Interfaces et Matériaux pour l'Energie, Université de Montpellier, 34095 Montpellier Cedex 5, France
*E-mail: sara.cavaliere@umontpellier.fr

9.1 Introduction

The development of clean and sustainable energy sources is one of the most impelling challenges of the 21st century to face fossil fuel depletion, carbon dioxide emissions and the consequent environmental pollution and global warming.[1] Amongst the existing electrochemical energy storage/conversion devices, fuel cells (FCs) play a key role for the electrification of transportation and for power supply in portable and stationary applications,[2] and water electrolysis (WE) for large-scale storage of renewable energy that can be converted by FCs.[3] Using hydrogen as energy carrier by coupling FCs and WE technologies may allow in the long term the worldwide objective of decarbonization of the energy sector to be reached.[4,5]

Soft Matter Series No. 7
Electrospinning: From Basic Research to Commercialization
Edited by Erich Kny, Kajal Ghosal and Sabu Thomas

Published by the Royal Society of Chemistry, www.rsc.org

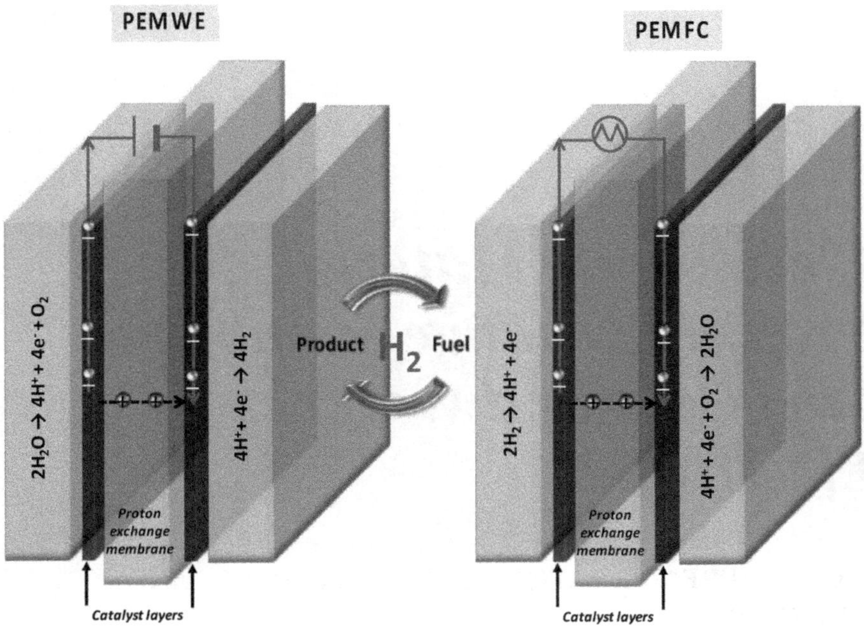

Figure 9.1 Schematic representations of a proton exchange membrane fuel cell (PEMFC) and a water electrolyzer (PEMWE).

FCs are devices directly converting the chemical energy of an oxidant and a fuel into electricity and heat. They are highly efficient[6] and versatile, as single cells can be combined in stacks to tune the output power for a wide range of applications. Their core is the membrane–electrode assembly (MEA), consisting of an ion conducting electrolyte sandwiched between two electrodes where the electrochemical reactions take place (Figure 9.1).

Amongst the numerous types of FC,[7] proton exchange membrane fuel cells (PEMFCs) working at relatively low temperature (80–120 °C) are considered the most promising, being already employed *e.g.* in commercialized vehicles. The reactions involved are hydrogen oxidation at the anode and oxygen reduction at the cathode, where the membrane allows proton transfer from the first to the latter; the fuel is thus hydrogen and the oxidant oxygen from the air or gaseous O_2. Alcohols can be also used as fuel in direct methanol or ethanol FCs, which will not be treated in this chapter. H_2-fed FCs are clean energy converters, producing only water. On the earth hydrogen is not present in its molecular structure (H_2) and so far it is mostly produced from hydrocarbons.[8] A promising and efficient method of hydrogen production is the electrochemical decomposition of water (WE). When using renewable power sources (*e.g.* wind, solar, biomass) to produce the electricity needed, this approach becomes completely sustainable and allows the storage of this clean but intermittent energy. Amongst the different existing types of devices[3] proton exchange membrane water electrolysers (PEMWEs) produce highly pure hydrogen with a high efficiency, and can be operated at high current densities.[9] In PEMWEs the electrochemical reactions are opposite to those occurring in PEMFCs: at the anode water is split into oxygen, protons and electrons by applying a voltage higher than the standard at ambient temperature and pressure (1.23 V/RHE). Protons and electrons are transferred to the cathode (through the electrolyte membrane and the external circuit, respectively), where they recombine to form molecular hydrogen. The core of the system is the membrane–electrode assembly as for PEMFCs (Figure 9.1).

In the past 50 years the development of PEMFCs and water electrolysers has witnessed great progress with applications of PEMFC in particular in the transportation sector. Nevertheless, some bottlenecks still slow down their full introduction into the market, mainly their durability and the high material costs.[2] Academic and industrial research is thus very active in the improvement of all core components of the cells: electrolyte membranes,[10] electrocatalysts[11] and their supports.[12] In general, the challenges concern the preparation of proton exchange membranes with high conductivity, elevated mechanical stability and low gas permeability, electrodes with zero or ultra-low amounts of noble metals (Pt, Ir) and corrosion-resistant electrocatalyst supports. In order to address them, novel materials with different chemical compositions are developed (*e.g.* novel proton-conducting (co)-polymers for membranes, electrocatalysts with partial or total replacement of noble metals, corrosion-resistant supports). A crucial role is also played by the material structure at the micro- and nanoscale, at which the key processes of proton and electron transfer and electrocatalysis take place. For this reason, research on PEMFC and PEMWE cores is focusing on composite nanostructured materials with targeted architectures. One of the techniques of choice to prepare these materials in a great range of compositions (polymers, inorganic or hybrid organic/inorganic materials) is electrospinning.[13,14] This simple, low-cost and up-scalable technique allows elaboration of micro/nanofibers with a great variety of morphologies (solid, hollow, porous, core-sheath) as well as their combination (each other or with other nanomaterials or polymers) in multi-scale assemblies that can present different geometries (non-woven, aligned or patterned fibers). It is thus possible to develop a panoply of materials based on electrospun fibers to be applied in the cores of PEMFC and PEMWE. Recent advances and future trends of electrospun materials in PEMFCs and electrolysers MEA are described below.

9.2 Nanocomposite Fibrous Ionomer Membranes

The electrolyte membrane in FCs and WE must be highly proton-conducting, electronically insulating, impermeable to reactants, as well as mechanically and chemically robust in operating conditions.[15] Furthermore, PEMWE membranes have to sustain higher pressures and avoid hydrogen permeability: low gas crossover is required to yield hydrogen with high purity and to work under a wide range of power inputs,[3] and is crucial for safety issues.

The state-of-the-art materials used with this purpose are perfluorosulfonic acid (PFSA) polymers including the long side-chain (LSC) Nafion®, the short side-chain (SSC) Aquivion® and 3M Corporation ionomers. Their structure comprises ionic clusters with hydrated sulfonic acid groups well percolated and phase separated from the hydrophobic backbone (polytetrafluoroethylene, PTFE).[16] The latter provides mechanical integrity and high stability in harsh electrochemical conditions, while the firsts ensure proton transport.[17] The effect of spatial organization of hydrophilic and hydrophobic domains on proton conduction and mechanical properties of ionomer membranes is fundamental.[18]

Proton conductivity of ionomer membranes is strongly related to the proportion of ionic domains in the polymer structure (expressed by the equivalent weight, E_W = mass of polymer per sulfonic acid group) and to their degree of hydration. Indeed, one of the main issues is the loss of conductivity at temperatures above 100 °C. Operating at such high temperature would present, thermodynamic, kinetic and engineering advantages. Higher membrane hydration and proton conductivity are obtained when the cell is pressurized, which is a clear advantage also for hydrogen storage purposes in PEMWE.[19] The development of mechanically robust proton exchange membranes that can sustain high operating temperature and pressure is an important challenge. This will improve the membrane lifetime, that in PEMFC is also compromised by mechanical failures due to swelling/shrinkage processes occurring during membrane hydration and dehydration under variable loads.[10] High dimensional stability membranes also allow reduction of their thickness, which lowers the electrical resistance (and facilitates water management in FC). To conclude the chemical stability of polymer membranes must also be considered, that is compromised by radical species forming during fuel cell operation: hydroxyl, hydroperoxyl and superoxide ($HO^{\cdot}$, $HOO^{\cdot}$, $O^{\cdot -}_2$) radicals can attack vulnerable polymer sites, leading to structure defragmentation, membrane thinning, and failure.

All these challenges have triggered extensive research on the development of improved proton exchange membranes. This includes the development of novel advanced ionomers (SSC PFSAs[20,21] or sulfonated polyaromatic polymers and their blends[22–33]) or the combination of them with strong acids/heterocycles/ionic liquids.[34–37] Besides those approaches related to the chemical composition of the membranes, the preparation of heterogeneous systems comprising mechanical[38] and/or chemical[39,40] reinforcement affecting the structure at the micro/nanoscale was demonstrated to be a promising strategy. The mechanical reinforcement in such a membrane design mitigates premature aging and failure of the material by limiting dimensional changes. Furthermore, these supports can contain radical scavengers, which prevent radical attack and the consequent ionomer decomposition.

Reinforcing substrates are usually based on thermostable and robust polymers or inorganic (nano)materials.[32,41–43] The main well-established membrane reinforcements used so far include expanded PTFE (ePTFE) sheet, for example from Gore® Fuel Cell Technologies,[38] PTFE microfibrils from Asahi Glass Corporation,[44] polysulfone/microglass fiber fleece[45] and laser-drilled polysulfone or polyimide from Giner Electrochemical Systems.[46] Although improvement of the mechanical and dimensional stability achieved with those materials, they present some disadvantages including a high processing cost, non-specific interface with the ionomer and low flexibility in architecture design. The introduction of electrospun nanofibers demonstrated to be a very effective and versatile strategy leading to nanocomposite reinforced membranes with targeted architectures and interfaces.[47–49] The possibility of preparing a variety of morphologies and compositions including further embedding of functional nanoparticles highly widen the range of novel multi-scale composite systems achievable. Control of the diameter of the fibers and their distribution in the matrix allows fine tuning of the interface and spatial organization between inert and conducting phases as well as pore interconnectivity. For these reasons, in recent years electrospun fibers have been thoroughly employed in proton exchange membrane fabrication and some of the key advances are described below. For the sake of clarity, this paragraph is organized into three sections according to materials design (ionomer electrospun or in a matrix) and composition (polymer, ionomer, inorganic electrospun materials), criteria also schematically depicted in Figure 9.2.

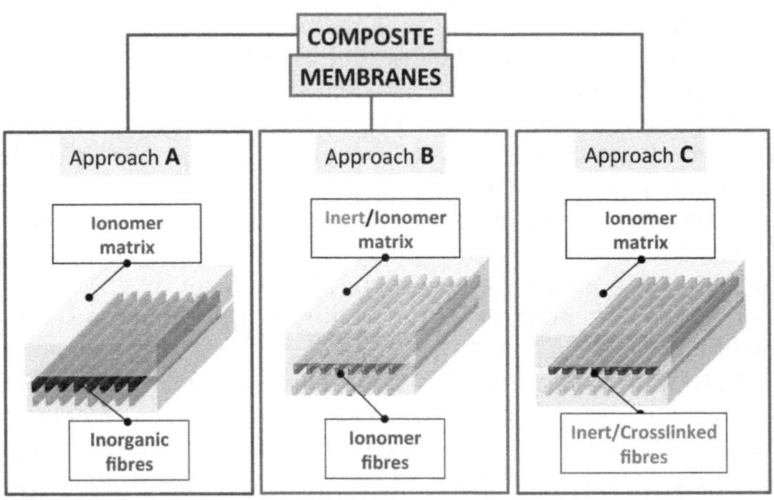

Figure 9.2 Types of membrane architectures based on electrospun materials. (a) Electrospun (proton conducting or inert) inorganic web embedded in a (proton conducting or inert) polymer matrix (Section 9.2.1). (b) Inert material/ionomer surrounding a 3D interconnected nanofiber web of a proton-conducting polymer (Section 9.2.2). (c) Proton-conducting polymer matrix surrounding an electrospun web of nanofibers of an inert or proton-conducting or crosslinked electrospun polymer (Section 9.2.3).

9.2.1 Composite Membranes with Electrospun Inorganic Materials Embedded in a Polymer/Ionomer Matrix

Inorganic materials such as metal oxides, zeolites, metal hydrogen phosphates and heteropolyacids have been incorporated in fuel cell[32,33,50,51] and WE composite membranes[52,53] to enhance mechanical strength, water retention, proton conductivity under low RH, thermal stability and to reduce fuel crossover. Some metal oxides such as SnO_2, TiO_2, CeO_2 or MnO_2 have also demonstrated radical scavenger properties.[15,54] The main challenge is to achieve their homogenous dispersion in the ionomer matrix, control of their morphology and their interface with the polymer matrix. Among the numerous strategies to incorporate inorganic materials into membranes,[55] electrospinning has emerged due to the enhanced control that this approach offers of composition and morphology of the fibers and interaction with the ionomer that finally affect their dispersion.

Hygroscopic metal oxide electrospun fibers and tubes based on SiO_2,[56–58] TiO_2,[59] and ZrO_2[60] were incorporated in fuel cell membranes to enhance their performance in dry and hot operating conditions. For instance, mesoporous TiO_2, CeO_2 and $ZrO_{1.95}$ nanotubes embedded into a Nafion® membrane increased water retention capability as well as water diffusion in the composite systems.[59] This resulted in low ohmic and mass transport resistance leading to higher PEMFC performance than with pristine Nafion® 212. The maximum power density value of a cell operating at 18% RH and 80 °C was 641 mW cm^{-2} for the MEA comprising the membrane with 1.5 wt% TiO_2 nanotubes *vs* 186 mW cm^{-2} obtained in an MEA with a Nafion® 212 membrane. The tubular morphology and the diameter of the electrospun materials, that inversely increase with their size, play a crucial role in water retention.[61] MEAs based on such hybrid membranes also demonstrated greater lifetime, which was ascribed to improved water back diffusion.[61,62] In a similar approach, sulfonated zirconia nanofibers were embedded into a crosslinked poly(2-acrylamido-2-methylpropane-sulfonic acid)[60] giving rise to membranes with the exceptionally high proton conductivity of 340 mS cm^{-1} at 100 °C. The authors suggested that the inorganic fiber web enhanced the anchoring of water molecules and facilitated the proton transport within the membrane.

Metal phosphate nanofibers have also been employed to reinforce PFSA membranes. Our group prepared zirconium phosphate/zirconium oxide (ZrP/ZrO_2) nanofibers by 'reactive' coaxial electrospinning and embedded them into Aquivion®.[63] ZrP prevented membrane dehydration at high temperature and assisted proton transport, and the high aspect ratio of the fibers provided an extended interaction with the ionomer matrix so increasing the mechanical properties of the composite system. Indeed, the membrane reinforced with ZrP/ZrO_2 demonstrated improved elastic modulus, yield point and proton conductivity in comparison to pristine Aquivion® membranes.[63] In a similar approach, composite membranes based on carbon nanotubes embedded in sulfonated poly (ether ether ketone) (sPEEK) electrospun fibers demonstrated improved mechanical stability and performance over recast Nafion® and sPEEK membranes.[64]

Another strategy is to embed or attach metal oxide nanoparticles to polymer nanofibers further embedded in an ionomer. Membranes based on SiO_2/sPEEK fibers impregnated with Nafion® demonstrated high water uptake capacity and significantly reduced membrane swelling in the membrane in-plane directions.[65]

In an original strategy, the hydrophobic/hydrophilic phase separation existing in Nafion® (see Section 9.2) has been mimicked in organic-inorganic hybrid membranes based on poly(vinylidene fluoride-hexafluoropropylene) (PVDF-HFP)/SO_3H-silica fiber networks.[66,67] For the optimal composition of 61 vol% of SiO_2, such membranes presented high proton conductivity and high elastic modulus above 80 °C, which was attributed to their microstructure comprising bundles of polymer fibers surrounded by functionalized silica domains.

Other fibrous metal oxides have been chosen for their radical scavenger properties and embedded into proton exchange membranes in order to mitigate chemical membrane degradation.[15] In particular, cerium oxide is one of the most efficient due to the fast kinetics of hydroxyl radical quenching. For instance, composite membranes based on mesoporous ceria nanotubes and Nafion® presented higher performance and durability in hot and dry operation conditions than a pristine PFSA membrane.[62] We developed layers of PFSA nanofibers embedding CeO_x that were incorporated into the MEA at the anode or cathode interface (Figure 9.3).[40] The lifetime of MEAs with such asymmetric composite membranes overcame by eight times that of an unmitigated MEA, especially when the PFSA/CeO_x layer was oriented to the anode side.[68] This result was ascribed to possible Ce^{3+} regeneration in a reductive environment and to the partial dissolution of CeO_x at the anode interface and migration of cerium ions through the membrane. In contrast, cerium species at the cathode side are likely leached, thus contributing over a shorter timeframe to membrane stabilization. Immobilizing free radical scavengers into electrospun nanofiber webs has demonstrated high potential in controlled release of these active species during fuel cell operation.[69,70]

9.2.2 Composite Membranes with Electrospun Ionomer Materials Embedded in a Polymer/Ionomer Matrix

Proton-conducting nanofibers can be prepared by electrospinning and embedded in an inert matrix that fills the voids. In this case, the nanofibers must be interconnected to allow a continuous proton conduction pathway through the membrane.

Electrospinning of ionomers may induce the alignment of their chains as well as influence their organization at the nano/microscale, thus affecting proton conductivity, water uptake and mechanical stability of the corresponding

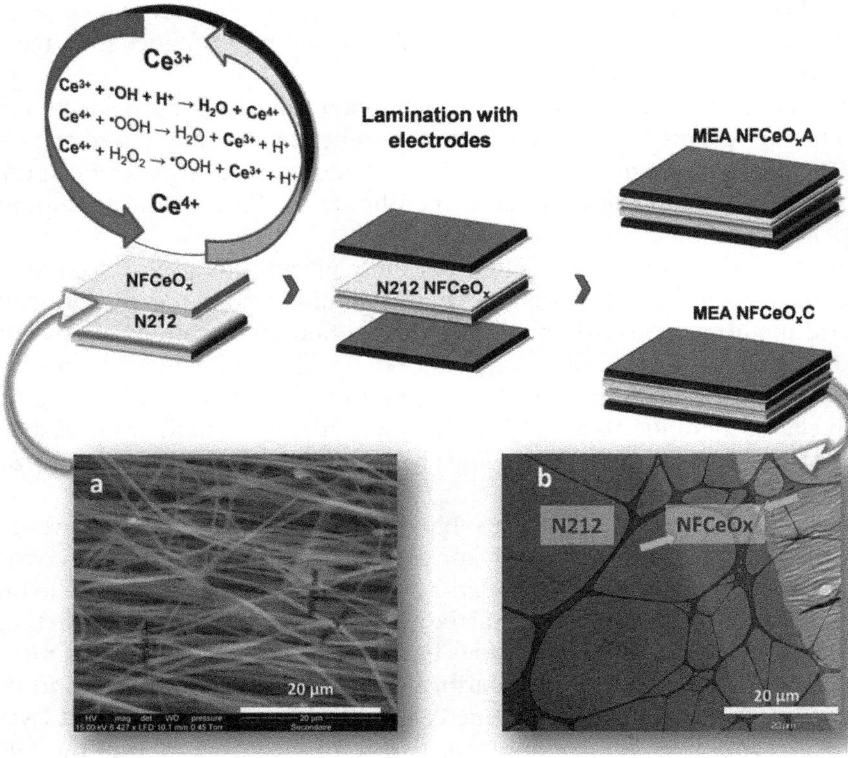

Figure 9.3 (Above) Schematic representation of composite membrane preparation and elaboration of MEAs. (Below) SEM micrograph of PFSA nanofibers (a) and TEM micrograph of the composite membrane cross-section after hot pressing (b). Reproduced from ref. 40 with permission from the Royal Society of Chemistry.

membrane.[71–75] In particular, the long-range ordered arrangements of polar groups in the polymer chains is likely to form proton-conducting channels and favor proton conductivity.[71,73–76] This phenomenon can be exacerbated in uniaxially aligned fibers. For instance, electrospun aligned sulfonated polyimide nanofibers demonstrated ultra-high single-fiber proton conductivity values >1 S cm^{-1} at 30–90 °C and 95% RH as well as improved mechanical properties in the axial direction.[73,74,77] The exceptional proton conductivity value of 1.5 S cm^{-1} was reported for a single Nafion® nanofiber with diameter of 400 nm.[71] The diameter of the ionomer fibers is crucial for proton transfer, as the physical confinement in thinner fibers may assist the alignment of the ionic domains in the longitudinal direction. Indeed, it has been observed that proton conductivity values of large (>2 µm diameter) Nafion® fibers are similar to that of bulk membrane (~100 mS cm^{-1}), and sharply increase as the diameter decreases (<1 µm).[72]

The electrospinning of PFSA ionomers is particularly challenging,[78,79] as they form aggregates or micelles when dispersed in aqueous and non-aqueous media.[80] Lacking adequate entanglement of their polymer chains, their dispersions give rise to an electrospray instead of electrospinning. Nevertheless, it is possible to prepare electrospun fibers by adding to the ionomer solution an inert carrier polymer increasing chain entanglement and viscosity. The most used is polyethylene oxide (PEO),[81–83] as well as polyacrylic acid (PAA),[84] polyvinylpyrrolidone (PVP),[85,86] polyvinyl alcohol (PVA)[83] and polyacrylonitrile (PAN).[87] The main disadvantage of this approach is a diminution of proton conductivity of the composite fibers due to the presence of non-conducting carrier that interacts with the ionomer and can interrupt the conductive pathways it formed, as was observed for the electrospinning of 5 wt% Nafion® dispersion with 16 wt% PEO.[82,83] To minimize the amount of carrier polymer needed, it is possible either to use high molecular weight carriers[71] or SSC and low equivalent weight ionomers.[88,89] Indeed, the amount of polymer carrier required for fiber formation was found to be significantly lower for SSC ionomers, which was attributed to greater inter-chain interactions than in LSC PFSA, leading to a more viscous dispersion. The electrospinning of LSC and SSC PFSA polymers with various equivalent weights (Nafion® EW 1100,[90] EW 1115,[91] 3M™ EW 825 and 733 [82]) was also investigated. To obtain homogeneous bead-free nanofibers of 3M PFSA 1 wt% of PEO (M_w = 300 kDa) and 10 wt% of ionomer dispersion were needed. Nevertheless, with higher molecular weight carrier (1000 kDa) only 0.3 wt% PEO was necessary. Another strategy employed to improve jet stability and PFSA fiber homogeneity is the use of solvents with high dielectric constant and dipole moment (*e.g.* dimethylacetamide, DMAc and dimethylformamide, DMF).[91]

Sulfonated polyaromatic hydrocarbon polymers can generally be electrospun without addition of a carrier because they form true solutions, by using adequate molecular weight, solvent and concentration.[92,93] Nanofibers of sulfonated polystyrene (sPS), sulfonated poly(ether ether ketone) (sPEEK), sulfonated polysulfone (sPSU), sulfonated poly(arylene ether sulfone) (sPAES), and sulfonated polyimides (sPI) have been obtained by electrospinning.[74,75,92,94–99]

Once the nanofiber web is obtained, dense composite membranes can be prepared by embedding it into a non-proton-conducting polymer matrix. The conventional method is based on impregnation/casting with a polymer (or prepolymer followed by crosslinking) to fill the voids of porous webs.[49] The interaction of the large surface area web with this matrix can affect crucial properties of the final membranes including ion conductivity, chemical, mechanical and thermal stability, fuel permeability. The characteristics of the composite systems are thus not a simple sum of the individual contributions of their components, as will be discussed in more detail in Section 9.2.3.

Pintauro *et al.*[82,90,95,100] prepared membranes based on PFSA or sulfonated aromatic polymer nanofibers impregnated with an inert crosslinkable monomer. Previously, the web was densified to increase the volume density of the nanofibers and weld them into a proton-conducting interconnected network. Mechanical properties and proton conductivity of the composite membranes were superior to those of monocomponent ionomer membranes, due to the inert robust polymer matrix and fiber web structure. Other membranes prepared with a similar approach and their characteristics are summarized in Table 9.1.[64,65,73,101–111]

The same group developed a single-step strategy to prepare dense composite fibrous membranes based on simultaneous electrospinning of charged and inert polymers, referred as 'dual electrospinning'.[91,112–115] The final composite is obtained by selectively melting the nanofiber web to be used as a matrix. Therefore, by using one dual electrospun web, two distinct membrane structures were prepared: (i) Nafion® nanofibers embedded in polyphenylsulfone (PPSU) and (ii) PPSU nanofibers embedded into Nafion®.[113] Both systems presented similar proton conductivity values directly related to the ionomer content, while the membrane type (i) demonstrated superior mechanical properties.

9.2.3 Composite Membranes with Electrospun Polymer Materials Embedded in an Ionomer Matrix

In another approach, composite nanofiber ionomer membranes are prepared by embedding a nanofiber polymer web into an ionomer matrix ensuring the proton conductivity. The electrospun material is thus a robust polymer mostly playing the role of mechanical reinforcement in an approach similar to that of e-PTFE/PFSA membranes (see Section 9.2).[38,116–118] The possibility of exploiting the synergy created by an extensive nanofiber/matrix interface drove the research towards the use of proton-conducting materials or polymers functionalized with acidic/basic moieties instead of inert polymers, dramatically increasing the possible materials associations (see Figure 9.2).

Reinforcing webs were prepared by electrospinning polysulfone (PS), polyimide (PI), polyvinylidene fluoride (PVDF), polystyrene (PS) and polyvinyl alcohol (PVA).[57,119–132] Their introduction in the form of fibers improved membrane mechanical properties in particular, even with no or weak (*e.g.* hydrophobic) interactions between them and the ionomer matrix. It is worth noting that when the same polymers (*e.g.* PVDF and Nafion®) were associated in blend membranes, they gave rise to poor mechanical strength and high cell resistance.[133–137] Our group developed Aquivion® membranes reinforced with electrospun PSU fiber webs and applied them for the first (and so far unique) time in WE. Dimensional and mechanical properties and performance in PEMWE were enhanced, while hydrogen crossover was reduced compared to non-modified Nafion® 212 and pristine Aquivion® membranes.[130]

With the aim of further improving membrane properties and durability, currently the challenge is the optimization of the interface between fibers and matrix. This can be applied for instance by chemical functionalization, exploiting the interactions between basic polymer nanofibers and acidic PFSA or sulfonated polyaromatic polymers, that may result in enhanced proton

Table 9.1 Overview of composite membranes comprising electrospun nanofibers of proton-conducting polymers embedded in a matrix of inert polymer.

Electrospun polymer	Matrix polymer	Membrane processing	Proton conductivity (mS cm^{-1})	Reference
sPI	sPI	Web impregnation	~100 (80 °C, 98% RH), 10 wt% nanofiber content	149
sPES	Nafion®	Web impregnation	87.5 (25 °C, 95% RH), 70 wt% nanofiber content	108
sPAES/sPOSS (60/40 w/w)	Norland Optical Adhesive (NOA) 63	Web impregnation	94 (30 °C, 80% RH), 70 vol% nanofiber content	101
sPAES	NOA 63	Web impregnation	86 (25 °C, 100% RH), 70 wt% nanofiber content	95
sPFEK/PES	—	Co-electrospinning layer by layer	61 (80 °C, 100% RH)	104
3M™ ionomer/PPSU	PPSU/3M™ ionomer	Dual electrospinning	93 (120 °C, 50% RH)	115
sPS	Nafion®	Web impregnation	180 (80 °C, 100% RH)	106
sPS/PEO (70/30 w/w)	Vinyl-terminated poly(dimethylsiloxane)	Web impregnation	100 (25 °C, 98% RH)	109
sPEEK/PVB (70/30 w/w)	sPEEK/PVA (65/35 w/w)	Web impregnation	13.5 (60 °C)	110
sPEEK/PVB (70/30 w/w)	sPEEK/PVA (65/35 w/w)	Web impregnation	38 (120 °C)	269
PVDF/PVA	Chitosan	Web impregnation	23 (25 °C, 100% RH)	270
sPPESK/ZCCH	—	Co-electrospinning followed by hot pressing	82 (160 °C)	97
sPPESK	sPPESK EW 580	Web impregnation	186.4 (80 °C, 100% RH)	145
sPEEK	Aquivion®	Web impregnation	110 (80 °C, 90% RH)	111
sPEK	PEK	Dual electrospinning	112 (80 °C, 100% RH)	112

hopping pathways.[138] In our group we have exploited such acid/base interaction in PBI/PFSA composite membranes. Electrospun poly[2,2′-(*m*-phenylene)-5,5′-dibenzimidazole] (PBI) webs were embedded into low E_w SSC Aquivion®. Membrane mechanical properties were improved compared to the pristine PFSA membranes: Young modulus increased from *ca* 40 MPa to *ca* 160 MPa at ambient conditions, while proton conductivity exceeded 30 mS cm^{-1} at 110 °C, 50% RH and 160 mS cm^{-1} at 80 °C, 95% RH. Furthermore, the outstanding durability of such composite membranes was demonstrated in fuel cell tests intended to accelerate chemical and mechanical degradation mechanisms (open-circuit-voltage hold testing at 85 °C and 13% RH, wet–dry cycling at OCV).[139] This superior stability was ascribed to the ionic crosslinking with proton transfer from the ionomer to the surface basic sites of PBI nanofibers, as well to the possible antioxidant properties of PBI (Section 9.2.1). Other composite membranes have exploited acid/base interaction between nanofibers and matrix, including sPEEK and chitosan[140] or polydopamine modified graphene oxide,[141] resulting in enhanced dimensional stability and proton conductivity. The effect of different types of interfacial fibers/polymer synergies (acid/acid, acid/base, acid/neutral, base/base, base/neutral) was investigated.[138] This systematic study confirmed that the acid/base interaction is the most effective in improving the membrane mechanical and proton transfer properties in hydrated and anhydrous conditions. Further proof of the decisive role of acid/base interactions is the improvement in tensile strength and proton conductivity when Aquivion® was reinforced by PS (weak hydrophobic interaction)[130] and 1,2,3-triazole functionalized PSU nanofibers[142] (strong ionic interaction).[143] The comparison of PSU/PFSA interface in dry conditions with (a) or without (b) functionalization is depicted in Figure 9.4, where the interaction is notably stronger in the presence of triazole moieties.

Electrospun polymer fibers were also functionalized with acidic groups potentially interacting with those of the ionomer matrix. Composite membranes based on Nafion®-functionalized PVDF nanofibers embedded into Nafion® presented higher elastic modulus (1840 MPa) and proton

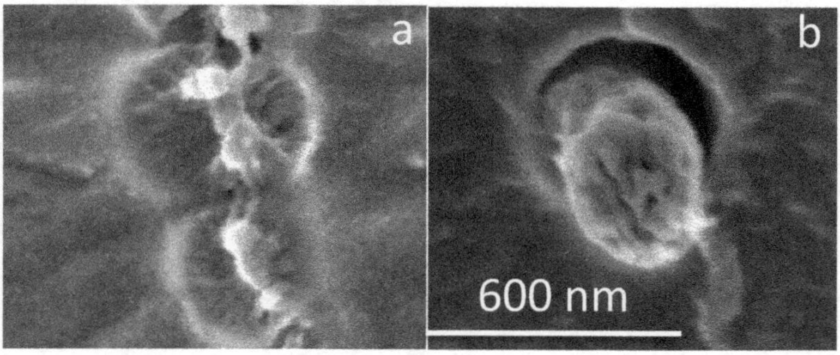

Figure 9.4 Details of SEM micrographs of PFSA membranes embedding bare (b) and 1,2,3-triazole functionalized PSU (a).[143]

conductivity values (60 mS cm^{-1}) than recast Nafion® (1280 MPa and 22 mS cm^{-1}). The power density in H_2/O_2 single cells for an MEA based on this composite membrane was higher than with MEAs comprising non-reinforced Nafion®.[122] These findings may be ascribed to the aggregation of ionomer chains on the nanofiber surface forming of proton-conducting channels. PBI nanofibers were also doped with phytic acid and embedded into Nafion®, enabling the preparation of ultra-thin (<5 μm) robust membranes.[144] They also were more conducting than the pristine recast Nafion® membrane in low humidity conditions (3.1 mS cm^{-1} *vs* 1.2 mS cm^{-1} at 80 °C and 40% RH). The MEA comprising the composite membrane outperformed the recast Nafion®-based MEA, especially at low relative humidity.

The approach described in this paragraph was not only extended to the incorporation of fibers of highly interacting polymers but also of ionomer fibers into ionomer matrix (see Table 9.1). Using the same ionomer in the fiber mat and in the matrix (*e.g.* sulfonated polyphthalazinone ether sulfone ketone,[145] sPEEK,[110,146] and sPI[74,77,147–149]) increased proton conductivity, mechanical, dimensional and thermal stability and gas barrier properties compared to the corresponding homogeneous cast membranes. This demonstrated the role of the membrane architecture at the nano/microscale driven by the incorporation of the nanofiber webs in increasing mechanical robustness and proton transfer. The dissolution of the nanofiber mat during impregnation with the same polymer was avoided by dissolving the latter either at high temperature[145] or in a non-protonated form[77] or using crosslinking reactions. In another approach, different proton-conducting polymers were used for the fibers and the matrix. Among them, sPEEK and sPI nanofibers were embedded in PFSA giving rise to composite membranes with improved mechanical and dimensional stability.[111,146,150] The advances in fuel cell and electrolyser membranes and the versatility of the electrospinning technique led also to the implementation of novel preparation processes, including dual electrospinning,[115] direct membrane deposition onto gas diffusion electrodes followed by inkjet-printing of Nafion®[151] and *in situ* polymerization.[152]

Besides ionic crosslinking, covalent crosslinking is widely used to increase mechanical strength and dimensional stability of polymer electrolyte membranes,[10] and has been used especially with sulfonated polyaromatic membranes. The drawback is that when ionic functions form the crosslinks, the proton conductivity can be diminished. The advantage of using crosslinking in electrospun fibers is the limitation of this unfavorable effect and of the fiber solubilization during impregnation. It can be performed *via* the application of heat, light, or pressure or the addition of bi-functional molecular crosslinkers. The latter can be used into the electrospinning solution[110,153,154] or react with the electrospun web in vapor or liquid form.[121,155,156] Using this strategy, to reduce their solubility and render possible their impregnation with Nafion®, electrospun PVA webs were exposed to glutaraldehyde vapor to induce crosslinking between their –OH side groups. This resulted in PVA/Nafion® membrane with improved mechanical properties.[121,157] Thermally induced crosslinking was also applied to sPEEK fibers to allow the use of highly functionalized thus proton-conducting sPEEK in composite membranes, demonstrating reduced swelling and increased tensile properties.[158,159]

9.3 Electrospun Electrocatalyst Supports

Performance and durability of PEMFC and PEMWE electrodes are strongly related to morphology, composition and structure of the electrocatalyst used and its support material. The limiting reactions for the two devices are the oxygen reduction reaction (ORR, at the PEMFC cathode) and the oxygen evolution reaction (OER, at the PEMWE anode), respectively. The conventional electrocatalyst to accelerate the sluggish rate of these reactions are Pt nanoparticles deposited onto support materials for PEMFC and unsupported IrO_2/RuO_2 for PEMWE. The support material needs to possess high electrical conductivity, high surface area to maximize catalyst dispersion and the three-phase boundary and porosity to allow facile access to the active sites, as well as high chemical and electrochemical stability in operating conditions.[160] Specific electronic interactions and adequate architectures leading to ultra-low electrocatalyst loading (as for noble metals such Pt)[161] can further improve activity and lifetime of the electrodes.

Commercial high surface area carbon blacks (*e.g.* Vulcan XC-72, Black Pearl 2000, Shawinigan and Ketjen Black) are conventionally used as the electrocatalyst support in PEMFC, as they mostly satisfy these requirements. However, the carbon corrosion occurring in fuel cell operating conditions according to reaction (eqn (9.1)) dramatically affects the electrode durability.[162,163]

$$C + 2H_2O \rightarrow CO_2 + 4H^+ + 4e^- \tag{9.1}$$

Indeed, during cell operation, high potentials can occur at both cathode and anode sides as a consequence of start-stop and fuel starvation.[164,165] These phenomena, along with high humidity and temperature,[166] can also promote the degradation of the metal nanocatalyst (*e.g.* aggregation, detachment and migration)[162,167,168] leading to a global loss of activity of the device.

In the case of WE, anode catalysts are often non-supported. A recent area of research focuses on the development of corrosion-resistant supports as a possible means of enhancing their activity and reducing precious metal (*e.g.* Ir) loading. Therefore, the long-term use of carbon supports is not possible, and titanium, metal carbides and oxides are used instead.[3]

These issues have pushed research towards novel corrosion-resistant support materials, with high conductivity and porosity. Beside the chemical composition, the support/electrode architecture plays a crucial role. In this regard, 1D nanomaterials, such as nanofibers and nanotubes, and in particular those prepared by electrospinning, have gained increased attention.[47,169] In the past 20 years, not only carbon-based, but also metal oxide, nitrides and carbide electrospun nanofiber catalyst supports have been developed[12,170] and are reviewed in the following paragraphs.

9.3.1 Carbon-based Nanofibrous Supports

In spite of its instability at high potential, carbon is still the material of choice as electrocatalyst support, and was investigated in the form of electrospun carbon nanofibers (CNFs). Because of high reproducibility and high yield, the most used preparation of these materials by electrospinning involves a solution of polyacrylonitrile (PAN) as a precursor, followed by thermal treatments. Other polymers (polyvinyl alcohol [PVA], polyvinylidene fluoride [PVDF], polyacrylic acid [PAA], polyimide [PI], polybenzimidazole [PBI]) were also reported as viable alternatives to PAN.[171] Typically, electrospun PAN nanofibers undergo a series of stabilization thermal treatments in air to allow cyclization, and then a final carbonization step in an inert atmosphere to achieve (at least partially) an sp^2 carbon structure. Other polymers, such as the aromatic PBI, do not need a multistep procedure and can be directly carbonized, avoiding time and energy-consuming processes,[171] although it is reported that CNF mats issued from the electrospinning of PBI are more fragile than the those based on PAN.[172] The type and concentration of the precursor as well the electrospinning conditions allow carbon nanofiber webs to be obtained, eventually self-standing, with promising connectivity and mechanical properties.[173–176] The morphological characteristics of carbon nanofiber webs allows a variety of platinum deposition methods to be applied. Polyol methods have been extensively used to synthesize Pt nanoparticles to be deposited onto CNFs.[177–179] H_2/O_2 PEMFC performance of so-prepared Pt/CNFs was superior to that of Pt/Vulcan XC-72R, which was attributed to the improvement of the triple-phase boundary due to the higher surface area of the fiber-based electrodes.[177] Furthermore, the shallow pores with rough surfaces of the CNF support increased Pt utilization from 35% (for Pt/carbon black) to 69%. The introduction of controlled porosity in the fibers can affect the dispersion and stability of Pt nanoparticles. That was demonstrated by Wang *et al.*,[179] who reported that platinum covered porous CNFs exhibited higher and more durable electrocatalytic activity than commercial catalyst and Vulcan XC72-supported platinum prepared with the same procedure. Single-cell PEMFC tests also demonstrated the better performance at high current densities due to the 3D connected architecture achieved with the electrospun materials, decreasing mass-transfer resistance, facilitating water drainage and reactant transport.

It is therefore evident that the preparation of self-standing CNF webs, potentially directly used in fuel cell electrodes or gas diffusion layers (GDLs) is a promising development for fuel cell application.[180,181] This approach may allow the complete exploitation of the porosity of the structure for water and gas transfer. Furthermore, electrospun self-standing webs represent an ideal substrate for the deposition of continuous thin film electrocatalysts instead of discrete nanoparticles,[182] which are known to maximize the utilization of platinum as well as the corrosion protection of the support, while minimizing the precious metal amount.[183] In our group we developed an overpotential deposition method to obtain thin Pt nanostructures on self-standing CNF webs that indeed presented enhanced activity and stability.[182] The use of electrospun materials with high aspect ratio and porous structure of the corresponding webs are expected to bring advanced properties when directly used in MEAs, in particular with regard to mechanical, mass transport and ionic and electronic transfer properties, with beneficial effects on the performance and lifetime of the resulting electrodes. Nevertheless, recent work has highlighted issues that need to be addressed for the full direct exploitation of nanofibrous electrodes obtained from self-standing electrospun mats, mainly their adequate thickness and porosity, their mechanical stability, and their interaction with the ionomer.[181,184]

9.3.2 Metal Oxide-based Nanofibrous Support

Besides carbon-fiber type materials,[185] metal oxides have been considered as electrocatalyst supports in particular because of their higher electrochemical stability and corrosion resistance at high voltage.[12,186,187] Thermochemical calculations also demonstrated the high stability of oxides of several transition and post-transition metals (Ti, Nb, Ta, W, Sb, Sn) at 80 °C, 1.0 V/SHE and pH = 0.[188] Furthermore, the strong metal–support interaction (SMSI) that may take place in catalyzed metal oxides can both promote the electrocatalyst activity and increase its durability.[189] Materials developed so far include TiO_x,[190–192] SnO_2,[193–196] SiO_2,[197] WO_x[198,199] and MoO_x.[200] The main issue with these materials is their poor electrical conductivity, which can be enhanced either by doping with aliovalent ions (*e.g.* Nb, V, Ta, Sb)[188,201,202] or by forming non-stoichiometric oxides by chemical reduction.[203,204]

In the past decade several metal oxide electrocatalyst supports were prepared by electrospinning,[205] in particular titanium dioxide catalyzed with Pt for ORR.[178,206–209] Nb-doped anatase[206] and rutile@sub-oxide,[178] and Ti_4O_7[208] nanofibers showed higher stability than commercial Pt/C after prolonged electrochemical cycling, keeping a higher percentage of their electrochemical surface area (ECSA). However, their ECSA and ORR activity was generally lower than those of reference carbon-based electrocatalysts. Many factors can contribute to these findings, from their lower electrical conductivity to the Pt nanoparticle encapsulation that may take place on metal oxides associated with the SMSI discussed above.[210,211] Another issue that occurred for the sub-oxide fibrous support[208] is the drop of ORR mass activity observed after cycling tests, due to the formation of an electronically insulating oxide on the nanofiber surface.

Pt/SnO_2 electrospun fibers demonstrated electroactivity towards the hydrogen oxidation reaction (HOR), but not ORR.[212,213] In contrast, Pt deposited onto tin oxide loose-tubes doped with Nb exhibited ORR activity as well as improved stability compared to Pt deposited onto carbon-based electrodes.[214–216] SEM analysis performed on the Nb–SnO_2 cathodes after the *in situ* accelerated stress tests confirmed that the tubular morphology was retained and Pt nanoparticles did not agglomerate nor detach from the support, in contrast to what occurred in the carbon-based cathode.[215] The same materials, when doped with Sb, were very highly electrically conductive, similar to those of carbon (1 S cm^{-1} at 20 °C), and provided higher performance in a fuel cell, while keeping high stability to degradation over time.[217]

In order to improve the dispersion of the Pt electrocatalyst on metal oxide nanofiber several approaches were developed, including *in situ* growth of the metal catalyst,[209,212] galvanic replacement,[218] and atomic layer deposition (ALD).[219]

A crucial issue concerning metal oxide nanofiber webs is their brittleness, which renders difficult the possibility of their direct use in MEAs. However, the use of purely inorganic Nb–TiO_2 supports for Pt ALD was reported.[219,220] The consequent electrochemical characterization demonstrated high stability with only 10% loss of initial activity after 30 000 potential cycles up to 1.0 V_{RHE}. Another strategy to circumvent the low mechanical properties of oxide webs is the preparation of composite materials with carbon nanofibers that can eventually be freestanding. This also gives the advantage of enhancing the support conductivity.

C/MO_x composite materials were prepared by electrospinning metal oxide precursors[218,221] or nanoparticles[222] with polymers followed by thermal treatment. For instance, Nb-doped TiO_2/carbon catalyzed with Pt nanocatalyst presented an ORR activity comparable to that of a commercial Pt/C with an increased durability.[221] H_2/O_2 fuel cell tests were also carried out with this composite material at the cathode side, yielding a peak power density of 0.34 W cm^{-2} at 75 °C.

Another two-step approach to achieving a 'core/sheath' nanocomposite comprised preparing CNF and their decoration by MO_x. Following this strategy, we selectively decorated electrospun Pt/CNF with tin dioxide using ALD.[223] The presence of SnO_2 enhanced the catalyst durability during voltage cycling and not inhibit the Pt active sites, with twice the ECSA retained after 10 000 cycles to 1.2 *V*/NHE (normal hydrogen electrode) compared to that observed with unprotected Pt/CNFs.

Metal oxide electrospun nanofibers were also applied in PEMWE cell components, both as supports for the anode and cathode sides. Porous Sb–SnO_2 fibers were used as iridium oxide support (Figure 9.5a and b) for the OER.[224] IrO_2 demonstrated a great enhancement of OER mass activity compared with that of the same catalyst unsupported or supported onto Sb–SnO_2 nanoparticles (Figure 9.5c and d) possibly due to the high porosity and electrical conductivity of the support. PEMWE at 80 °C at 0.45 A cm^{-2} demonstrated the stability of the nanofibrous catalyst up to around 650 h.

Sb–SnO_2 has also been used as support for cobalt hexacyanoferrate,[225] demonstrating similar activity for OER as other non-noble metal electrocatalysts.[226] Porous perovskite $La_{0.6}Sr_{0.4}Co_{0.6}Fe_{0.4}O_{3-\delta}$ (LSCF) nanofibers obtained by electrospinning were also used for OER electrocatalysis, presenting a lower potential (647 mV/Ag/AgCl) at 10 mA cm^{-2} compared to a powder-type commercial LSCF (786 mV) and conventional IrO_2 catalysts (660 mV).[227] Furthermore, they demonstrated higher durability compared to the other electrocatalysts, with an almost unvaried potential at 10 mA cm^{-2} for more than 3 h.

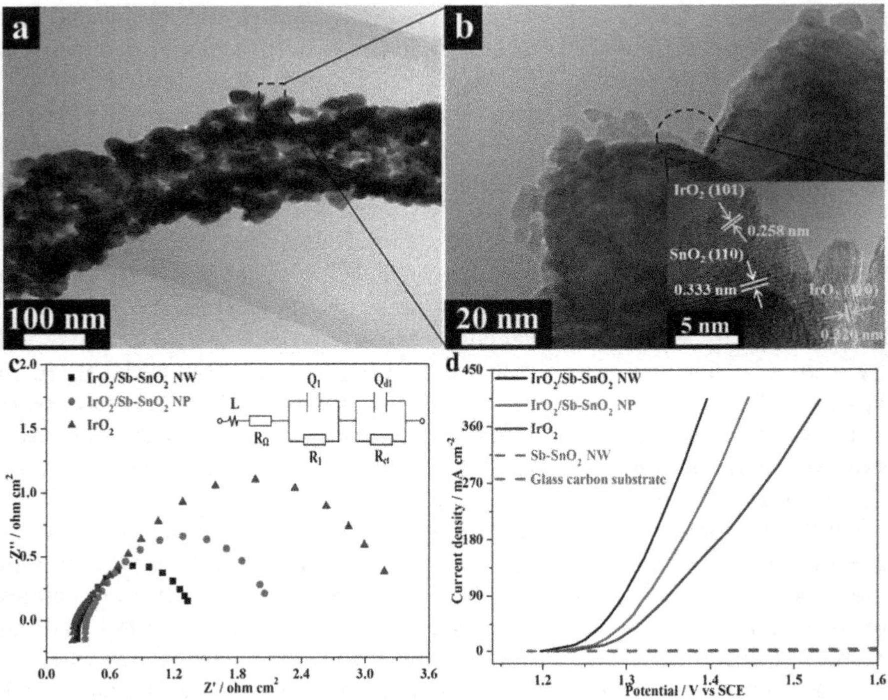

Figure 9.5 (a) Low and (b) high magnification TEM images (inset is the corresponding HRTEM) of IrO_2/Sb–SnO_2 nanofibers. (c) Nyquist diagram of three types of prepared catalysts measured at 1.25 *V*/SCE during oxygen evolution in 0.5 M H_2SO_4 at 25 °C [the inset is an equivalent circuit]. (d) Steady-state polarization curves of the three types of prepared catalysts in 0.5 M H_2SO_4 at 25 °C at a scan rate of 1 mV s^{-1}. Reproduced from ref. 224 with permission from the Royal Society of Chemistry.

Other recent work describes the use of electrospun oxide fibers for hydrogen evolution reaction (HER) catalysis, for example $Ca_3Co_4O_9$ and $MgCo_3O_5$ nanofibers that gave low HER Tafel slopes (76 and 98 mV/decade, respectively),[228] and WO_{3-x} nanoplates grown onto CNFs that provided low HER overpotential, large cathodic current density, and high exchange current density.[229]

9.3.3 Nitride and Carbide Nanofibrous Supports

Transition metal nitrides and carbides are another class of alternative support materials, with high stability and electrical conductivity, which can act both as catalyst support and as catalysts themselves.[170,230] Little work is reported so far on electrospun metal nitrides. TiN nanofibers prepared by electrospinning and catalyzed with platinum nanoparticles[231] led to ORR performance comparable to conventional Pt/C catalysts and higher stability in accelerated stress testing, which was ascribed as due to the stronger interaction and consequent stabilization of the nanocatalyst with the support as well as to the extended connectivity and high inertness of TiN. Carbon nitride particles also demonstrated ORR activity and were incorporated into electrospun PAN fibers.[232]

In a study from Liu *et al.*,[233] carbon nanofibers embedding TiC crystallites and surface-decorated with platinum nanoparticles presented enhanced ORR onset potential and current density, reaching mass activities as high as 300 A/g_{Pt}. These findings were attributed to the high specific surface area of the fibrous support and the synergic effect (SMSI) of TiC and Pt on electrocatalysis.

Niobium carbide/carbon porous nanotubes prepared in our group by electrospinning and catalyzed with Pt nanoparticles exhibited also improved stability over high potential cycling.[234,235] They presented comparable initial mass activity and retained 31% of the initial ECSA after 10 000 cycles between 0.6 and 1.4 *V*/NHE, while the Pt/C reference only retained 5%. The presence of a non-conducting niobium oxide surface layer did not hinder the electroactivity, instead it probably played a role in enhancing the corrosion resistance of the carbide support. The preparation of an MEA based on a cathode using Pt/NbC-C led to a maximum power density of 0.43 W cm^{-2} in a single fuel cell comparable with other non-carbon supports.

9.4 Self-supported Electrocatalysts

The preparation of self-supported electrocatalysts by electrospinning holds promise for enhancing the gas accessibility and the charge transfer in PEMFC electrodes. Nevertheless, this research area is more limited compared to that on support materials, probably due to the difficulty in achieving diameters comparable to conventional nanocatalysts and freestanding electrodes. Nevertheless, metal (Cu, Ni, Pt), or alloy (Pt–Fe) nanofibers were prepared,[223,236–239] And in general, presented higher ORR catalytic activity and stability than conventional Pt/C towards oxygen reduction.[238]

The drive for electrospun noble metal-free self-supported catalyst led to the exploration for suitable alternative materials with catalytic properties. On the one hand, mention must be made here of the metal oxides and carbides that were already reported in Sections 9.3.2 and 9.3.3 as catalyst supports, but can also present ORR electroactivity in several media.[240] Due to the similarity of its electronic structure to that of platinum, tungsten carbide was the earliest carbide to be assessed as electrocatalyst.[241] Electrospun carbon-coated WC nanofibers demonstrated catalytic activity towards the ORR and stability that were further improved by NH_3 post-treatment at high temperature.[242] These authors considered that a carbon coating could play a role in the ORR electrocatalysis. Concerning metal oxides, WO_3 nanofibers demonstrated activity towards the HOR, which was further improved by the addition of carbon black.[243]

On the other hand, CNFs can be used not only as a support for metal electrocatalysts, but also as catalyst themselves.[244–246] Indeed, carbon nanostructures doped with heteroatoms (mostly nitrogen) are able to form porous materials active towards the oxygen reduction reaction. The preparation of these structures involves the necessity of fine control over nanostructure and dimensions of the pores and the active sites.[247–252] It is also possible to couple these structures together with non-noble metals (*e.g.* Fe and Co) to build ORR active fibers.[253–259] In an original approach, bimetallic Fe–Co nanoparticles were used in PAN-based CNF, promoting the growth of carbon nanotubes during high temperature treatment in NH_3 atmosphere. The Fe/Co-based materials formed demonstrated higher durability with electroactivity similar to that of commercial Pt/C, which was attributed to the high stability of the carbons with high graphitization degree. One appealing strategy to incorporate metal active sites is the electrospinning of carbon precursors with iron salts or metal-organic frameworks.[254,260] These materials show promising activity towards oxygen reduction, as well as greater tolerance to methanol poisoning than conventional Pt/C catalysts.[260] When used in a single-cell test as cathodes, they delivered exceptional volumetric activities of 3.3 A cm^{-3} at 0.9 V while presenting improved durability in operating conditions.[254]

9.5 Electrospun Electrodes

Together with the use of self-standing webs as electrocatalyst supports, electrospinning was recognized as a promising approach to prepare dimensionally ordered electrode architectures. The fabrication of electrode nanofibers comprising ionomer and electrocatalyst demonstrated several interesting features potentially improving the continuous pathways for proton and/or electron transfer within catalyst layers. Non-woven nanofibrous cathodes prepared by electrospinning of a catalyst/ionomer mixture (commercial Pt/C catalyst, Nafion® ionomer and poly(acrylic acid) [PAA] binder) exhibited high performance in an H_2/air fuel cell with a Pt loading as low as 0.1 mg Pt/cm^{-2}.[261] These results were attributed to the highly uniform distribution of Pt/C electrocatalyst and Nafion® in the nanofiber web. Furthermore, the highly porous and interconnected architecture of electrospun cathode facilitated proton transport and oxygen diffusion to catalyst sites, and water transport within the electrode. Compared to conventional painted electrodes, electrospun inks exhibited outstanding stability in accelerated durability tests.[261,262] To further improve the durability over potential cycling, PVDF was added in several percentages to the electrospinning suspension formulation to enhance the hydrophobicity of the cathode. Indeed, reducing the water content at the electrocatalyst surface effectively mitigated carbon corrosion.[263] In a similar approach, in order to improve the water management in the electrode layers when operating at high current densities, 5 wt% PTFE was added to the electrode ink, based on Pt/C, Nafion® and PAA.[264] Coupling of electrospinning with other techniques such as electrospray[265] and underpotential deposition was also proposed.[266]

9.6 Conclusion

This overview of polymer, inorganic and hybrid materials resulting from electrospinning and applied in all the core components of PEMFC and PEMWE clearly demonstrates the versatility of the technique, whether in terms of chemical composition, morphologies, architectures or associations. Furthermore, the rational assembly of electrospun nanofibers with ionomers or electrocatalysts to build up composite 3D hierarchical materials for membranes and electrodes allows control over properties and interfaces. Great advances have been realized for fuel cell devices, and are emerging for WE.

Amongst the challenges remaining to address, the direct integration of a catalyzed electrospun support web into an MEA is of great interest and to date little explored, still requiring optimization of the porosity and of the fiber interconnection to ensure efficient proton, electron and reactant transport. Furthermore, this kind of self-standing electrode is currently achievable only for carbon, as electrospun ceramic supports are particularly brittle. As one means of circumventing this issue, and also to extend this application to PEMWE, the preparation of twisted yarns[267] and ribbons[268] can be investigated.

To decrease the precious metal loading in PEMFC and PEMWE, the combination of electrospinning with techniques allowing thin film deposition (*e.g.* ALD and overpotential electrodeposition) or the preparation of support fiber bearing non-noble metal active sites (*e.g.* Fe, Co) are valuable approaches under investigation.

Composite nanofiber membranes are showing promise of becoming a viable alternative to conventional reinforced membranes for FCs. Furthermore, the specific architectures developed notably with reinforcement across the membrane thickness are of great interest for WE where the industry targets very high pressure differentials. The main challenge remains the control of the fiber/matrix interface that so far is performed mainly by ionically crosslinking the electrospun and matrix components. The implementation of covalent crosslinking by surface modification of the fibers (*e.g.* by ALD, reactive plasma, electron beam) is a promising area for future investigations. To increase membrane chemical stability, another advantage of electrospinning is the possibility to prepare radical scavenger porous layers that can be selectively located in the MEA.

The final goal of an all-fibrous electrospun membrane–electrode assembly is approaching, with great promise for PEMFC and PEMWE but also other sister technologies rapidly emerging such as redox flow systems. The possibility of up-scale and industrialization of the electrospinning technique is further proof of the real impact it can have on hydrogen FCs and electrolysers.

While the commercialization of nanofiber-based materials for PEMFC and PEMWE is not yet a reality, non-woven nanofiber webs have already been introduced in various marketed technologies, *e.g.* filtration membranes based on Ultra-Web® technology,[271] separators for high performance lithium ion batteries.[272] For fuel cell and electrolyser applications, and as emphasized throughout this chapter, the fibers are part of composite systems: membranes or electrodes. Techniques for membrane reinforcement by ePTFE or woven fabrics are already in use, with opportunity for the implementation of electrospun nanofiber webs into these existing processes. For instance, membranes with reinforcing electrospun nanofiber mats are being up-scaled in the European project VOLUMETRIQ,[273] opening up exciting possibilities in terms of composition and architecture, thereby leading to a new generation of nanocomposite electrolytes for energy conversion devices.

Acknowledgements

The author would like to acknowledge funding received from the European Research Council under the European Union's Seventh Framework Programme (FP/2007–2013)/ERC Grant Agreement no. 306682.

References

1. E. P. Giannelis, *Adv. Mater.*, 1996, **8**, 29–35.
2. R. Borup, J. Meyers, B. Pivovar, Y. S. Kim, R. Mukundan, N. Garland, D. Myers, M. Wilson, F. Garzon, D. Wood, P. Zelenay, K. More, K. Stroh, T. Zawodzinski, J. Boncella, J. E. McGrath, M. Inaba, K. Miyatake, M. Hori, K. Ota, Z. Ogumi, S. Miyata, A. Nishikata, Z. Siroma, Y. Uchimoto, K. Yasuda, K.-I. Kimijima and N. Iwashita, *Chem. Rev.*, 2007, **107**, 3904–3951.
3. M. Carmo, D. L. Fritz, J. Mergel and D. Stolten, *Int. J. Hydrogen Energy*, 2013, **38**, 4901–4934.
4. A. Züttel, A. Borgschulte and L. Schlapbach, *Hydrogen Future Energy Carrier*, Wiley-VCH Verlag GmbH & Co. KGaA, Weinheim, Germany, 2008.
5. S. Sharma and G. Sib Krishna, *Renewable Sustainable Energy Rev.*, 2015, **43**, 1151–1158.
6. A. Alaswad, A. Baroutaji, H. Achour, J. Carton, A. Al Makky and A. G. Olabi, *Int. J. Hydrogen Energy*, 2016, 1–10.
7. O. Z. Sharaf and M. F. Orhan, *Renewable Sustainable Energy Rev.*, 2014, **32**, 810–853.
8. I. Dincer and C. Acar, *Int. J. Hydrogen Energy*, 2015, **40**, 11094–11111.
9. P. Millet, R. Ngameni, S. A. Grigoriev, N. Mbemba, F. Brisset, A. Ranjbari and C. Etiévant, *Int. J. Hydrogen Energy*, 2010, **35**, 5043–5052.
10. S. Subianto, M. Pica, M. Casciola, P. Cojocaru, L. Merlo, G. Hards and D. J. Jones, *J. Power Sources*, 2013, **233**, 216–230.
11. A. Rabis, P. Rodriguez and T. Schmidt, *ACS Catal.*, 2012, **2**, 864–890.
12. L. Du, Y. Shao, J. Sun, G. Yin, J. Liu and Y. Wang, *Nano Energy*, 2016, **29**, 314–322.
13. J. H. Wendorff, S. Agarwal and A. Greiner, *Electrospinning: Materials, Processing and Applications*, Wiley VCH, Weinheim, Germany, 2012.

14. S. Subianto, D. Cornu and S. Cavaliere, in *Electrospinning for Advanced Energy and Environmental Applications*, ed. S. Cavaliere, CRC Press, Boca Raton, 2015, pp. 1–27.
15. M. Zatoń, J. Rozière and D. Jones, *Sustainable Energy Fuels*, 2017, 409–438.
16. Q. Berrod, S. Lyonnard, A. Guillermo, J. Ollivier, B. Frick, A. Manseri, B. Améduri and G. Gébel, *Macromolecules*, 2015, **48**, 6166–6176.
17. K. A. Mauritz and R. B. Moore, *Chem. Rev.*, 2004, **104**, 4535–4585.
18. K. D. Kreuer and G. Portale, *Adv. Funct. Mater.*, 2013, **23**, 5390–5397.
19. A. Ursua, L. M. Gandia and P. Sanchis, in *Proceedings of the IEEE*, 2012, vol. 100, pp. 410–426.
20. A. Ghielmi, P. Vaccarono, C. Troglia and V. Arcella, *J. Power Sources*, 2005, **145**, 108–115.
21. D. J. Jones, *Mater. Matters*, 2015, **10**, 88–99.
22. T. Higashihara, K. Matsumoto and M. Ueda, *Polymer*, 2009, **50**, 5341–5357.
23. M. A. Hickner, H. Ghassemi, Y. S. Kim, B. R. Einsla and J. E. McGrath, *Chem. Rev.*, 2004, **104**, 4587–4611.
24. F. A. De Bruijn, V. A. T. Dam and G. J. M. Janssen, *Fuel Cells*, 2008, **8**, 3–22.
25. J. Peron, E. Ruiz, D. J. Jones and J. Rozière, *J. Membr. Sci.*, 2008, **314**, 247–256.
26. F. Ng, D. J. Jones, J. Rozière, B. Bauer, M. Schuster and M. Jeske, *J. Membr. Sci.*, 2010, **362**, 184–191.
27. T. J. Peckham and S. Holdcroft, *Adv. Mater.*, 2010, **22**, 4667–4690.
28. C. H. Park, C. H. Lee, M. D. Guiver and Y. M. Lee, *Prog. Polym. Sci.*, 2011, **36**, 1443–1498.
29. D. S. Kim, A. Labouriau, M. D. Guiver and Y. S. Kim, *Chem. Mater.*, 2011, **23**, 3795–3797.
30. B. Bae, T. Yoda, K. Miyatake, H. Uchida and M. Watanabe, *Angew. Chem., Int. Ed.*, 2010, **49**, 317–320.
31. M. Tanaka, M. Koike, K. Miyatake and M. Watanabe, *Macromolecules*, 2010, **43**, 2657–2659.
32. D. J. Jones and J. Rozière, in *Fuel Cells I - Advances in Polymer Science*, Springer Berlin Heidelberg, Berlin, Heidelberg, 2008, vol. 215, pp. 219–264.
33. J. A. Kerres, *J. Membr. Sci.*, 2001, **185**, 3–27.
34. J.-T. Wang, R. F. Savinell, J. Wainright, M. Litt and H. Yu, *Electrochim. Acta*, 1996, **41**, 193–197.
35. K. D. Kreuer, A. Fuchs, M. Ise, M. Spaeth and J. Maier, *Electrochim. Acta*, 1998, **43**, 1281–1288.
36. Q. Li, R. He, J. O. Jensen and N. J. Bjerrum, *Chem. Mater.*, 2003, **15**, 4896–4915.
37. R. Sood, C. Iojoiu, E. Espuche, F. Gouanvé, H. Mendil-Jakani, S. Lyonnard, F. Gouanvé, H. Mendil-Jakani and S. Lyonnard, *J. Membr. Sci.*, 2015, **495**, 445–456.

38. B. Bahar, A. R. Hobson, J. A. Kolde and D. Zuckerbrod, U.S. Patent, US 5547551, 1996.
39. P. Trogadas, J. Parrondo and V. Ramani, *ACS Appl. Mater. Interfaces*, 2012, **4**, 5098–5102.
40. M. Zatoń, J. Rozière and D. J. Jones, *J. Mater. Chem. A*, 2017, **5**, 5390–5401.
41. G. Alberti, M. Casciola, D. Capitani, A. Donnadio, R. Narducci, M. Pica and M. Sganappa, *Electrochim. Acta*, 2007, **52**, 8125–8132.
42. V. Di Noto, R. Gliubizzi, E. Negro, M. Vittadello and G. Pace, *Electrochim. Acta*, 2007, **53**, 1618–1627.
43. C. Laberty-Robert, K. Vallé, F. Pereira and C. Sanchez, *Chem. Soc. Rev.*, 2011, **40**, 961–1005.
44. Y. Higuchi, I. Terada, H. Shimoda and S. Hommura, U.S. Patent, US 2001/0026883 A1, 2001.
45. S. Haufe and U. Stimming, *J. Membr. Sci.*, 2001, **185**, 95–103.
46. C. Mittelsteadt and J. Willey, in *7th International Energy Conversion Engineering Conference*, American Institute of Aeronautics and Astronautics, Reston, Virigina, 2009, pp. 1–14.
47. S. Cavaliere, S. Subianto, I. Savych, D. J. Jones and J. Rozière, *Energy Environ. Sci.*, 2011, **4**, 4761.
48. V. Thavasi, G. Singh and S. Ramakrishna, *Energy Environ. Sci.*, 2008, **1**, 205–221.
49. R. Sood, S. Cavaliere, D. J. Jones and J. Rozière, *Nano Energy*, 2016, **26**, 729–745.
50. G. Alberti and M. Casciola, *Annu. Rev. Mater. Res.*, 2003, **33**, 129–154.
51. V. Neburchilov, J. Martin, H. Wang and J. Zhang, *J. Power Sources*, 2007, **169**, 221–238.
52. V. Antonucci, A. Di Blasi, V. Baglio, R. Ornelas, F. Matteucci, J. Ledesma-Garcia, L. G. Arriaga and A. S. Aricò, *Electrochim. Acta*, 2008, **53**, 7350–7356.
53. V. Baglio, R. Ornelas, F. Matteucci, F. Martina, G. Ciccarella, I. Zama, L. G. Arriaga, V. Antonucci and A. S. Aricò, *Fuel Cells*, 2009, **9**, 247–252.
54. M. P. Rodgers, L. J. Bonville, H. R. Kunz, D. K. Slattery and J. M. Fenton, *Chem. Rev.*, 2012, **112**, 6075–6103.
55. M. Rezakazemi, M. Sadrzadeh, T. Mohammadi and T. Matsuura, in *Organic-inorganic Composite Polymer Electrolyte Membranes: Preparation, Properties, and Fuel Cell Applications*, ed. D. Inamuddin, A. Mohammad and A. M. Asiri, Springer International Publishing, Cham, 2017, pp. 311–325.
56. H. Wang, X. Li, X. Zhuang, B. Cheng, W. Wang, W. Kang, L. Shi and H. Li, *J. Power Sources*, 2017, **340**, 201–209.
57. H. H.-J. Lee, J.-H. J. Kim, J.-H. J. Won, J.-M. Lim, Y. T. Hong and S.-Y. Lee, *ACS Appl. Mater. Interfaces*, 2013, **5**, 5034–5043.
58. J.-R. Lee, N.-Y. Kim, M.-S. Lee and S.-Y. Lee, *J. Membr. Sci.*, 2011, **367**, 265–272.
59. K. Ketpang, K. Lee and S. Shanmugam, *ACS Appl. Mater. Interfaces*, 2014, **6**, 16734–16744.

60. Y. Yao, B. Guo, L. Ji, K.-H. Jung, Z. Lin, M. Alcoutlabi, H. Hamouda and X. Zhang, *Electrochem. Commun.*, 2011, **13**, 1005–1008.
61. K. Ketpang, B. Son, D. Lee and S. Shanmugam, *J. Membr. Sci.*, 2015, **488**, 154–165.
62. K. Ketpang, K. Oh, S.-C. Lim and S. Shanmugam, *J. Power Sources*, 2016, **329**, 441–449.
63. S. Subianto, A. Donnadio, S. Cavaliere, M. Pica, M. Casciola, D. J. Jones and J. Rozière, *J. Mater. Chem. A*, 2014, **2**, 13359.
64. J. Choi, C. Lee, S. Hawkins, C. Huynh, J. Park, Y. Jeon, Y. Truong, I. L. Kyratzis, Y.-G. Shul and R. Caruso, *RSC Adv.*, 2014, **4**, 32787–32790.
65. C. Lee, S. M. Jo, J. Choi, K.-Y. Baek, Y. B. Truong, I. L. Kyratzis and Y.-G. Shul, *J. Mater. Sci.*, 2013, **48**, 3665–3671.
66. L. Dos Santos, S. Rose, O. Sel, M. Maréchal, H. Perrot and C. Laberty-Robert, *J. Membr. Sci.*, 2016, **513**, 12–19.
67. V. Maneeratana, J. D. Bass, T. Azaïs, A. Patissier and K. Vallé, *Adv. Funct. Mater.*, 2013, **23**, 2872–2880.
68. M. Zatoń, D. Jones and J. Roziere, *ECS Trans.*, 2014, **61**, 15–23.
69. J. Li, K. Wang and Y. Yang, Patent, CA 2950775, 2017.
70. M. Breitwieser, C. Klose, A. Hartmann, A. Büchler, M. Klingele, S. Vierrath, R. Zengerle and S. Thiele, *Adv. Energy Mater.*, 2017, **7**, 1602100.
71. B. Dong, L. Gwee, D. Salas-De La Cruz, K. I. Winey and Y. A. Elabd, *Nano Lett.*, 2010, **10**, 3785–3790.
72. C. Pan, H. Wu, C. Wang, B. Wang, L. Zhang, Z. Cheng, P. Hu, W. Pan, Z. Zhou, X. Yang and J. Zhu, *Adv. Mater.*, 2008, **20**, 1644–1648.
73. T. Tamura and H. Kawakami, *Nano Lett.*, 2010, **10**, 1324–1328.
74. R. Takemori, G. Ito, M. Tanaka and H. Kawakami, *RSC Adv.*, 2014, **4**, 20005–20009.
75. X. Li, X. Hao, D. Xu, G. Zhang, S. Zhong, H. Na and D. Wang, *J. Membr. Sci.*, 2006, **281**, 1–6.
76. C. Pan, H. Wu, C. Wang, B. Wang, L. Zhang, Z. Cheng, P. Hu, W. Pan, Z. Zhou, X. Yang and J. Zhu, *Adv. Mater.*, 2008, **20**, 1644–1648.
77. T. Tamura, R. Takemori and H. Kawakami, *J. Power Sources*, 2012, **217**, 135–141.
78. E. H. Sanders, K. A. McGrady, G. E. Wnek, C. a. Edmondson, J. M. Mueller, J. J. Fontanella, S. Suarez and S. G. Greenbaum, *J. Power Sources*, 2004, **129**, 55–61.
79. J. D. Snyder and Y. A. Elabd, *J. Power Sources*, 2009, **186**, 385–392.
80. S. Jiang, K.-Q. Q. Xia and G. G. Xu, *Macromolecules*, 2001, **34**, 7783–7788.
81. T. Song, Z. Chen, H. He, Y. Liu, Y. Liu and S. Ramakrishna, *J. Appl. Polym. Sci.*, 2015, **132**, 41755.
82. J. Choi, K. Lee, R. Wycisk, P. N. Pintauro and P. T. Mather, *J. Mater. Chem.*, 2010, **20**, 6282–6290.
83. A. Laforgue, L. Robitaille, A. Mokrini and A. Ajji, *Macromol. Mater. Eng.*, 2007, **292**, 1229–1236.
84. H. Chen, J. D. Snyder and Y. A. Elabd, *Macromolecules*, 2008, **41**, 128–135.

85. J. Zhao, W. Z. Yuan, A. Xu, F. Ai, Y. Lu and Y. Zhang, *React. Funct. Polym.*, 2011, **71**, 1102–1109.
86. R. Bajon, S. Balaji and S. M. Guo, *J. Fuel Cell Sci. Technol.*, 2009, **6**, 31004–31006.
87. D. K. Sharma, J. Shen and F. Li, *RSC Adv.*, 2014, **4**, 39110.
88. S. Subianto, S. Cavaliere, D. J. Jones and J. Rozière, *J. Polym. Sci., Part A: Polym. Chem.*, 2013, **51**, 118–128.
89. S. Subianto, S. Cavaliere, D. J. Jones and J. Rozière, *ECS Trans.*, 2011, **41**, 1517–1520.
90. K. M. Lee, J. Choi, R. Wycisk, P. N. Pintauro and P. Mather, *ECS Trans.*, 2009, **25**, 1451–1458.
91. J. B. Ballengee and P. N. Pintauro, *J. Electrochem. Soc.*, 2011, **158**, B568–B572.
92. C. Subramanian, R. a. Weiss and M. T. Shaw, *Polymer*, 2010, **51**, 1983–1989.
93. S. Shenoy, W. Bates, H. Frisch and G. Wnek, *Polymer*, 2005, **46**, 3372–3384.
94. S. Roddecha, Z. Dong, Y. Wu and M. Anthamatten, *J. Membr. Sci.*, 2012, **389**, 478–485.
95. J. Choi, K. M. K. Lee, R. Wycisk, P. N. Pintauro and P. T. Mather, *Macromolecules*, 2008, **41**, 4569–4572.
96. P. P. Lu, Z. L. Xu, H. Yang, Y. M. Wei and H. T. Xu, *J. Polym. Res.*, 2012, **19**, 9854.
97. B. Wu, J. Pan, L. Ge, L. Wu, H. Wang and T. Xu, *Sci. Rep.*, 2014, **4**, 1–7.
98. R. Boonpoo-nga, M. Sriring, P. Nasomjai and S. Martwiset, *ScienceAsia*, 2014, **40**, 232.
99. M. Sadrjahani, A. A. Gharehaghaji and M. Javanbakht, *Polym. Eng. Sci.*, 2017, **57**, 789–796.
100. J. Choi, R. Wycisk, W. Zhang, P. N. Pintauro, K. M. Lee and P. T. Mather, *ChemSusChem*, 2010, **3**, 1245–1248.
101. J. Choi, K. M. Lee, R. Wycisk, P. N. Pintauro and P. T. Mather, *J. Electrochem. Soc.*, 2010, **157**, B914.
102. L. Wang, J. Zhu, J. Zheng, S. Zhang and L. Dou, *RSC Adv.*, 2014, **4**, 25195.
103. J. H. Won, H. J. Lee, J. M. Lim, J. H. Kim, Y. T. Hong and S. Y. Lee, *J. Membr. Sci.*, 2014, **450**, 235–241.
104. W. Liu, S. J. Wang, M. Xiao, D. M. Han and Y. Z. Meng, *Chem. Commun.*, 2012, **48**, 3415.
105. Y. Yao, B. Guo, L. Ji, K. H. Jung, Z. Lin, M. Alcoutlabi, H. Hamouda and X. Zhang, *Electrochem. Commun.*, 2011, **13**, 1005–1008.
106. Y. Yao, L. Ji, Z. Lin, Y. Li, M. Alcoutlabi, H. Hamouda and X. Zhang, *ACS Appl. Mater. Interfaces*, 2011, **3**, 3732–3737.
107. M. M. Hasani-Sadrabadi, I. Shabani, M. Soleimani and H. Moaddel, *J. Power Sources*, 2011, **196**, 4599–4603.
108. I. Shabani, M. M. Hasani-Sadrabadi, V. Haddadi-Asl and M. Soleimani, *J. Membr. Sci.*, 2011, **368**, 233–240.

109. C. Subramanian, R. A. Weiss and M. T. Shaw, *Ind. Eng. Chem. Res.*, 2013, **52**, 15088–15093.
110. S. Mollá and V. Compañ, *J. Membr. Sci.*, 2015, **492**, 123–136.
111. C. Boaretti, L. Pasquini, R. Sood, S. Giancola, A. Donnadio, M. Roso, M. Modesti and S. Cavaliere, *J. Membr. Sci*, 2018, **545**, 66–74.
112. M. Oroujzadeh, S. Mehdipour-Ataei and M. Esfandeh, *Chem. Eng. J.*, 2015, **269**, 212–220.
113. J. B. Ballengee and P. N. Pintauro, *Macromolecules*, 2011, **44**, 7307–7314.
114. J. B. Ballengee, G. M. Haugen, S. J. Hamrock and P. N. Pintauro, *J. Electrochem. Soc.*, 2013, **160**, F429–F435.
115. J. B. Ballengee and P. N. Pintauro, *J. Membr. Sci.*, 2013, **442**, 187–195.
116. K. M. Nouel and P. S. Fedkiw, *Electrochim. Acta*, 1998, **43**, 2381–2387.
117. F. Liu, B. Yi, D. Xing, J. Yu and H. Zhang, *J. Membr. Sci.*, 2003, **212**, 213–223.
118. H.-L. Lin, T. L. Yu, L.-N. Huang, L.-C. Chen, K.-S. Shen and G.-B. Jung, *J. Power Sources*, 2005, **150**, 11–19.
119. W. G. Jang, J. Hou and H. S. Byun, *Desalin. Water Treat.*, 2011, **34**, 315–320.
120. S. W. Choi, Y.-Z. Fu, Y. R. Ahn, S. M. Jo and A. Manthiram, *J. Power Sources*, 2008, **180**, 167–171.
121. S. Mollà and V. Compan, *J. Membr. Sci.*, 2011, **372**, 191–200.
122. H.-Y. Li and Y.-L. Liu, *J. Mater. Chem. A*, 2014, **2**, 3783.
123. S. Shahgaldi, M. Ghasemi, W. R. Wan Daud, Z. Yaakob, M. Sedighi, J. Alam and A. F. Ismail, *Fuel Process. Technol.*, 2014, **124**, 290–295.
124. H. Y. Li, Y. Y. Lee, J. Y. Lai and Y. L. Liu, *J. Membr. Sci.*, 2014, **466**, 238–245.
125. M. M. Mannarino, D. S. Liu, P. T. Hammond and G. C. Rutledge, *ACS Appl. Mater. Interfaces*, 2013, **5**, 8155–8164.
126. D. M. Yu, S. Yoon, T. H. Kim, J. Y. Lee, J. Lee and Y. T. Hong, *J. Membr. Sci.*, 2013, **446**, 212–219.
127. C. Nah, K. U. Jeong, Y. S. Lee, S. H. Lee, M. M. A. Kader, H. K. Lee and J. H. Ahn, *Polym. Int.*, 2013, **62**, 375–381.
128. D. M. Yu, S. Yoon, T. H. Kim, J. Y. Lee, J. Lee and Y. T. Hong, *Macromol. Res.*, 2014, **22**, 79–84.
129. J. Jin, R. Hao, X. He and G. Li, *Int. J. Hydrogen Energy*, 2015, 1–7.
130. S. Giancola, M. Zaton, A. Reyes Carmona, S. Cavaliere, M. Dupont, J. Rozière and D. Jones, submitted.
131. J.-M. Lim, J.-H. Won, H.-J. Lee, Y. T. Hong, M.-S. Lee, C. H. Ko and S.-Y. Lee, *J. Mater. Chem.*, 2012, **22**, 18550.
132. R. Wycisk, J. Chisholm, J. Lee, J. Lin and P. N. Pintauro, *J. Power Sources*, 2006, **163**, 9–17.
133. K. Y. Cho, H. Y. Jung, N. S. Choi, S. J. Sung, J. K. Park, J. H. Choi and Y. E. Sung, *Solid State Ionics*, 2005, **176**, 3027–3030.
134. H. J. J. Kim, H. J. J. Kim, Y. G. Shul and H. S. Han, *J. Power Sources*, 2004, **135**, 66–71.

135. M. M. Nasef, N. A. Zubir, A. F. Ismail, K. Z. M. Dahlan, H. Saidi and M. Khayet, *J. Power Sources*, 2006, **156**, 200–210.
136. K. Y. Cho, H. Y. Jung, K. a. Sung, W. K. Kim, S. J. Sung, J. K. Park, J. H. Choi and Y. E. Sung, *J. Power Sources*, 2006, **159**, 524–528.
137. M. K. Song, Y. T. Kim, J. M. Fenton, H. Russell Kunz and H. W. Rhee, *J. Power Sources*, 2003, **117**, 14–21.
138. Y. He, H. Zhang, Y. Li, J. Wang, L. Ma, W. Zhang and J. Liu, *J. Mater. Chem. A*, 2015, **3**, 21832–21841.
139. D. Jones, J. Rozière, S. Cavaliere, S. Subianto and S. Burton, WO2016020668, 2016.
140. J. Wang, Y. He, L. Zhao, Y. Li, S. Cao, B. Zhang and H. Zhang, *J. Membr. Sci.*, 2015, **482**, 1–12.
141. Y. He, J. Wang, H. Zhang, T. Zhang, B. Zhang, S. Cao and J. Liu, *J. Mater. Chem. A*, 2014, **2**, 9548–9558.
142. R. Sood, A. Donnadio, S. Giancola, A. Kreisz, D. J. Jones and S. Cavaliere, *ACS Appl. Mater. Interfaces*, 2016, **8**, 16897–16906.
143. R. Sood, A. Donnadio, S. Giancola, A. Kreisz, D. J. Jones, J. Rozière and S. Cavaliere, *Meet. Abstr.*, 2016, 2605.
144. M. Tanaka, Y. Takeda, T. Wakiya, Y. Wakamoto, K. Harigaya, T. Ito, T. Tarao and H. Kawakami, *J. Power Sources*, 2017, **342**, 125–134.
145. S. Zhang, G. He, X. Gong, X. Zhu, X. Wu, X. Sun, X. Zhao and H. Li, *J. Membr. Sci.*, 2015, **493**, 58–65.
146. X. Xu, L. Li, H. Wang, X. Li and X. Zhuang, *RSC Adv.*, 2015, **5**, 4934–4940.
147. R. Takemori and H. Kawakami, *J. Power Sources*, 2010, **195**, 5957–5961.
148. Y. Ding, H. Hou, Y. Zhao, Z. Zhu and H. Fong, *Prog. Polym. Sci.*, 2016, **61**, 67–103.
149. T. Tamura and H. Kawakami, *Nano Lett.*, 2010, **10**, 1324–1328.
150. T. Makinouchi, M. Tanaka and H. Kawakami, *J. Membr. Sci.*, 2017, **530**, 65–72.
151. M. Breitwieser, C. Klose, M. Klingele, A. Hartmann, J. Erben, H. Cho, J. Kerres, R. Zengerle and S. Thiele, *J. Power Sources*, 2017, **337**, 137–144.
152. I. Javakhishvili, I. Dimitrov, O. Tynelius, J. Hales, K. Jankova and S. Hvilsted, *Macromol. Mater. Eng.*, 2017, **302**, 1600410.
153. S. Mollá and V. Compañ, *J. Power Sources*, 2011, **196**, 2699–2708.
154. H.-L. Lin and S.-H. Wang, *J. Membr. Sci.*, 2014, **452**, 253–262.
155. E. Quartarone and P. Mustarelli, *Energy Environ. Sci.*, 2012, **5**, 6436.
156. S.-H. Yun, J.-J. Woo, S.-J. Seo, L. Wu, D. Wu, T. Xu and S.-H. Moon, *J. Membr. Sci.*, 2011, **367**, 296–305.
157. H.-L. Lin, S.-H. Wang, C.-K. Chiu, T. L. Yu, L.-C. Chen, C.-C. Huang, T.-H. Cheng and J.-M. Lin, *J. Membr. Sci.*, 2010, **365**, 114–122.
158. M. L. Di Vona, E. Sgreccia, S. Licoccia, G. Alberti, L. Tortet and P. Knauth, *J. Phys. Chem. B*, 2009, **113**, 7505–7512.
159. B. Maranesi, H. Hou, R. Polini, E. Sgreccia, G. Alberti, R. Narducci, P. Knauth and M. L. Di Vona, *Fuel Cells*, 2013, **13**, 107–117.
160. Y. Shao, J. Liu, Y. Wang and Y. Lin, *J. Mater. Chem.*, 2009, **19**, 46–59.
161. S. G. Chalk and J. F. Miller, *J. Power Sources*, 2006, **159**, 73–80.

162. Y. Shao, J. Wang, R. Kou, M. Engelhard, J. Liu, Y. Wang and Y. Lin, *Electrochim. Acta*, 2009, **54**, 3109–3114.
163. B. J. Eastwood, P. a. Christensen, R. D. Armstrong and N. R. Bates, *J. Solid State Electrochem.*, 1999, **3**, 179–186.
164. A. Taniguchi, T. Akita, K. Yasuda and Y. Miyazaki, *J. Power Sources*, 2004, **130**, 42–49.
165. C. Reiser, L. Bregoli and T. Patterson, *Electrochem. Solid-State Lett.*, 2005, **8**, A273.
166. N. Yousfi-Steiner, P. Moçotéguy, D. Candusso and D. Hissel, *J. Power Sources*, 2009, **194**, 130–145.
167. X. Yu and S. Ye, *J. Power Sources*, 2007, **172**, 145–154.
168. J. Wu, X. Z. Yuan, J. J. Martin, H. Wang, J. Zhang, J. Shen, S. Wu and W. Merida, *J. Power Sources*, 2008, **184**, 104–119.
169. *Electrospinning for Advanced Energy and Environmental Applications*, ed. S. Cavaliere, CRC Press, 2015.
170. Y. Liu, T. G. Kelly, J. G. Chen and W. E. Mustain, *ACS Catal.*, 2013, **3**, 1184–1194.
171. M. Inagaki, Y. Yang and F. Kang, *Adv. Mater.*, 2012, **24**, 2547–2566.
172. C. Kim, Y.-J. Kim and Y.-A. Kim, *Solid State Commun.*, 2004, **132**, 567–571.
173. C. Lai, Z. Zhou, L. Zhang, X. Wang, Q. Zhou, Y. Zhao, Y. Wang, X. F. Wu, Z. Zhu and H. Fong, *J. Power Sources*, 2014, **247**, 134–141.
174. Q. Guo, D. Zhao, S. Liu, S. Chen, M. Hanif and H. Hou, *Electrochim. Acta*, 2014, **138**, 318–324.
175. A.-F. Che, V. Germain, M. Cretin, D. Cornu, C. Innocent and S. Tingry, *New J. Chem.*, 2011, **35**, 2848–2853.
176. S. N. Arshad, M. Naraghi and I. Chasiotis, *Carbon*, 2011, **49**, 1710–1719.
177. J. J.-H. Park, Y.-W. Y. Ju, S.-H. S. Park, H. H.-R. Jung, K.-S. Yang and W.-J. Lee, *J. Appl. Electrochem.*, 2009, **39**, 1229–1236.
178. I. Savych, J. Bernard D'Arbigny, S. Subianto, S. Cavaliere, D. J. Jones and J. Rozière, *J. Power Sources*, 2014, **257**, 147–155.
179. Y. Wang, J. Jin, S. Yang, G. Li and J. Qiao, *Electrochim. Acta*, 2015, **177**, 181–189.
180. D. Todd and W. Mérida, *J. Power Sources*, 2015, **273**, 312–316.
181. M. K. Kayarkatte, Ö. Delikaya and C. Roth, *ChemElectroChem*, 2017, **4**, 404–411.
182. G. Ercolano, F. Farina, S. Cavaliere, D. J. Jones and J. Rozière, *J. Mater. Chem. A*, 2017, **5**, 3974–3980.
183. M. K. Debe, *Nature*, 2012, **486**, 43–51.
184. G. Ercolano, F. Farina, S. Cavaliere, D. J. Jones and J. Rozière, *ECS Trans.*, 2017, **80**, 757–762.
185. M. Mathias, R. Makharia, H. Gasteiger, J. J. Conley, T. J. Fuller, C. S. Gittleman, S. S. Kocha, D. P. Miller, C. K. Mittelsteadt, T. Xie, S. G. Yan and P. T. Yu, *Electrochem. Soc. Interface*, 2005, 24–35.
186. Y. Shao, Y. Cheng, W. Duan, W. Wang, Y. Lin, Y. Wang and J. Liu, *ACS Catal.*, 2015, **5**, 7288–7298.
187. Y.-J. Wang, D. P. Wilkinson and J. Zhang, *Chem. Rev.*, 2011, **111**, 7625–7651.

188. K. Sasaki, F. Takasaki and Z. Noda, *ECS Trans.*, 2010, **33**, 473–482.
189. L. Timperman and N. Alonso-Vante, *Electrocatalysis*, 2011, **2**, 181–191.
190. Y.-J. Wang, D. P. Wilkinson, A. Guest, V. Neburchilov, R. Baker, F. Nan, G. a. Botton and J. Zhang, *J. Power Sources*, 2013, **221**, 232–241.
191. S.-Y. Huang, P. Ganesan and B. N. Popov, *ACS Catal.*, 2012, **2**, 825–831.
192. L. Chevallier, A. Bauer, S. Cavaliere, R. Hui, J. Rozière and D. J. Jones, *ACS Appl. Mater. Interfaces*, 2012, **4**, 1752–1759.
193. K. Kanda, Z. Noda, Y. Nagamatsu, T. Higashi, S. Taniguchi, S. M. Lyth, A. Hayashi and K. Sasaki, *ECS Electrochem. Lett.*, 2014, **3**, F15–F18.
194. M. Dou, M. Hou, D. Liang, W. Lu, Z. Shao and B. Yi, *Electrochim. Acta*, 2013, **92**, 468–473.
195. N. R. Elezovic, B. M. Babic, V. R. Radmilovic and N. V. Krstajic, *J. Electrochem. Soc.*, 2013, **160**, F1151–F1158.
196. K. Kakinuma, Y. Chino, Y. Senoo, M. Uchida, T. Kamino, H. Uchida, S. Deki and M. Watanabe, *Electrochim. Acta*, 2013, **110**, 316–324.
197. B. Seger, A. Kongkanand, K. Vinodgopal and P. V. Kamat, *J. Electroanal. Chem.*, 2008, **621**, 198–204.
198. M. S. Saha, M. N. Banis, Y. Zhang, R. Li, X. Sun, M. Cai and F. T. Wagner, *J. Power Sources*, 2009, **192**, 330–335.
199. P. J. Kulesza, K. Miecznikowski, B. Baranowska, M. Skunik, A. Kolary-Zurowska, A. Lewera, K. Karnicka, M. Chojak, I. Rutkowska, S. Fiechter, P. Bogdanoff, I. Dorbandt, G. Zehl, R. Hiesgen, E. Dirk, K. S. Nagabhushana and H. Boennemann, *J. Appl. Electrochem.*, 2007, **37**, 1439–1446.
200. N. R. Elezovic, B. M. Babic, V. R. Radmilovic, L. M. Vracar and N. V. Krstajic, *Electrochim. Acta*, 2009, **54**, 2404–2409.
201. N. Elezović, B. Babić, L. Gajić-Krstajić, V. Radmilović, N. V. Krstajić and L. J. Vračar, *J. Power Sources*, 2010, **195**, 3961–3968.
202. H. Chhina, S. Campbell and O. Kesler, *J. Electrochem. Soc.*, 2009, **156**, B1232.
203. T. Ioroi, Z. Siroma, N. Fujiwara, S. Yamazaki and K. Yasuda, *Electrochem. Commun.*, 2005, **7**, 183–188.
204. S. Siracusano, V. Baglio, C. D'Urso, V. Antonucci and a. S. Aricò, *Electrochim. Acta*, 2009, **54**, 6292–6299.
205. S. Subianto, S. Giancola, G. Ercolano, Y. Nabil, D. Jones, J. Rozière and S. Cavaliere, in *Electrospinning for Advanced Energy and Environmental Applications*, ed. S. Cavaliere, CRC Press, Boca Raton, 2015, vol. 1, pp. 29–60.
206. A. Bauer, K. Lee, C. Song, Y. Xie, J. Zhang and R. Hui, *J. Power Sources*, 2010, **195**, 3105–3110.
207. A. Bauer, L. Chevallier, R. Hui, S. Cavaliere, J. Zhang, D. Jones and J. Rozière, *Electrochim. Acta*, 2012, **77**, 1–7.
208. K. Senevirathne, R. Hui, S. Campbell, S. Ye and J. Zhang, *Electrochim. Acta*, 2012, **59**, 538–547.
209. S. Cavaliere, S. Subianto, L. Chevallier, D. J. Jones and J. Rozière, *Chem. Commun.*, 2011, **47**, 6834–6836.
210. V. A. de la Pena O'Shea, M. C. Alvarez Galvan, A. E. Platero Prats, J. Campos-Martin and J. Fierro, *Chem. Commun.*, 2011, **47**, 7131–7133.

211. S. J. Tauster, *Acc. Chem. Res.*, 1987, **20**, 389–394.
212. S. Sago, A. B. Suryamas, G. M. Anilkumar, T. Ogi and K. Okuyama, *Mater. Lett.*, 2013, **105**, 202–205.
213. A. A. B. Suryamas, G. M. G. Anilkumar, S. Sago, T. Ogi and K. Okuyama, *Catal. Commun.*, 2013, **33**, 11–14.
214. S. Cavaliere, S. Subianto, I. Savych, M. Tillard, D. J. Jones, J. Roziére and D. J. Jones, *J. Phys. Chem. C*, 2013, **117**, 18298–18307.
215. I. Savych, S. Subianto, Y. Nabil, S. Cavaliere, D. Jones and J. Rozière, *Phys. Chem. Chem. Phys.*, 2015, **17**, 16970–16976.
216. G. Cognard, G. Ozouf, C. Beauger, I. Jiménez-Morales, S. Cavaliere, D. Jones, J. Rozière, M. Chatenet and F. Maillard, *Electrocatalysis*, 2017, **8**, 51–58.
217. S. Cavaliere, I. Jiménez-Morales, G. Ercolano, I. Savych, D. Jones and J. Rozière, *ChemElectroChem*, 2015, **2**, 1966–1973.
218. K. Zhang, C. Feng, B. He, H. Dong, W. Dai, H. Lu and X. Zhang, *J. Electroanal. Chem.*, 2016, **781**, 198–203.
219. Q. Du, J. Wu and H. Yang, *ACS Catal.*, 2013, **4**, 144–151.
220. CATAPULT project (Novel catalyst structures employing Pt at ultra-low and zero loadings for automotive MEAs), FCH-JU grant no 325268.
221. A. Bauer, R. Hui, A. Ignaszak, J. Zhang and D. J. Jones, *J. Power Sources*, 2012, **210**, 15–20.
222. Y. Ito, T. Takeuchi, T. Tsujiguchi, M. A. Abdelkareem and N. Nakagawa, *J. Power Sources*, 2013, **242**, 280–288.
223. C. Marichy, G. Ercolano, G. Caputo, M. G. Willinger, D. Jones, J. Rozière, N. Pinna and S. Cavaliere, *J. Mater. Chem. A*, 2016, **4**, 969–975.
224. G. Liu, J. Xu, Y. Wang and X. Wang, *J. Mater. Chem. A*, 2015, **3**, 20791–20800.
225. F. Jaouen, B. Rodriguez-Garcia, A. Reyes-Carmona, I. Jiménez-Morales, M. Blasco-Ahicart, S. Cavaliere, D. Jones and J.-R. Galan-Mascaros, in *68th Annual Meeting of the International Society of Electrochemistry*, 2017.
226. M. Huynh, T. Ozel, C. Liu, E. C. Lau and D. G. Nocera, *Chem. Sci.*, 2017, **8**, 4779–4794.
227. D. Zhen, B. Zhao, H.-C. Shin, Y. Bu, Y. Ding, G. He and M. Liu, *Adv. Mater. Interfaces*, 2017, **4**, 1700146.
228. C. Wu, J. Li, D. Zhang, B. Yang, L. Li, T. Zhou, C. Zhang, G. Yang and Y. Shan, *Int. J. Hydrogen Energy*, 2016, **41**, 13915–13922.
229. J. Chen, D. Yu, W. Liao, M. Zheng, L. Xiao, H. Zhu, M. Zhang, M. Du and J. Yao, *ACS Appl. Mater. Interfaces*, 2016, **8**, 18132–18139.
230. G. García, O. Guillén-Villafuerte, J. L. Rodríguez, M. C. Arévalo and E. Pastor, *Int. J. Hydrogen Energy*, 2016, **41**, 19664–19673.
231. H. Kim, M. K. Cho, J. A. Kwon, Y. H. Jeong, K. J. Lee, N. Y. Kim, M. J. Kim, S. J. Yoo, J. H. Jang, H.-J. Kim, S. W. Nam, D. Lim, E. Cho, K. Lee and J. Y. Kim, *Nanoscale*, 2015, **7**, 18429–18434.
232. A. Jindal, D. K. Gautam and S. Basu, *J. Electroanal. Chem.*, 2016, **775**, 198–204.
233. H. Liu, F. Wang, Y. Zhao and H. Fong, *Nanoscale*, 2013, **5**, 3643–3647.
234. S. Cavaliere, J. Rozière, D. Jones and Y. Nabil-Moreau, WO2017001843, 2017.

235. Y. Nabil, S. Cavaliere, I. A. Harkness, J. D. B. Sharman, D. J. Jones and J. Rozière, *J. Power Sources*, 2017, **363**, 20–26.
236. M. T. Sung, M. H. Chang and M. H. Ho, *J. Power Sources*, 2014, **249**, 320–326.
237. H. J. Kim, Y. S. Kim, M. H. Seo, S. M. Choi, J. Cho, G. W. Huber and W. B. Kim, *Electrochem. Commun.*, 2010, **12**, 32–35.
238. J. L. Shui, C. Chen and J. C. M. Li, *Adv. Funct. Mater.*, 2011, **21**, 3357–3362.
239. H. Kim, Y. Kim, M. Seo, S. Choi and W. Kim, *Electrochem. Commun.*, 2009, **11**, 446–449.
240. Y. N. Regmi, G. R. Waetzig, K. D. Duffee, S. M. Schmuecker, J. M. Thode and B. M. Leonard, *J. Mater. Chem. A*, 2015, **3**, 10085–10091.
241. S. Sharma and B. G. Pollet, *J. Power Sources*, 2012, **208**, 96–119.
242. X. S. Zhou, Y. J. Qiu, J. Yu, J. Yin and S. Gao, *Int. J. Hydrogen Energy*, 2011, **36**, 7398–7404.
243. X. Zhou, Y. Qiu, J. Yu, J. Yin and X. Bai, *J. Mater. Sci.*, 2012, **47**, 6607–6613.
244. Y. Qiu, J. Yin, H. Hou, J. Yu and X. Zuo, *Electrochim. Acta*, 2013, **96**, 225–229.
245. J. Yin, Y. Qiu and J. Yu, *Electrochem. Commun.*, 2013, **30**, 1–4.
246. S. Uhm, B. Jeong and J. Lee, *Electrochim. Acta*, 2011, **56**, 9186–9190.
247. Q. Shi, Y. Lei, Y. Wang, H. Wang, L. Jiang, H. Yuan, D. Fang, B. Wang, N. Wu and Y. Gou, *Curr. Appl. Phys.*, 2015, 1–9.
248. Y. Wang, J. Jin, S. Yang, G. Li and J. Jiang, *Int. J. Hydrogen Energy*, 2016, **41**, 11174–11184.
249. S. Wang, C. Dai, J. Li, L. Zhao, Z. Ren, Y. Ren, Y. Qiu and J. Yu, *Int. J. Hydrogen Energy*, 2015, **40**, 4673–4682.
250. M. E. M. Buan, N. Muthuswamy, J. C. Walmsley, D. Chen and M. Rønning, *Carbon*, 2016, **101**, 191–202.
251. J. Yin, Y. Qiu and J. Yu, *J. Electroanal. Chem.*, 2013, **702**, 56–59.
252. G. Panomsuwan, N. Saito and T. Ishizaki, *ACS Appl. Mater. Interfaces*, 2016, **8**, 6962–6971.
253. W. Yang, Y. Zhang, C. Liu and J. Jia, *J. Power Sources*, 2015, **274**, 595–603.
254. J. Shui, C. Chen, L. Grabstanowicz, D. Zhao and D.-J. Liu, *Proc. Natl. Acad. Sci. U. S. A.*, 2015, **112**, 10629–10634.
255. X. Yan, L. Gan, Y. C. Lin, L. Bai, T. Wang, X. Wang, J. Luo and J. Zhu, *Small*, 2014, **10**, 4072–4079.
256. N. Wu, Y. Wang, Y. Lei, B. Wang, C. Han, Y. Gou, Q. Shi and D. Fang, *Sci. Rep.*, 2015, **5**, 17396.
257. G. Ren, X. Lu, Y. Li, Y. Zhu, L. Dai and L. Jiang, *ACS Appl. Mater. Interfaces*, 2016, **8**, 4118–4125.
258. B. Jeong, D. Shin, M. Choun, S. Maurya, J. Baik, B. S. Mun, S. H. Moon, D. Su and J. Lee, *J. Phys. Chem. C*, 2016, **120**, 7705–7714.
259. Q. Liu, S. Cao, Y. Qiu and L. Zhao, *Mater. Sci. Eng. B, Solid-State Mater. Adv. Technol.*, 2017, **223**, 159–166.
260. C. Liu, J. Wang, J. Li, J. Liu, C. Wang, X. Sun, J. Shen, W. Han and L. Wang, *J. Mater. Chem. A*, 2017, **5**, 1211–1220.
261. W. Zhang and P. N. Pintauro, *ChemSusChem*, 2011, **4**, 1753–1757.

262. M. Brodt, T. Han, N. Dale, E. Niangar, R. Wycisk and P. Pintauro, *J. Electrochem. Soc.*, 2015, **162**, F84–F91.
263. M. Brodt, R. Wycisk, N. Dale and P. Pintauro, *J. Electrochem. Soc.*, 2016, **163**, F401–F410.
264. S. Hong, B. Y. Ming Hou, Y. Xiao and Z. Shao, *Energy Technol.*, 2017, **5**, 1457–1463.
265. X. Wang, F. W. Richey, K. H. Wujcik and Y. A. Elabd, *J. Power Sources*, 2014, **264**, 42–48.
266. S. Hong, M. Hou, H. Zhang, Y. Jiang, Z. Shao and B. Yi, *Electrochim. Acta*, 2017, **245**, 403–409.
267. A. F. Lotus, E. T. Bender, E. A. Evans, R. D. Ramsier, D. H. Reneker and G. G. Chase, *J. Appl. Phys.*, 2008, **103**, 24910.
268. S. Huang, H. Wu, M. Zhou, C. Zhao, Z. Yu, Z. Ruan and W. Pan, *NPG Asia Mater.*, 2014, **6**, e86.
269. J. L. Reyes-Rodriguez, O. Solorza-Feria, A. García-Bernabé, E. Giménez, O. Sahuquillo and V. Compañ, *RSC Adv.*, 2016, **6**, 56986–56999.
270. C. Gong, H. Liu, B. Zhang, G. Wang, F. Cheng, G. Zheng, S. Wen, Z. Xue and X. Xie, *J. Membr. Sci.*, 2017, **535**, 113–121.
271. https://www.donaldson.com/en-be/industrial-dust-fume-mist/technical-articles/ultra-web-media-technology/.
272. http://www.dupont.com/products-and-services/membranes-films/products/energain/press-releases/energain-separators-lithium.html.
273. http://www.volumetriq.eu.

CHAPTER 10

Electrospinning in the Packaging Industry

S. TORRES-GINER, M. BUSOLO, A. CHERPINSKI AND J. M. LAGARON*

Novel Materials and Nanotechnology Group, Institute of Agrochemistry and Food Technology (IATA), Spanish Council for Scientific Research (CSIC), Calle Catedrático Agustín Escardino Benlloch 7, 46980, Paterna, Spain
*E-mail: lagaron@iata.csic.es

10.1 Introduction to Electrohydrodynamic Processing

Electrohydrodynamic processing (EHDP) has emerged as a straightforward, versatile and simple approach for the manufacturing of ultrathin, *i.e.* both submicron and nano-sized, polymer structures with high specific surface areas and porosities for a wide range of applications in industry.[1–4] This technology is based on the application of electrostatic forces to polymers in the liquid phase (*e.g.* solution or molten polymer with sufficient molecular entanglement), for which neither temperature or pressure are required during processing.[5–7] The main reason for the substantial interest in EHDP technology, including electrospinning and electrospraying, is its multitude of possibilities due to the extensive tunability and multi-functionality of the processed nanostructured materials, with independence of the type of polymers or material selection, which could be certainly scaled up to different industries.[7–12]

Soft Matter Series No. 7
Electrospinning: From Basic Research to Commercialization
Edited by Erich Kny, Kajal Ghosal and Sabu Thomas

Published by the Royal Society of Chemistry, www.rsc.org

Electrospun and electrosprayed materials have been reported to be obtained from different synthetic polymers,[13–16] naturally occurring polymers,[5,17] biodegradable polymers,[18,19] and their blends,[20] including proteins[9,12,21] and other food hydrocolloids.[5,12,22] In addition, materials processed by EHDP can be produced using polymer formulations containing metallic and/or inorganic fillers.[8,11,23,24] Over the years, different polymer-based compositions have been investigated in detail in the fields of drug delivery,[2,20,25–27] protection and/or controlled release of bioactive ingredients,[2,15,21,22,28–30] food packaging,[31–37] tissue engineering,[12,25,38] textiles,[12,39] filtration,[12,40,41] and nanodevices,[24,42,43] among other cutting-edge applications.[8,44]

The invention of using electrostatic forces for obtaining synthetic filaments was patented by Formhals[45] in 1934, when cellulose acetate fibers were obtained from acetone solutions. Since then, electrospinning continued being studied to refine the earlier approach and have overcome some drawbacks related to incomplete drying of the processed materials and, thus, a second patent was published in 1940.[46] In the 1960s, Taylor performed fundamental studies on the shape of the polymer droplet that is produced in the tip of the conducting tubes during the application of electric forces.[47] These studies, including the stability of the conical drop from polymer solutions and the jet-forming process, showed that the jets were ejected from the vertices of the cone, which was later renamed as the 'Taylor cone'.[19,46] In subsequent years, researchers focused on the study of the relationship between structural features and process parameters, investigating the electrospinnability of different polymer–solvent systems and examining the potential applications of the resultant electrospun nanofiber mats.[26,46] Indeed, the volume of articles published in international peer-reviewed journals related to electrospinning increased from less than 10 publications per year, in the period 1986 to 1999, to over 3000 articles in 2016. This increasing trend clearly shows the potential the researchers have found in this technology and also the way it has tremendously impacted many different science and engineering disciplines.

A basic electrospinning setup consists of a high-voltage power supply, a metallic spinneret or needle, a pump, and a conductive collector (Figure 10.1a).[1,11,48,49] Some specific collector configurations can also be based on rotating devices as the collector in order to mechanically stretch the nanofibers, thus aligning them during processing (Figure 10.1b).[17,50–53] In this process, the polymer liquid is extruded to the needle tip from a reservoir (*e.g.* syringe or tank) at a constant flow rate. The drop hemispherical surface in the needle tip stretches due to the electric field, *i.e.* surface charge repulsion, forming the previously described Taylor cone.[1,4,26] When the electrical forces overcome the surface tension of the liquid at a critical value, an expanding charged polymer jet is ejected.[49] Besides electrostatic forces, the jet is subjected to dragging forces, coulombic repulsion forces, surface tension, and viscoelastic forces.[9] The polymer jet continues thinning, *i.e.* rapid whipping, until total evaporation of solvent and the nanofibers are deposited on the collector with an opposing or zero polarity in the form of non-wovens.[19]

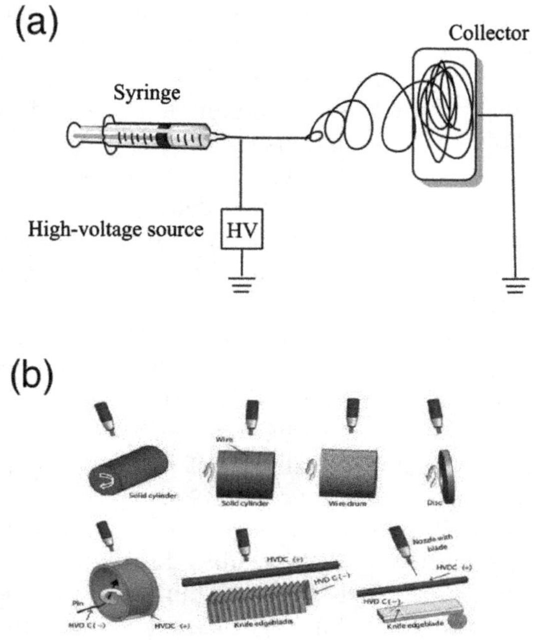

Figure 10.1 (a) Schematic diagram of a basic electrospinning setup. Reprinted from *Journal of Controlled Release*, **185**, X. Hu, S. Liu, G. Zhou, Y. Huang, Z. Hie, X. Jing, Electrospinning of polymeric nanofibers for drug delivery applications, 12–21, Copyright 2014, with permission from Elsevier.[48] (b) Representation of different rotating collectors used for electrospinning. Adapted from previous research.[52]

The thickness, density, and morphology of the electrospun and electrosprayed materials can be controlled by altering governing factors such as polymer and solution properties, electrospinning conditions, and environment variables (Table 10.1).[6,17–19,46,54] Among them, polymer concentration, molecular weight (M_W), and electrical conductivity are the dominant factors on the materials morphology.[19] Furthermore, the properties of the substrate, the flow-rate, and electric field strength and geometry play a major role in nanofibers and beads formation.[12] The choice of the EHDP conditions and polymer–solvent system will depend on the required properties of the processed materials and it is not possible to make general recommendations about the best processing conditions because the dominant factors will be given by the polymer–solvent system.[12] Empirical knowledge is also crucial to adjust correctly the process parameters and select the right environmental conditions, although these can also be optimized by applying theoretical models. Normally, the materials diameter decreases as the applied voltage and tip-to-collector distance increases. At very high voltages, the Taylor cone starts to destabilize causing the raise of the standard deviation of fibers diameter thickness while too low voltages can cause undesirable bead

Table 10.1 Main parameters affecting the morphology of electrospun materials.

Polymer properties	Solubility Molecular weight Melting point Glass transition temperature Crystallization rate
Solution parameters	Viscosity Polymer concentration Surface tension Solvent properties (*e.g.* boiling point) Electrical conductivity
Processing conditions	Applied voltage Tip-to-collector distance Collector configuration Flow-rate Needle inner diameter
Environmental factors	Temperature Relative humidity Atmospheric pressure

formation.[2,11] The charge density decreases as distance increases, and the fibers become thicker by overlapping and fusion caused by the bending and twisting as well as by the inadequate solvent evaporation. Besides the polymer nature, the relative humidity can induce porosity and affect the fiber diameter by controlling the solidification process of the charged jet, while temperature also influences the evaporation rate of solvent and the viscosity of the polymer solution.[5,25]

The setting of parameters provides diverse electrospun morphologies, including submicron nanofibers and beads. Electrospun fibers include beaded, smooth or ribbon-like fibers, but other sophisticated structures that can broaden their applications can be also obtained (Figure 10.2).[3] Starting from the original basic single needle arrangement, other electrospinning configurations have been developed to overcome limitations in polymer processing as well as for obtaining products with enhanced technical performance.[55] These new arrangements, together with the selection of the electrospinning parameters, have allowed application-targeted products to be produced, typically patterned fibers,[14,26,56] fiber yarns,[14,57] multicomponent fibers,[13,58] and tailored deposition area of the electrospun nanofibers mat.[50,59,60] Electrospinning setups comprising rotating devices or unique parts as collectors, counter electrode arrays, and special designs of polymer solution dispensing devices have been developed for these purposes and allow researchers to prepare different fiber-based morphologies.[21]

According to the state of the material to be processed, electrospinning methods can be classified as melt and solution electrospinning.[55] Solution electrospinning includes single and/or blending, emulsion, and coaxial configurations and it is suitable for most soluble polymers and particularly for those that are thermally unstable and degrade while melting. The solvent

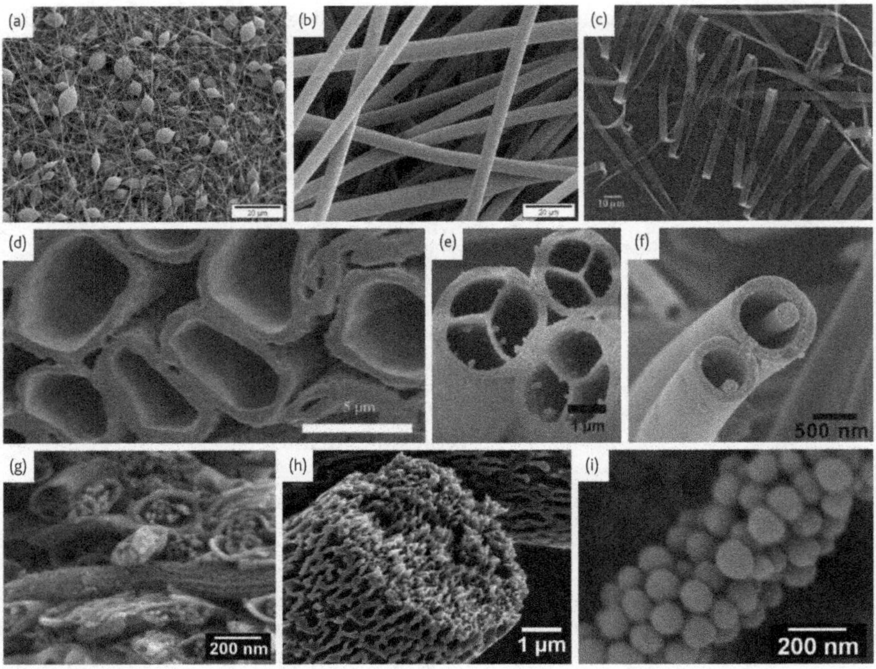

Figure 10.2 Different fiber-like morphologies of electrospun polymer materials: (a) beaded; (b) smooth; (c) ribbon; (d) hollow; (e) multichannel tubular; (f) nanowire-in-microtube; (g) multi-core cable-like; (h, i) porous. Reprinted from *Materials Today*, **13**, B. Ding, M. Wang, X. Wang, J. Yu, G. Sun, Electrospun nanomaterials for ultrasensitive sensors, 16–27, Copyright 2010, with permission from Elsevier.[3]

can be removed either by a dry process, *i.e.* through evaporation of the solvent during the jet fly or blowing hot air on the extruded filament, or a wet process, *i.e.* the fibers are extruded in a viscous coagulation bath consisting of a liquid that is miscible with the spinning solvent but not with the polymer.[55] Melt electrospinning involves the use of a molten polymer instead of a solution, which allows the issues related to solvent accumulation, costs, low solubility, and toxicity to be overcome, making this technique attractive from an environmental and productivity perspective.[5,12,26] Compared to solution electrospinning, melt electrospinning is strongly governed by flow rate and viscosity of the polymer,[5] where different petroleum-derived polymers and biopolymers have been processed.[5,12,61] In emulsion electrospinning, the solution is instead replaced by a water-in-oil (w/o) or oil-in-water (o/w) emulsion, which has gained special interest in fabricating core–shell structured nanofibers.[15,62,63] For instance, a bioactive agent, generally included in the aqueous phase, can be well incorporated in the w/o emulsions being protected from the solvent and available for encapsulation into the core rather than uniformly dispersed inside the fibers.[15] Core-shell fibers obtained by emulsion electrospinning can exhibit a higher controlled and sustained release capacity for the incorporated drugs or biomolecules than fibers obtained by single solution electrospinning.[63] This extended release capacity is related to the fact that the encapsulated agent needs to pass through the shell fiber matrix before reaching the medium.[15] Besides this, core–shell fibers exhibit improved thermal conductivity and mechanical strength.[23] Emulsion electrospinning has been used, for instance, in the encapsulation of anti-inflammatory drugs,[20] highly volatile organic compounds,[64] and oxygen-sensitive fish oil.[65] Finally, coaxial electrospinning is an innovative extension of solution electrospinning, also reported as a method for producing core–shell structures but starting from two different polymer solutions.[23,27,66,67] The coaxial arrangement enhances production stability due to its spinneret consisting of two concentrically arranged nozzles dispensing two different polymer solutions through inner/outer capillaries.[2,67] Therefore, nanofibers with a core–shell structure can be simultaneously formed from the Taylor cone.[68] The fact that the core and shell solutions are phase-separated makes the coaxial electrospinning an ideal technique, not only for suitable encapsulation and controlled release of active and sensitive agents,[66,67,69] but also for creating products with tailor-made properties for different technical fields.[23,70,71]

10.2 Applications of Electrohydrodynamic Processing in Packaging

The versatility and extraordinary technical features of the electrospun/electrosprayed materials have been explored over recent years for the development of novel packaging applications.[72] Various recent studies have shown promising results for the development of functional multilayer packaging structures involving transparent nanostructured coatings and interlayers by incorporating functional substances in polymer matrices.[6,7,34–37,73–75] Additionally, the electrospinning technology has been proposed to be a suitable platform for the development of active packaging technologies, such as antimicrobials, antioxidants, and oxygen scavengers, the encapsulation and controlled released of bioactive ingredients, and, more recently, the generation of smart packaging systems with the capacity to notify consumers about the quality of the packaged materials.[44] The following subsections describe in detail the different potential applications of EHDP in the packaging area.

10.2.1 Coatings

Food packaging materials have been used during the past decade not only for containing and keeping fresh foodstuffs during shelf life all the way up to consumption but also to guarantee safety by minimizing food losses and preserving quality.[76] Previous research works have added a new insight into the potential of the so-called 'electrospinning coating technology' to develop packaging structures of higher quality and safety, showing a high capacity to extend the shelf life of food products. To convert the electrospun mats

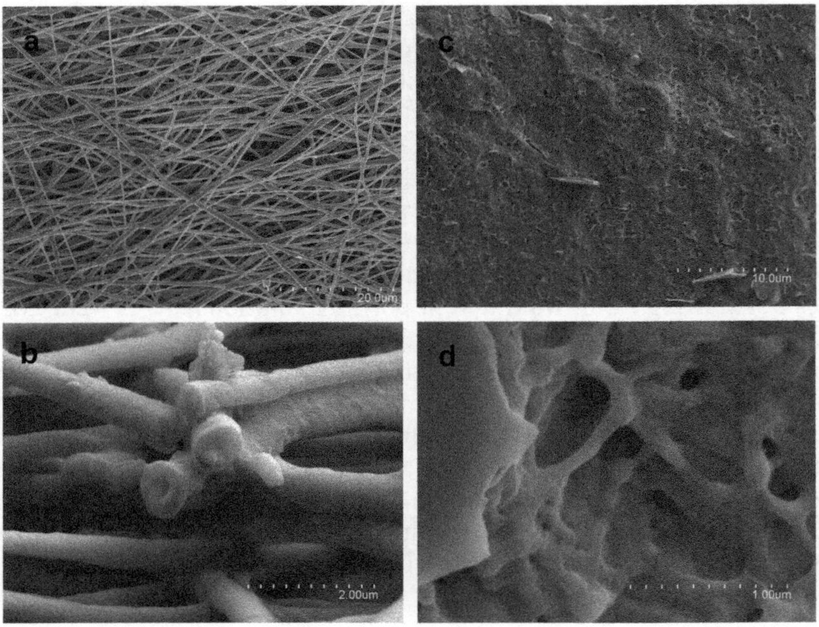

Figure 10.3 Scanning electron microscopy (SEM) images of electrospun poly(3-hydroxybutyrate) (PHB) submicron fibers, before annealing. (a) Surface view with scale marker of 20 μm and (b) cross-sectional view with scale marker of 2 μm. Resultant electrospun film after annealing at 160 °C. (c) Cross-sectional view with scale marker of 10 μm and (d) detail of the fiber coalescence with scale marker of 1 μm. Adapted from previous research work.[77]

into continuous films and, therefore, to increase their application interest in packaging, these can be annealed, *i.e.* thermally post-processed at temperatures below the melting point of the polymer. For instance, electrospun submicron poly(3-hydroxybutyrate) (PHB) fibers were recently treated at 160 °C in a hot press without pressure and then cooled slowly to obtain homogeneous transparent films.[77] Figure 10.3 shows that the annealing process applied to the electrospun PHB mat (Figure 10.3a) led to coalescence of the submicron fibers and, hence, to a homogeneous and continuous film with virtually little or no porosity (Figure 10.3b). The resultant electrospun PHB films exhibited similar rigidity and barrier properties to oxygen, water vapor, and limonene vapors than conventional compression-molded PHB films, but with enhanced elongation at break and toughness. As a result, multilayer films produced by this electrospinning technique have many potential applications, including the design of barrier and functional coatings for fiber- and plastic-based food packaging. For instance, wheat gluten films have been coated by thermally post-processed electrospun PHB and poly(3-hydroxybutyrate-*co*-3-valerate) (PHBV) ultrathin fibers.[78,79] In the case of the electrospun PHB films, in another study it was interestingly found that the storage time at dry conditions did not significantly modify the degree of crystallinity of the homopolyester.[80] In particular, the optical and barrier properties of electrospun PHB films stored for 3 months remained nearly constant although the stored multilayer films were stiffer and less deformable.

More recently, for the first time, a fiber-based packaging material was coated by annealed electrospun ultrathin fibers made of different biopolymers.[81] The resultant mono- and multilayer structures self-adhered to the paper substrate, inducing a significant improvement of the paper barrier properties to water and limonene vapors when built with sufficient thickness. Figure 10.4 gives an example of the fiber-based packaging material developed by the electrospinning coating technology. The optical micrograph included in Figure 10.4a displays the top view of a neat electrospun PHB film indicating that, after the annealing process, the regularly introduced electrospun PHB nanofibers yielded totally transparent films. Figure 10.4b displays the top view of one of the coated paper showing that, due to the opaque nature of the uncoated paper substrate, the resultant packaging materials were also opaque but with a glossier surface finish provided by the electrospun coating. Through lateral cryo-fracture of a portion of this multilayer fiber-based

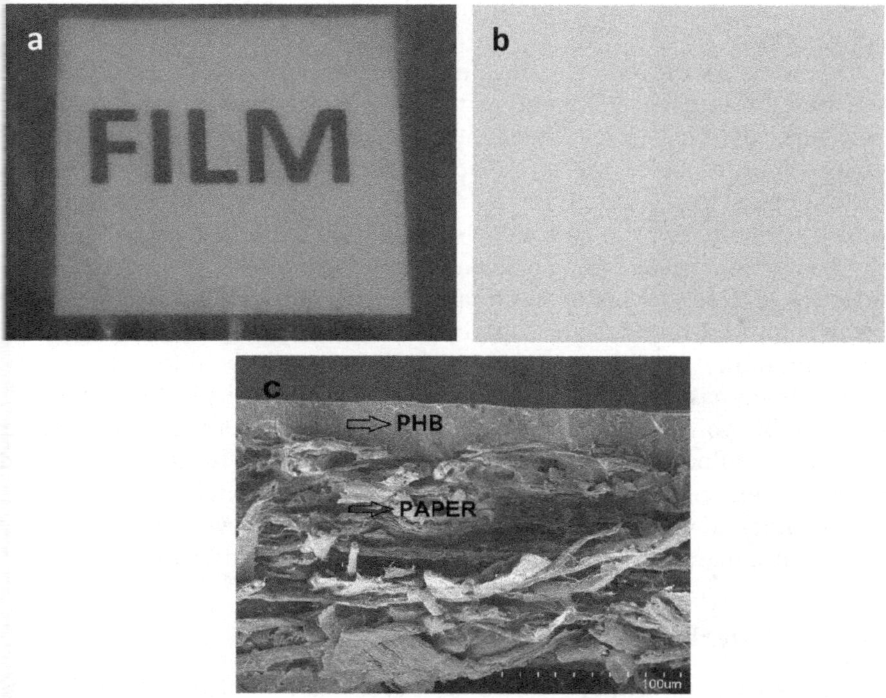

Figure 10.4 Optical images of the top views of (a) an electrospun poly(3-hydroxybutyrate) (PHB) film and (b) a fiber-based packaging coated by an electrospun PHB film. (c) A scanning electron microscopy (SEM) image of its cross-section with a scale marker of 100 μm. Based on previous research work.[81]

packaging structure, which is shown in the scanning electron microscopy (SEM) image of Figure 10.4c, one can observe that the electrospun PHB coating self-adhered to the paper during post-processing due to the high surface-to-volume ratio of the nanostructured PHB fibers. The electrospinning coating technology described here is also currently being explored for packaging materials based on nanopapers to develop multilayers with outstanding barrier properties to oxygen even at high moisture conditions.[82]

As opposed to the above, thin electrospun coatings can be added as outer layers onto both sides of hydrophilic thermoplastic films to develop highly permeable biodegradable multilayer systems to protect fresh products.[75] In this type of multilayer structures, proteins and polysaccharides (*e.g.* wheat gluten or starch) have to be firstly plasticized to become thermoplastic materials. These can then form continuous packaging films that can be useful for containing some fresh produce (*e.g.* mushroom and strawberry) with high respiration rates such as high oxygen and carbon dioxide permeability values. However, the barrier properties of these bioplastic films can be detrimentally altered by the presence of plasticizers and they could be disintegrated or even dissolved at high relative humidity conditions. Therefore, the presence of electrospun hydrophobic coatings on both sides of the bioplastic films can avoid these limitations and certainly increase their packaging applications.

The use of EHDP, both electrospinning and electrospraying, is also one of the most advantageous alternatives to create packaging materials with repellent surfaces. This technology can produce polymer nanostructures with various morphologies and size distributions, which introduce one degree of roughness to the materials. By tuning the EHDP parameters, applying post-treatments steps and/or using some additives in the polymer solution, the roughness needed for achieving superhydrophobicity is created. For instance, different types of fluorinated and nonfluorinated polymers have been employed to fabricate superhydrophobic fibers, and although fluorine-containing polymers are not suitable for use in solution electrospinning in many cases, approaches to increase their solubility by blending them with soluble polymers have been carried out.[83] To further reduce the surface energy, some fluorine-based chemicals (*e.g.* fluoroalkylsilanes) can be added, however most of them are not permitted for food contact materials. As a result, other novel approaches have been performed with nonfluorinated surface modifiers such as stearic acid-based modifiers and nanoparticles.[84,85]

10.2.2 Interlayers

The enhancement of biopolymers' properties can be also tackled through the use of the so-called 'electrospinning biolayering technology'. This produces high barrier structures to oxygen and other gases, even at high relative humidity, by combining layers of different nanofiber-based mats made of biopolyesters with layers based on proteins and polysaccharides. In the first research work of our group, the oxygen barrier properties of polylactide (PLA) films made by wet casting were improved over 70% by electrospun ultrathin zein fibers laminated in a sandwich-type structure.[86,87] Later, in a series of research works, Fabra *et al.*[88–90] demonstrated that the use of electrospun zein, whether pure or in blends with other hydrocolloids, and pullulan nanofibers as interlayers in films made of polyhydroxyalkanoates (PHAs) provided fully bio-based films with minimum changes in the mechanical and optical properties but enhanced barrier performance. In particular, the addition of a zein interlayer improved the water and oxygen permeability values of both PHB and PHBV films, reaching permeability values similar to those obtained for polyethylene terephthalate (PET).[89] The SEM micrographs gathered in Figure 10.5 show some examples of multilayer structures obtained when different biopolymer interlayers were introduced into PHBV films.

As summarized in a previous review,[35] the electrospinning biolayering technology presents significant advances and advantages *versus* other conventional technologies for the development of multilayer biopolymer-based structures of interest in packaging applications. These can be summarized as follows:

- There is virtually no need for tie layers. The nanofibers-based electrospun interlayers made of biopolymers have demonstrated to have the ability of acting as natural adhesives. Indeed, due to the lack of natural adhesives, this currently represents a challenge for the development of multilayer structures fully based on natural materials that do not compromise the biodegradability and/or renewability of the packaging materials.
- Development of tailor-made packaging structures. Alternating layers of different biopolymers, including their blends and/or composites, can be electrospun as intermediate or even coating barrier layers. One can also tightly control the thickness of the layers depending on the specific requirements in terms of barrier performance.
- Improvement of optical properties of the packaging materials. As the interlayers are nanofabricated, no significant changes in transparency or optical properties are habitually observed when developing the multilayer material. This is an important feature in the case of packaging materials for food applications, especially in flexible packaging.
- Overall enhancement of barrier performance. The nanostructure of the biomaterials and the flexibility to combine different biopolymers and/or to introduce nanofillers during fabrication allow a more efficient barrier effect than expected by thicker individual layers.
- Potentially, there is no effect on the overall mechanical properties. The layers are in nanostructured form and, therefore, do not affect flexibility and toughness and can even improve crack resistance with no need for compatibilizers.

The electrospinning biolayering technology has also been applied to develop moisture resistant protein-based multilayer films, showing excellent adhesion between layers that is, in fact, the key factor for the resulting

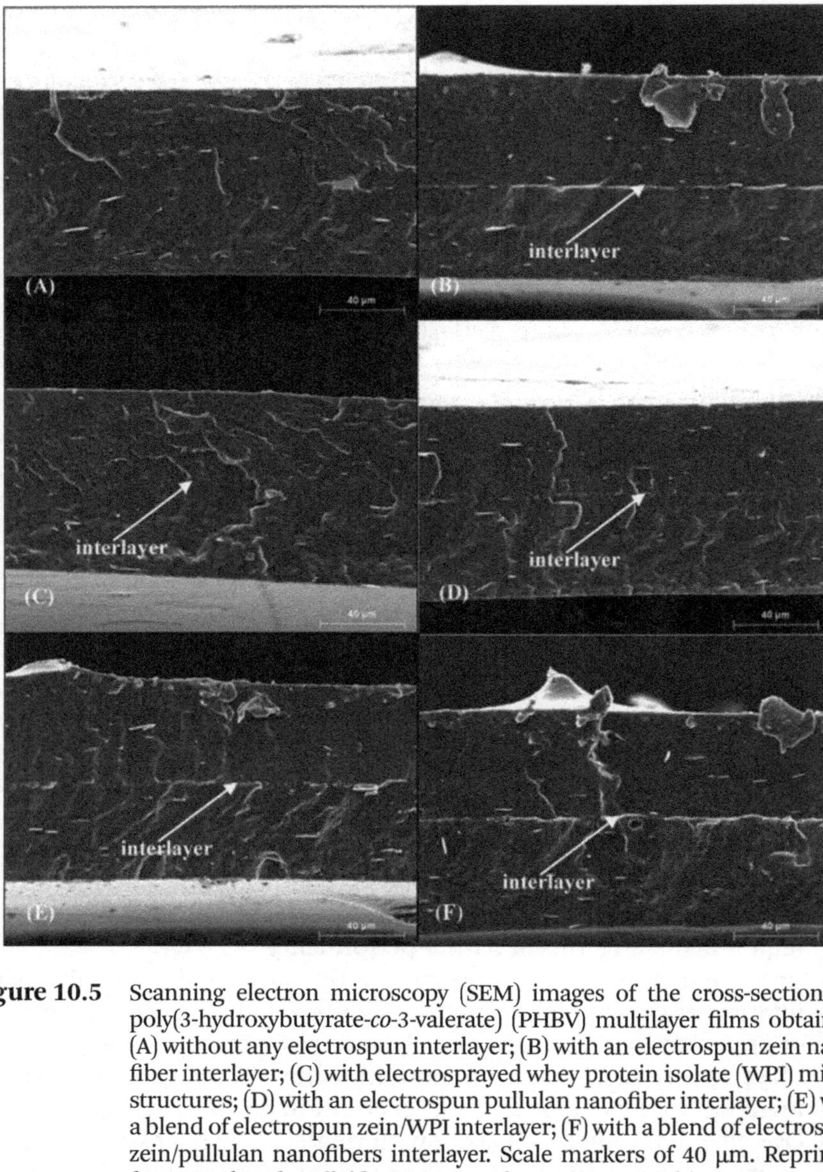

Figure 10.5 Scanning electron microscopy (SEM) images of the cross-sections of poly(3-hydroxybutyrate-*co*-3-valerate) (PHBV) multilayer films obtained: (A) without any electrospun interlayer; (B) with an electrospun zein nanofiber interlayer; (C) with electrosprayed whey protein isolate (WPI) microstructures; (D) with an electrospun pullulan nanofiber interlayer; (E) with a blend of electrospun zein/WPI interlayer; (F) with a blend of electrospun zein/pullulan nanofibers interlayer. Scale markers of 40 μm. Reprinted from *Food Hydrocolloids*, **39**, M. J. Fabra, A. Lopez-Rubio, J. M. Lagaron, On the use of different hydrocolloids as electrospun adhesive interlayers to enhance the barrier properties of polyhydroxyalkanoates of interest in fully renewable food packaging concepts, 77–84, Copyright 2014, with permission from Elsevier.[88]

improved barrier properties.[78] In addition to the improved barrier properties, it was observed that the stretchability of the developed multilayer systems remained constant over the whole storage time. Furthermore, the electrospinning technology allows the incorporation of nano-reinforcing agents into the electrospun ultrathin fibers, such as nanoclays[91] and bacterial cellulose nanowhiskers (BCNWs).[36] Therefore, the resultant hybrid electrospun layers can further improve the barrier properties, among other physical properties, when these are properly applied in packaging materials.

10.2.3 Active, Bioactive, and Smart Functionalities

Active packaging is one of the most innovative food concepts, including oxygen scavengers, moisture absorbers, flavor and odor absorbers/releasers, and antimicrobials systems, that has been introduced in response to the continuous changes in consumer demands and market trends.[92] This packaging concept involves different solutions depending on the quality attribute to be preserved, including components that release or absorb substances, from or into the packaged food, and the surrounding environment, thus extending the shelf life of foods, preventing and/or indicating the migration of contaminants while maintaining or even improving their nutritional quality and ensuring food safety.[93] In this sense, electrospun layers, especially in the form of coatings, can be used as carriers of active ingredients such as antimicrobial or antioxidant compounds to reduce, inhibit, and stop the growth of microorganisms or oxidation/rancidity, respectively.

From the above, a layer deposition of electrospun active ultrathin fibers or capsules on films and sheets that would be exposed to the environmental conditions can be used to introduce new biocide functionalities in the packaging material.[94] Based on this concept, electrospun or electrosprayed coatings made of either intrinsically antimicrobial polymers (*e.g.* chitosan[95,96] and its blends[97]) or functionalized polymers with biocide substances, including essential oils,[87] antibiotics,[27] and metal nanoparticles (MNPs),[73] have shown high antimicrobial activity against both Gram-positive and Gram-negative bacteria. Apart from offering efficient protection to the packaged food from microbial growth, it can be considered that the application of the electrospinning technology will serve to avoid the technical difficulties that would be encountered by the packaging industries in the development of such new active structures by means of conventional technologies.[75] In addition to antimicrobials, other research works have reported the suitability of the electrospinning process in order to develop antioxidant multilayer packaging structures with properties to inhibit oxidation processes. For example, α-tocopherol, as a model antioxidant, was encapsulated in three different hydrocolloid matrices, namely whey protein isolate (WPI), soy protein isolate (SPI), and zein, and these were directly applied as coatings onto a thermoplastic wheat gluten film.[37] It was observed that the antioxidant activity of the active compound was preserved during the encapsulation process up to 95% and the resultant bilayer films protected the antioxidant from degradation during retorting.

A more innovative concept of packaging structure having an active role in food preservation is the development of active materials with heat management properties that can control thermal variations during storage and distribution of food. The development of packaging structures with thermal energy storage capacity can be achieved through the incorporation of phase change materials (PCMs) into the electrospun/electrosprayed materials.[34,98] These pioneering works have demonstrated the potential of EHDP to develop efficient temperature buffering materials based on PLA, poly(ε-caprolactone) (PCL), and polystyrene (PS) containing mainly paraffins as PCMs for a thermal buffering effect of relevance in packaging applications as well as in the walls of refrigeration equipment. The resultant PCM-containing packaging materials can be designed to play an active role to maintain the food temperature within desired limits, preventing temperature variations along the cold chain that otherwise could have negative effects on food due to crystal ice growth, acceleration of chemical reactions, and/or microorganism growth, and result in a reduction of quality and may shorten the shelf life of the food products.

One of the most promising applications of the EHDP technology is undoubtedly related to the development of novel materials containing integrated bioactive components.[44] These components, when properly built into packages, will have the ability of being released into the food to transform conventional foods into functional ones with potential health benefits for the consumer and counteracting some of the industrial problems encountered when developing novel nutraceuticals.[99] However, to the best of our knowledge, there are only a few works that explore this concept for packaging applications, which seeks to equally contribute to the functional foods area. For instance, recent results from our research group have already demonstrated the enormous potential of bio-based electrospun materials to protect and release in a controlled way different interesting bioactive compounds,[29,30,100,101] where a large number of food hydrocolloids with potential applications in packaging have been used as encapsulation matrices. This opens up and encourages next research in the field of the conceptual approach of bioactive packaging[102] where a food package or coating is given the unique role of enhancing food impact over the consumer's health.

Finally, there are also some previous works that have incorporated electrospun nanolayers within packaging structures but with the aim of developing smart packaging technologies. As an example, the use of electrically conductive fibers filled with carbon black nanoparticles can serve to create labels or tags that allow monitoring the conditions and quality of the packaged products (*e.g.* food freshness) from the production line to the end user.[103] This includes relevant information for the consumer such as temperature, pH or different gases, chemical contaminants, and pathogens presence. Additionally, such smart labels can include components that range from bar codes to radio frequency transmitters, *i.e.* radio frequency identification (RFID) devices.[104] These conductive films can be achieved by the above-described electrospinning coating technology when applied to different polymer and biopolymer matrices containing conductive carbon black nanoparticles such as graphene nanoplatelets (GNPs) and carbon nanotubes (CNTs). For instance, in recent years, GNPs have been successfully nanodispersed in electrospun fibers of poly(vinylpyrrolidone) (PVP),[105] PS and polyvinyl chloride (PVC),[106] and poly(vinyl acetate) (PVAC),[107] among others.

10.3 Industrial Upscaling

The low cost and high quality of nanofibers mass production is a key factor for supporting the growing research activities and applications of the electrospun nanofiber mats in the packaging industry.[108] However, the scale-up production of nanofibers through a traditional single-jet electrospinning setup is not economically or technically feasible because of its low productivity (typically 0.01–0.10 g h^{-1}).[10,54] Specific technological improvements have been introduced not only for enhancing the quality of nanofibers and for tuning their properties according to the requirements of different applications but also in view of large-scale production.[108] The efforts for enhancing the productivity have been directed towards increasing the number of jets in multiple needles arrangements and needleless systems. Multi-nozzle techniques comprising needle-based arrangements have been proven in the uniaxial configuration (both linear and circular), co-, and tri-axial geometry, *etc.*[55] Furthermore, several types of needleless systems have been developed for increasing productivity, such as those involving conductive parts that can be agitated or rotated (*e.g.* cylinder, discs, wires, and balls), roller electrospinning, bubble electrospinning, and coil electrospinning.[4,16]

The simplest and direct way to increase productivity is the use of multi-jet needle-based systems that can also increase the covered area and the thickness of the resultant electrospun mats.[75] In order to fulfil the rising demands related to productivity in the packaging industry and also take advantage of the technical benefits of the electrospinning machines, some companies offer different devices and accessories involving single- and multi-nozzle spinnerets that target both academic research and industrial applications (Table 10.2). For academia and small-scale production, some key requirements are demanded by customers, including low prices, multi-functionality, flexibility for carrying out different research projects, and compactness. Instead, industrial applications require not only high throughputs, but also process monitoring, specialized customer support, and in-line quality control systems. Most electrospinning companies largely satisfy these demands.

Some research and development activities carried out at commercially available or high-efficient electrospinning designs have been already reported in the packaging field. For instance, Chalco-Sandoval *et al.*[34] used a Fluidnatek® pilot plant apparatus from Bioinicia S.L.[109] (Valencia, Spain), which was equipped with a 20-needle injector tool and a climate control and roll-to-roll collector, to manufacture electrospun PCM-containing PCL mats (Figure 10.6). More recently, Torres-Giner *et al.*[30] also used a Fluidnatek® LE500 pilot line, which was equipped with 24 emitter multi-nozzle injector and static collector, to nanoencapsulate *Aloe vera* in both synthetic polymer nanofibers, *i.e.* polyvinylpyrrolidone (PVP) and polyvinyl alcohol (PVOH), and naturally occurring polymers, *i.e.* barley starch (BS), maltodextrin, and whey protein concentrate (WPC). Kim *et al.*[108] also used a proprietary designed cylinder-type electrospinning design for mass production of smooth and non-beaded polyurethane (PU) nanofibers. The authors reported the use of

Table 10.2 Some manufacturers of electrospinning machines with single- and multi-nozzle spinnerets.

Company	Website	Products
Bioinicia (Valencia, Spain)	www.bioinicia.com	• LE™ series: lab devices, pilot production, and mass production machines • R&D and manufacturing services
Electrospunra	www.electrospunra.com	• Pilot-line equipment • Custom equipment manufacturing • Consulting and research
E-Spin Nanotech (Kanpur, India)	www.espinnanotech.com	• Super ES series pilot-line machines
IME Technologies	www.imetechnologies.com	• Electrospinning machines for biomedical applications • R&D and manufacturing services
Inovenso (Istanbul, Turkey)	www.inovenso.com	• Nanospinner series: Lab-scale devices, pilot-line and industrial machines • Consulting services
Linari Nanotech (Pisa, Italia)	www.linariobiomedical.com	• R&D devices, industrial systems
NaBond (Shenzen, China)	www.electrospinning.com	• TL-series: lab setup, pilot production, and mass production machines

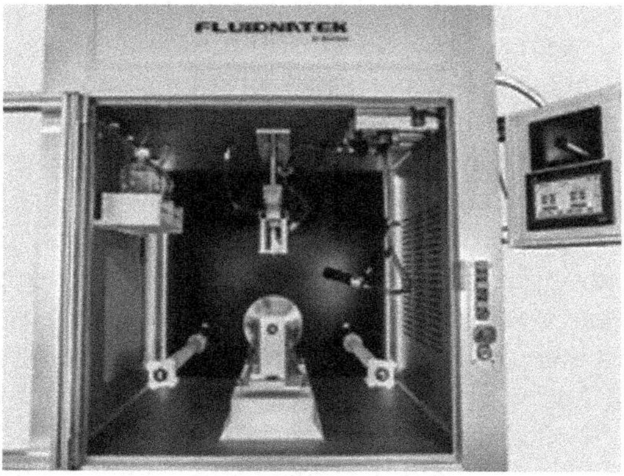

Figure 10.6 Fluidnatek® pilot line LE500. Courtesy of Bioinicia S.L.[109] (Valencia, Spain).

an upward 1 m length cylinder-type nozzle with 120 multi-nozzles installed into the cylinder, which is represented in Figure 10.7.

One of the main drawbacks related to the use of multi-jet systems is the alteration of the electric field profile by the presence of jets nearby.[55] For instance, Theron *et al.*[54] found that the interacting jets in a multiple arrangement show the characteristic instabilities of electrospinning though these were reduced by the introduction of more jets. In the case of a large number of jets, the distance between the nozzles and the collector must be increased in order to compensate the driven instabilities and obtain a larger collection area.

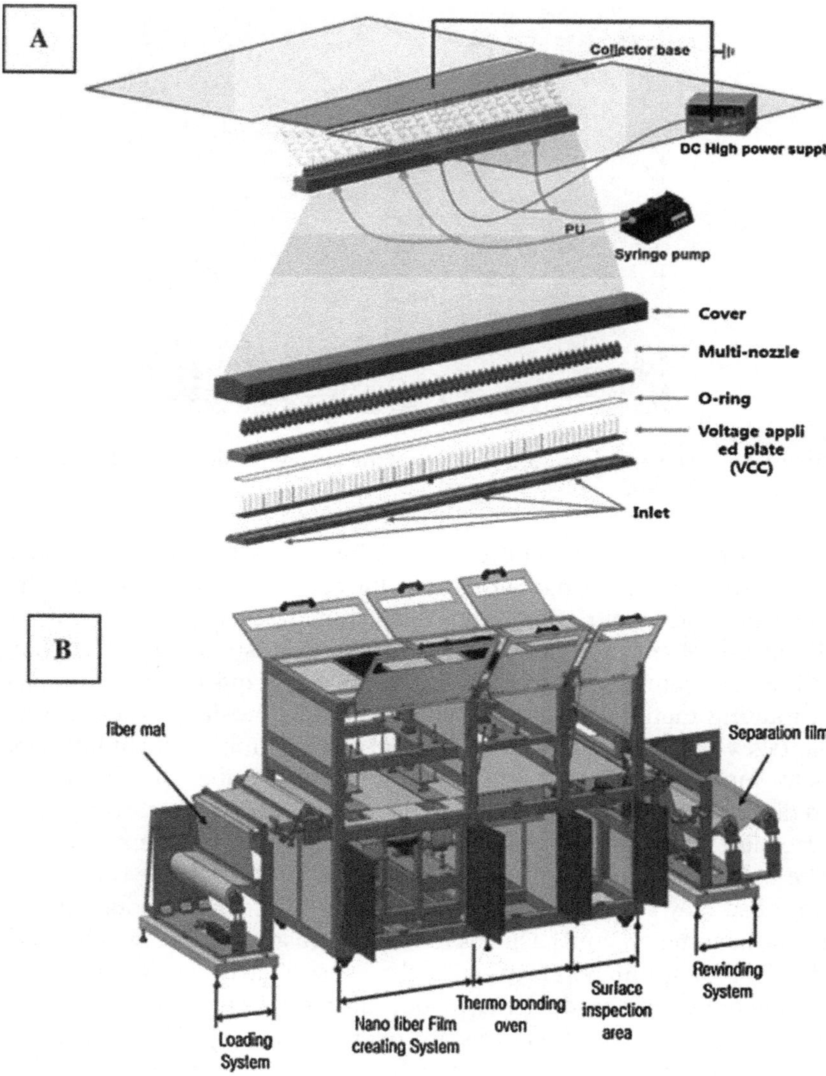

Figure 10.7 Schematic diagram of: (a) the components of a multi-nozzle electrospinning system and (b) a nanofiber mass production system. Reprinted from *Journal of Industrial and Engineering Chemistry*, **31**, I. G. Kim, J.-H. Lee, A. R. Unnithan, C.-H. Park, C. S. Kim, A comprehensive electric field analysis of cylinder-type multinozzle electrospinning system for mass production of nanofibers, 251–256, Copyright 2015, with permission from Elsevier.[108]

In another work, Varesano *et al.*[110] used a 9-nozzle spinning head pilot line apparatus from Nuova ROJ Electrotex SRL[111] (Biella, Italy) to electrospin polyethylene oxide (PEO) by using a circular multi-needle injector with 37 needles distributed in a hexagonal pattern. The authors observed some jet perturbations that were overcome by adjusting the voltage applied as a function of the

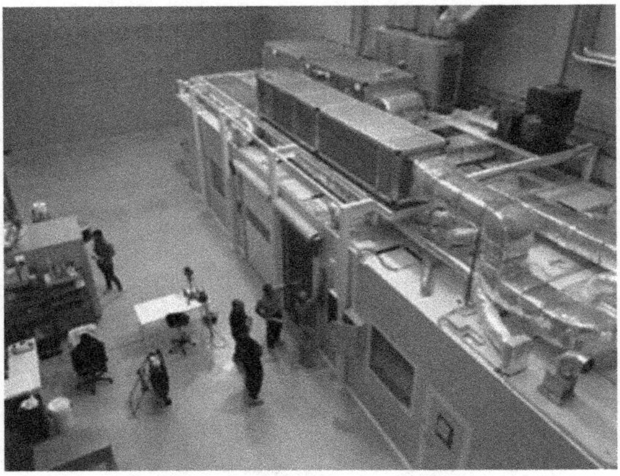

Figure 10.8 The electrospinning manufacturing site of Bioinicia S.L.[109] (Valencia, Spain).

weight of the substrate. In a novel outlook for potential industrial broadening of electrospinning in the packaging area, the coupling of a twin-screw extruder as the front end of an electrospinning machine was assessed.[14] In this work, PCL was processed *via* this hybrid extrusion-electrospinning process using a multi-nozzle spinneret between 3 and 20 needles, and fibers were collected in a rotating mandrel. This approach showed to provide several promising advantages in comparison to conventional electrospinning, including capabilities for melting and pressurization/pumping, continuous process, mixing of ingredients, and the incorporation of particles or even nanoparticles into the electrospun fibers without observing agglomeration or sedimentation.

The use of multi-needle systems can significantly increase the productivity because they are easy to arrange, however they could be not sufficiently efficient in some practical aspects due to the obstruction of needles and the difficulties for spinneret cleaning.[10,55,112] These drawbacks can undoubtedly scale down the fiber productivity, also resulting in a poor fibers quality. Overcoming these drawbacks requires a proper engineering design of the injector and very stringent process controls. The company Bioinicia S.L.[109] has designed and built two robust EHDP plants, one for materials based on electrospun fibers and another one for electrosprayed particles or capsules at a rate of 5 kg h^{-1}. The former electrospinning plant (Figure 10.8) is now both good manufacturing practice (GMP) and ISO13485 certified and is currently producing over three continuous shifts non-woven mats at high production rates. Bioinicia S.L.[109] is also currently building a plant for electrospun coatings and interlayers for packaging applications aiming at 0.2 gsm electrospun fibers deposition at a roll-to-roll production rates of up to 500 m of substrate per min within a H2020 EU project called YPACK.

The design of injectors, however, is not limited to the use of needles and different needleless configurations can be used according to the application requirements, allowing to overcome the disadvantages of multi-needle spinnerets. Alternative spinneret designs to conventional multi-nozzle design include stationary spinnerets (*e.g.* corona,[112] bubble spinneret,[113] stationary wire,[114] and magnetic liquid[115]) and rotating spinnerets (*e.g.* cylinder and disc[116] and spiral wire coil or cone[10]). Some needleless electrospinning systems are commercially available showing a significant throughput. For instance, Nanospider™ Electrospinning Technology from Elmarco[117] (Liberec, Czech Republic) is a patented needle-free technique, which is based on the use of either a wire that is constantly covered by the polymer solution or a rotating cylinder. SNC Nanofibers[118] (Cape Town, South Africa) has also patented a high-throughput nanofibers production platform called *Ball ElectroSpinning Technology* (SNC BEST™). Unlike the traditional needle-based electrospinning process, where the number of jets is limited by the number of installed needles, the free-surface spinning nature of SNC BEST™ allows the jets to self-arrange and -optimize. The disadvantages of needleless technologies are mainly related to limitations in the solvents that can be used, production instabilities with some polymers, and less control on fiber size and size distribution.

10.4 Current Limitations and Future Scope

Recent years have witnessed a tremendous expansion of research and technology developments in the electrospinning field, of which some of them have resulted in a significant application development in the packaging field. In particular, significant advances in the development of multilayer structures obtained by means of electrospun coatings and interlayers of biopolymers will certainly provide a more solid ground towards increasing the technical and economic competitiveness of renewable polymers in the packaging industry. However, there is still a long way to go, not only in the materials development and large-scale processing parcels, but also regarding the widespread commercialization of the resultant packaging materials and their full characterization in terms of food contact application and environmental impact.

Considering the high technical advantages, versatility, and applications of the electrospun nanofibers in many industrial sectors it is expected that their market and, hence, that of the electrospinning machines, both at pilot and industrial scales, will significantly grow in the next years. For this, some significant issues (*e.g.* low productivity, insufficient process stability, and high costs) still need to be addressed *via* research efforts and technology transferences from academia to representative companies, including both nanofibers and machines manufacturers. This will certainly allow the required improvements of EHDP to be completely scaled up for mass production, in turn allowing the technique to benefit different commodity areas including packaging. In the next few years, therefore, some of these developments, which are currently at different stages of product definition and refining, are expected to reach full market introduction.

Acknowledgements

The authors acknowledge support received from the Spanish Ministry of Economy and Competitiveness (MINECO) project AGL2015-63855-C2-1-R and the Brazilian Council for Scientific and Technological Development (CNPq) of the Brazilian Government for the predoctoral grant 205955/2014-2.

References

1. X. Wang, B. Ding, G. Sun, M. Wang and J. Yu, *Prog. Mater. Sci.*, 2013, **58**, 1173–1243.
2. S. Chakraborty, I. C. Liao, A. Adler and K. W. Leong, *Adv. Drug Delivery Rev.*, 2009, **61**, 1043–1054.
3. B. Ding, M. Wang, X. Wang, J. Yu and G. Sun, *Mater. Today*, 2010, **13**, 16–27.
4. R. K. Srivastava, in *Electrospun Nanofibers*, Woodhead Publishing, Duxford, UK, 2017, pp. 399–447.
5. N. Bhardwaj, S. Kundu, N. Bhardwaj and S. C. Kundu, *Biotechnol. Adv.*, 2010, **28**, 325–347.
6. J. Anu Bhushani and C. Anandharamakrishnan, *Trends Food Sci. Technol.*, 2014, **38**, 21–33.
7. M. Noruzi, *J. Sci. Food Agric.*, 2016, 4663–4678.
8. W.-E. Teo and S. Ramakrishna, *Compos. Sci. Technol.*, 2009, **69**, 1804–1817.
9. B. Ghorani and N. Tucker, *Food Hydrocolloids*, 2015, **51**, 227–240.
10. M. Yu, R.-H. Dong, X. Yan, G.-F. Yu, M.-H. You, X. Ning and Y.-Z. Long, *Macromol. Mater. Eng.*, 2017, **302**, 1700002.
11. S. T. Aruna, L. S. Balaji, S. S. Kumar and B. S. Prakash, *Renewable Sustainable Energy Rev.*, 2017, **67**, 673–682.
12. A. Greiner and J. H. Wendorff, *Angew. Chem., Int. Ed. Engl.*, 2007, **46**, 5670–5703.
13. Y. Lin, Y. Yao, X. Yang, N. Wei, X. Li, P. Gong, R. Li and D. Wu, *J. Appl. Polym. Sci.*, 2008, **107**, 909–917.
14. S. Senturk Ozer, D. Kalyon, H. Gevgilili and D. Ward, *Polym. Eng. Sci.*, 2013, **53**, 1463–1474.
15. X. Wang, Y. Yuan, X. Huang and T. Yue, *J. Appl. Polym. Sci.*, 2015, **132**, 41811.
16. Y. Liu, L. Zhang, X.-F. Sun, J. Liu, J. Fan and D.-W. Huang, *Mater. Lett.*, 2015, **141**, 153–156.
17. A. Rogina, *Appl. Surf. Sci.*, 2014, **296**, 221–230.
18. D. Kai, S. S. Liow and X. J. Loh, *Mater. Sci. Eng., C*, 2014, **45**, 659–670.
19. S. H. Tan, R. Inai, M. Kotaki and S. Ramakrishna, *Polymer*, 2005, **46**, 6128–6134.
20. A. O. Basar, S. Castro, S. Torres-Giner, J. M. Lagaron and H. Turkoglu Sasmazel, *Mater. Sci. Eng., C*, 2017, **81**, 459–468.

21. A. Fernandez, S. Torres-Giner and J. M. Lagaron, *Food Hydrocolloids*, 2009, **23**, 1427–1432.
22. A. López-Rubio, E. Sanchez, S. Wilkanowicz, Y. Sanz and J. M. Lagaron, *Food Hydrocolloids*, 2012, **28**, 159–167.
23. H. Qu, S. Wei and Z. Guo, *J. Mater. Chem. A*, 2013, **1**, 11513–11528.
24. X. Wu, X. Li, C. Zhu, P. Li, H. Yu, Z. Guo and J. Shu, *Mater. Today Energy*, 2016, **1–2**, 17–23.
25. A. Haider, S. Haider and I.-K. Kang, *Arabian Journal of Chemistry*, 2015, DOI: 10.1016/j.arabjc.2015.11.015.
26. T. J. Sill and H. A. von Recum, *Biomaterials*, 2008, **29**, 1989–2006.
27. S. Torres-Giner, A. Martinez-Abad, J. V. Gimeno-Alcañiz, M. J. Ocio and J. M. Lagaron, *Adv. Eng. Mater.*, 2012, **14**, B112–B122.
28. R. H. de Freitas Zômpero, A. López-Rubio, S. C. de Pinho, J. M. Lagaron and L. G. de la Torre, *Colloids Surf., B*, 2015, **134**, 475–482.
29. S. Torres-Giner, A. Martinez-Abad, M. J. Ocio and J. M. Lagaron, *J. Food Sci.*, 2010, **75**, N69–N79.
30. S. Torres-Giner, S. Wilkanowicz, B. Melendez-Rodriguez and J. M. Lagaron, *J. Agric. Food Chem.*, 2017, **65**, 4439–4448.
31. N. Alvarado, J. Romero, A. Torres, C. López de Dicastillo, A. Rojas, M. J. Galotto and A. Guarda, *J. Food Eng.*, 2018, **217**, 1–10.
32. M. Arkoun, F. Daigle, M. C. Heuzey and A. Ajji, *Molecules*, 2017, **22**, 585.
33. Z. Aytac, N. O. S. Keskin, T. Tekinay and T. Uyar, *J. Appl. Polym. Sci.*, 2017, **134**, 44858.
34. W. Chalco-Sandoval, M. J. Fabra, A. López-Rubio and J. M. Lagaron, *J. Food Eng.*, 2017, **192**, 122–128.
35. M. J. Fabra, M. A. Busolo, A. Lopez-Rubio and J. M. Lagaron, *Trends Food Sci. Technol.*, 2013, **31**, 79–87.
36. M. J. Fabra, A. López-Rubio, J. Ambrosio-Martín and J. M. Lagaron, *Food Hydrocolloids*, 2016, **61**, 261–268.
37. M. J. Fabra, A. López-Rubio and J. M. Lagaron, *Food Hydrocolloids*, 2016, **55**, 11–18.
38. T. Chae and F. Ko, in *Electrospun Nanofibers*, Woodhead Publishing, Duxford, UK, 2017, pp. 521–550.
39. M. Gorji, R. Bagherzadeh and H. Fashandi, in *Electrospun Nanofibers*, Woodhead Publishing, Duxford, UK, 2017, pp. 571–598.
40. X. Qin and S. Subianto, in *Electrospun Nanofibers*, Woodhead Publishing, Duxford, UK, 2017, pp. 449–466.
41. S. Zhang, H. Liu, F. Zuo, X. Yin, J. Yu and B. Ding, *Small*, 2017, **13**, 1603151.
42. X. Cui, H. M. Zhang and T. F. Wu, *Spectrochim. Acta, Part A*, 2011, **79**, 1998–2002.
43. Y. Wang, M. Shahid, J. Cheng, H. Nishijima and W. Pan, *Nanotechnology*, 2017, 28.
44. S. Torres-Giner, R. Pérez-Masiá and J. M. Lagaron, *Polym. Eng. Sci.*, 2016, **56**, 500–527.

45. F. Anton, Process and apparatus for preparing artificial threads, US1975504A, 1934.
46. T. Subbiah, G. S. Bhat, R. W. Tock, S. Parameswaran and S. S. Ramkumar, *J. Appl. Polym. Sci.*, 2005, **96**, 557–569.
47. F. R. S. Sir Geoffrey Taylor, *Electrically driven jets, Proc. R. Soc. London, Ser. A*, 1969, **313**, 453–475.
48. X. Hu, S. Liu, G. Zhou, Y. Huang, Z. Xie and X. Jing, *J. Controlled Release*, 2014, **185**, 12–21.
49. S. Torres-Giner, E. Gimenez and J. M. Lagaron, *Food Hydrocolloids*, 2008, **22**, 601–614.
50. R. Sahay, V. Thavasi and S. Ramakrishna, *J. Nanomater.*, 2011, **2011**, 17.
51. J. H. Wendorff, S. Agarwal and A. Greiner, *Electrospinning : Materials, Processing, and Applications (1)*, Wiley-VCH, Hoboken, NJ, USA, 2012.
52. R. Sahay, V. Thavasi and S. Ramakrishna, *J. Nanomater.*, 2011, 317673.
53. S. Torres-Giner, J. V. Gimeno-Alcañiz, M. J. Ocio and J. M. Lagaron, *J. Appl. Polym. Sci.*, 2011, **122**, 914–925.
54. S. A. Theron, A. L. Yarin, E. Zussman and E. Kroll, *Polymer*, 2005, **46**, 2889–2899.
55. L. Persano, A. Camposeo, C. Tekmen and D. Pisignano, *Macromol. Mater. Eng.*, 2013, **298**, 504–520.
56. N. Bhattarai, D. Edmondson, O. Veiseh, F. A. Matsen and M. Zhang, *Biomaterials*, 2005, **26**, 6176–6184.
57. L. Tian, C. Zhao, J. Li and Z. Pan, *Text. Res. J.*, 2015, **85**, 621–631.
58. Z. Sun, E. Zussman, A. L. Yarin, J. H. Wendorff and A. Greiner, *Adv. Mater.*, 2003, **15**, 1929–1932.
59. Y. Ying, J. Zhidong, L. Qiang, H. Lei and G. Zhicheng, *J. Phys.: Conf. Ser.*, 2008, **142**, 012027.
60. G. H. Kim, *J. Polym. Sci., Part B: Polym. Phys.*, 2006, **44**, 1426–1433.
61. R. Nayak, R. Padhye and L. Arnold, in *Electrospun Nanofibers*, Woodhead Publishing, Duxford, UK, 2017, pp. 11–40.
62. C. Wang and M. Wang, *Adv. Mater. Res.*, 2012, **410**, 118–121.
63. X. Xu, X. Zhuang, X. Chen, X. Wang, L. Yang and X. Jing, *Macromol. Rapid Commun.*, 2006, **27**, 1637–1642.
64. A. Camerlo, A. M. Bühlmann-Popa, C. Vebert-Nardin, R. M. Rossi and G. Fortunato, *J. Mater. Sci.*, 2014, **49**, 8154–8162.
65. P. J. García-Moreno, K. Stephansen, J. van der Kruijs, A. Guadix, E. M. Guadix, I. S. Chronakis and C. Jacobsen, *J. Food Eng.*, 2016, **183**, 39–49.
66. A. Repanas, W. Wolkers, O. Gryshkov, P. Kalozoumis, M. Mueller, H. Zernetsch, S. Korossis and B. Glasmacher, *Clin. Exp. Pharmacol.*, 2015, **5**, 192.
67. H. Jiang, L. Wang and K. Zhu, *J. Controlled Release*, 2014, **193**, 296–303.
68. F. Li, Y. Zhao and Y. Song, in *Nanofibers*, ed. A. Kumar, InTech, Rijeka, 2010, ch. 22, DOI: 10.5772/8166.

69. A. López-Rubio, E. Sanchez, Y. Sanz and J. M. Lagaron, *Biomacromolecules*, 2009, **10**, 2823–2829.
70. X. Qin, in *Electrospun Nanofibers*, Woodhead Publishing, Duxford, UK, 2017, pp. 41–71.
71. C. Li, Q. Li, X. Ni, G. Liu, W. Cheng and G. Han, *Materials*, 2017, **10**, 572.
72. S. Torres-Giner, in *Multifunctional and Nanoreinforced Polymers for Food Packaging*, Woodhead Publishing Limited, Cambridge, UK, 2011, pp. 108–125.
73. J. L. Castro-Mayorga, M. J. Fabra, A. M. Pourrahimi, R. T. Olsson and J. M. Lagaron, *Food Bioprod. Process.*, 2017, **101**, 32–44.
74. M. A. Cerqueira, M. J. Fabra, J. L. Castro-Mayorga, A. I. Bourbon, L. M. Pastrana, A. A. Vicente and J. M. Lagaron, *Food Bioprocess Technol.*, 2016, **9**, 1874–1884.
75. Y. Echegoyen, M. J. Fabra, J. L. Castro-Mayorga, A. Cherpinski and J. M. Lagaron, *Trends Food Sci. Technol.*, 2017, **60**, 71–79.
76. C. E. Realini and B. Marcos, *Meat Sci.*, 2014, **98**, 404–419.
77. A. Cherpinski, S. Torres-Giner, L. Cabedo and J. M. Lagaron, *Food Addit. Contam., Part A*, 2017, **34**, 1817–1830.
78. M. J. Fabra, A. Lopez-Rubio and J. M. Lagaron, *Food Hydrocolloids*, 2015, **44**, 292–299.
79. M. J. Fabra, A. López-Rubio and J. M. Lagaron, *Food Bioprocess Technol.*, 2015, **8**, 2330–2340.
80. M. J. Fabra, G. Sánchez, A. López-Rubio and J. M. Lagaron, *LWT–Food Sci. Technol.*, 2014, **59**, 760–767.
81. A. Cherpinski, S. Torres-Giner, L. Cabedo, J. A. Méndez and J. M. Lagaron, *J. Appl. Polym. Sci.*, 2017, 45501.
82. A. Cherpinski, S. Torres-Giner, J. Vartiainen, M. S. Peresin, P. Lahtinen and J. M. Lagaron, *Cellulose*, 2018, **25**, 1291–1307.
83. K. Acatay, E. Simsek, C. Ow-Yang and Y. Z. Menceloglu, *Angew. Chem., Int. Ed.*, 2004, **43**, 5210–5213.
84. A. Nakajima, K. Hashimoto and T. Watanabe, *Monatsh. Chem. Chem. Mon.*, 2001, **132**, 31–41.
85. D. Öner and T. J. McCarthy, *Langmuir*, 2000, **16**, 7777–7782.
86. M. A. Busolo, S. Torres-Giner and J. M. Lagaron, Enhancing the gas barrier properties of polylactic acid by means of electrospun ultrathin zein fibers, *67th Annual Technical Conference - ANTEC™ of the Society of Plastics Engineers (SPE)*, Chicago, USA, 2009, **5**, 2763–2767.
87. S. Torres-Giner, A. Martinez-Abad and J. M. Lagaron, *J. Appl. Polym. Sci.*, 2014, **131**, 9270–9276.
88. M. J. Fabra, A. López-Rubio and J. M. Lagaron, *Food Hydrocolloids*, 2014, **39**, 77–84.
89. M. J. Fabra, A. Lopez-Rubio and J. M. Lagaron, *J. Food Eng.*, 2014, **127**, 1–9.
90. M. J. Fabra, A. Lopez-Rubio and J. M. Lagaron, *Food Hydrocolloids*, 2013, **32**, 106–114.

91. S. Torres-Giner and J. M. Lagaron, *J. Appl. Polym. Sci.*, 2010, **118**, 778–789.
92. A. L. Brody, E. R. Strupinsky and L. R. Kline, in *Active Packaging for Food Applications*, CRC Press, Boca Ratón, Florida, USA, 2001.
93. D. A. Pereira de Abreu, J. M. Cruz and P. Paseiro Losada, *Food Rev. Int.*, 2012, **28**, 146–187.
94. S. Torres-Giner, in *Antimicrobial Polymers*, John Wiley & Sons, Inc., Hoboken, New Jersey, USA, 2011, pp. 261–285.
95. S. Torres-Giner, M. J. Ocio and J. M. Lagaron, *Eng. Life Sci.*, 2008, **8**, 303–314.
96. S. Sreekumar, P. Lemke, B. M. Moerschbacher, S. Torres-Giner and J. M. Lagaron, *Food Addit. Contam., Part A*, 2017, **34**, 1795–1806.
97. S. Torres-Giner, M. J. Ocio and J. M. Lagaron, *Carbohydr. Polym.*, 2009, **77**, 261–266.
98. W. Chalco-Sandoval, M. J. Fabra, A. López-Rubio and J. M. Lagaron, *J. Food Eng.*, 2015, **164**, 55–62.
99. J. M. Lagaron and A. Lopez-Rubio, *Trends Food Sci. Technol.*, 2011, **22**, 611–617.
100. M. Aceituno-Medina, S. Mendoza, B. A. Rodríguez, J. M. Lagaron and A. López-Rubio, *J. Funct. Foods*, 2015, **12**, 332–341.
101. L. G. Gómez-Mascaraque, J. M. Lagarón and A. López-Rubio, *Food Hydrocolloids*, 2015, **49**, 42–52.
102. A. Lopez-Rubio, R. Gavara and J. M. Lagaron, *Trends Food Sci. Technol.*, 2006, **17**, 567–575.
103. J. H. Han, in *Innovations in Food Packaging*, Elsevier Ltd., London, UK, 2013, pp. 3–12.
104. Y. Echegoyen, in *Nanotechnologies in Food and Agriculture*, 2015, pp. 141–166.
105. N. Zhu, W. Liu, M. Xue, Z. Xie, D. Zhao, M. Zhang, J. Chen and T. Cao, *Electrochim. Acta*, 2010, **55**, 5813–5818.
106. R. Asmatulu, M. Ceylan and N. Nuraje, *Langmuir*, 2011, **27**, 504–507.
107. Q. Bao, H. Zhang, J.-x. Yang, S. Wang, D. Y. Tang, R. Jose, S. Ramakrishna, C. T. Lim and K. P. Loh, *Adv. Funct. Mater.*, 2010, **20**, 782–791.
108. I. G. Kim, J.-H. Lee, A. R. Unnithan, C.-H. Park and C. S. Kim, *J. Ind. Eng. Chem.*, 2015, **31**, 251–256.
109. Bioinicia S.L. http://bioinicia.com.
110. A. Varesano, F. Rombaldoni, G. Mazzuchetti, C. Tonin and R. Comotto, *Polym. Int.*, 2010, **59**, 1606–1615.
111. Nuova ROJ Electrotex SRL, http://www.roj.com.
112. K. Molnar and Z. K. Nagy, *Eur. Polym. J.*, 2016, **74**, 279–286.
113. K. Vadodaria and G. Stylios, *Polym. Polym. Compos.*, 2016, **24**, 265–272.
114. C. T. Hsieh, C. W. Lou, Y. J. Pan, C. L. Huang, J. H. Lin, Z. I. Lin, Y. S. Chen and K. C. Chiang, *Fibers Polym.*, 2016, **17**, 1217–1226.
115. A. L. Yarin and E. Zussman, *Polymer*, 2004, **45**, 2977–2980.
116. C. Zhao, Y. Lu and Z. Pan, *Text. Res. J.*, 2017, **87**, 1685–1695.
117. Elmarco, www.//elmarco.com.
118. Stellenbosch Nanofiber Company, www.//sncfibers.com.

Subject Index